METHODS IN MOLECULAR BIOLOGY

Series Editor
John M. Walker
School of Life Sciences
University of Hertfordshire
Hatfield, Hertfordshire, AL10 9AB, UK

For further volumes:
http://www.springer.com/series/7651

Hematopoietic Stem Cell Protocols

Third Edition

Edited by

Kevin D. Bunting and Cheng-Kui Qu

Department of Pediatrics, Aflac Cancer and Blood Disorders Center, Emory University School of Medicine, Atlanta, GA, USA

 Humana Press

Editors
Kevin D. Bunting
Department of Pediatrics
Aflac Cancer and Blood Disorders Center
Emory University School of Medicine
Atlanta, GA, USA

Cheng-Kui Qu
Department of Pediatrics
Aflac Cancer and Blood Disorders Center
Emory University School of Medicine
Atlanta, GA, USA

ISSN 1064-3745   ISSN 1940-6029 (electronic)
ISBN 978-1-4939-1132-5   ISBN 978-1-4939-1133-2 (eBook)
DOI 10.1007/978-1-4939-1133-2
Springer New York Heidelberg Dordrecht London

Library of Congress Control Number: 2014942724

Cover photograph: The cover photo shows the bone marrow vasculature (red), the bone (blue), and a single injected KLS cell (green) localized near the vaculature in the mouse calvarium that was capatured by non-invasive intravital microscopic imaging. The cover photo is provided by Dr. Charles P. Lin (MGH, Harvard, Boston, MA).

Printed on acid-free paper

Humana Press is a brand of Springer
Springer is part of Springer Science+Business Media (www.springer.com)

Preface

The hematopoietic stem cell (HSC) field has continued to grow and become more sophisticated with each passing year. Methods for isolation of HSC and progenitor subsets and a variety of single cell and molecular technologies have led to a greater understanding of the heterogeneity within the phenotypically defined HSC subsets. The first two editions of *Hematopoietic Stem Cell Protocols* are very thorough resources in the user-friendly *Methods in Molecular Medicine* format. This is the third edition of *Hematopoietic Stem Cell Protocols* and aims to provide timely protocols in the HSC field while continuing the tradition. This edition teaches major new technologies that have advanced the state of the art in the field and promise to drive research in many unexpected and exciting ways in the future.

The first chapter is an overview that ties the other chapters together into a larger picture of the HSC landscape.

The methods chapters include similar categories to the last editions but with entirely new approaches and innovative insights. The author list is made up of leading researchers who have made major contributions toward these technical advances and who have provided a much needed resource for new stem cell investigators.

We thank all of the contributors for their time and effort. It has been a pleasure for both co-editors to work with the contributing authors on this project. To provide updates in the fast paced HSC field is an important endeavor. We have specifically focused on the HSC population and did not focus directly on the stem cell-associated niche which comprises a new field in itself and deserving of a separate edition. We are delighted to have organized this information into a comprehensive and essential resource. It is our hope that this resource will be a critical addition to all laboratories seeking to isolate and characterize HSCs for research and for therapeutic applications.

Atlanta, GA, USA   *Kevin D. Bunting, Ph.D.*
                                                    Cheng-Kui Qu, M.D., Ph.D.

Contents

Contributors

JENNIFER E. ADAIR • *Fred Hutchinson Cancer Research Center, University of Washington School of Medicine, Seattle, WA, USA*

BYUNGWOOK AHN • *Department of Pediatrics, Aflac Cancer and Blood Disorders Center, Emory University School of Medicine, Atlanta, GA, USA*

DAVID R. ARCHER • *Division of Hem/Onc/BMT, Department of Pediatrics, Aflac Cancer and Blood Disorders Center, Emory University School of Medicine, Atlanta, GA, USA*

KEN-EDWIN ARYEE • *Program in Molecular Medicine, University of Massachusetts Medical School, Worcester, MA, USA*

BRIAN C. BEARD • *Fred Hutchinson Cancer Research Center, Seattle, WA, USA*

AMBER H. BEGTRUP • *Division of Human Genetics, Department of Pediatrics, College of Medicine, Cincinnati Children's Research Hospital Medical Center, Cincinnati, OH, USA*

HEATH L. BRADLEY • *Division of Hem/Onc/BMT, Department of Pediatrics, Aflac Cancer and Blood Disorders Center, Emory University School of Medicine, Atlanta, GA, USA*

MICHAEL A. BREHM • *Program of Molecular Medicine, University of Massachusetts Medical School, Worcester, MA, USA*

CHRISTIAN BRENDEL • *Boston Children's Hospital and Dana Farber Cancer Institute, Harvard Medical School, Boston, MA, USA*

KEVIN D. BUNTING • *Department of Pediatrics, Aflac Cancer and Blood Disorders Center, Emory University School of Medicine, Atlanta, GA, USA*

LEONID V. BYSTRYKH • *Laboratory of Ageing Biology and Stem Cells, European Research Institute for the Biology of Ageing, University Medical Centre Groningen, University of Groningen, Groningen, The Netherlands*

HUIMIN CAO • *CMSE, Commonwealth Scientific and Industrial Research Organization (CSIRO), Clayton, Australia*

MARTA CAPALA • *Department of Experimental Hematology, University Medical Center Groningen, University of Groningen, Groningen, The Netherlands*

MARCO CARRETTA • *Department of Experimental Hematology, University Medical Center Groningen, University of Groningen, Groningen, The Netherlands*

BRAHMANANDA REDDY CHITTETI • *Indiana University School of Medicine, Indianapolis, IN, USA*

JOHN P. CHUTE • *Division of Hem/Onc, Department of Medicine, University of California Los Angeles, Los Angeles, USA*

MARIOARA F. CIUCULESCU • *Boston Children's Hospital and Dana Farber Cancer Institute, Harvard Medical School, Boston, MA, USA*

PHUONG L. DOAN • *Division of Hematologic Malignancies and Cellular Therapy, Duke University Medical Center, Durham, NC, USA*

LEI DONG • *Division of Hem/Onc/BMT, Department of Pediatrics, Aflac Cancer and Blood Disorders Center, Emory, University School of Medicine, Atlanta, GA, USA*

TARIQ ENVER • *Stem Cell Laboratory, University College London Cancer Institute, London, UK*

DEBORAH L. FRENCH • *Department of Pathology and Laboratory Medicine and Center for Cellular and Molecular Therapeutics, Children's Hospital of Philadelphia, Philadelphia, PA, USA*

PAUL GADUE • *Department of Pathology and Laboratory Medicine and Center for Cellular and Molecular Therapeutics, Children's Hospital of Philadelphia, Philadelphia, PA, USA*

GERALDDE HAAN • *Laboratory of Ageing Biology and Stem Cells, European Research Institute for the Biology of Ageing, University Medical Centre Groningen, University of Groningen, Groningen, The Netherlands*

CHAD E. HARRIS • *Boston Children's Hospital and Dana Farber Cancer Institute, Harvard Medical School, Boston, MA, USA*

XI C. HE • *Stowers Institute for Medical Research, Kansas City, MO, USA*

JONATHAN HOGGATT • *Department of Stem Cell and Regenerative Biology, Harvard University, Cambridge, MA, USA*

WENHUO HU • *Human Oncology and Pathogenesis Program, Memorial Sloan Kettering Cancer Center, New York, NY, USA*

HANS-PETER KIEM • *Departments of Medicine and Pathology, Program in Transplantation Biology, Fred Hutchinson Cancer Research Center, University of Washington School of Medicine, Seattle, WA, USA*

FATIH KOCABAS • *Department of Education, and Texas Institute of Biotechnology, Education and Research, North American University, Houston, TX, USA*

WILBUR A. LAM • *Division of Hem/Onc/BMT, Department of Pediatrics, Aflac Cancer and Blood Disorders Center, Emory University School of Medicine, Atlanta, GA, USA*

ZHENRUI LI • *Department of Pathology and Laboratory Medicine, University of Kansas Medical Center, Kansas City, KS, USA*

LINHENG LI • *Investigator, Stowers Institute for Medical Research, Kansas City, MO, USA; Department of Pathology and Laboratory Medicine, University of Kansas Medical Center, Kansas City, KS, USA*

CHARLES P. LIN • *Center for Systems Biology and Wellman Center for Photomedicine, Massachusetts General Hospital and Harvard Medical School, Boston, MA, USA*

JASON A. MILLS • *Department of Pathology and Laboratory Medicine and Center for Cellular and Molecular Therapeutics, Children's Hospital of Philadelphia, Philadelphia, PA, USA*

YOHEI MORITA • *Division of Stem Cell Therapy, Center for Stem Cell Biology and Regenerative Medicine, The Institute of Medical Science, The University of Tokyo, Tokyo, Japan*

HIROMITSU NAKAUCHI • *Division of Stem Cell Therapy, Center for Stem Cell Biology and Regenerative Medicine, The Institute of Medical Science, The University of Tokyo, Tokyo, Japan*

SUSAN K. NILSSON • *CMSE, Commonwealth Scientific and Industrial Research Organization (CSIRO), Clayton, Australia*

PRASUNA PALURU • *Department of Pathology and Laboratory Medicine and Center for Cellular and Molecular Therapeutics, Children's Hospital of Philadelphia, Philadelphia, PA, USA*

CHRISTOPHER Y. PARK • *Human Oncology and Pathogenesis Program, Departments of Pathology and Laboratory Medicine, Memorial Sloan Kettering Cancer Center, New York, NY, USA*

LOUIS M. PELUS • *Department of Microbiology and Immunology and the Walther Oncology Center, Indiana University School of Medicine, and the Walther Cancer Institute, Indianapolis, IN, USA*

CRISTINA PINA • *Stem Cell Laboratory, University College London Cancer Institute, London, UK*

DEBORAH PRITCHETT • *ProteinSimple, Richmond, VA, USA*

CHENG-KUI QU • *Department of Pediatrics, Aflac Cancer and Blood Disorders Center, Emory University School of Medicine, Atlanta, GA, USA*

NALINI RAGHAVACHARI • *Division of Geriatrics and Clinical Gerontology, National Institute on Aging, Bethesda, MD, USA*

JASON ROSS • *Sanford Consortium for Regenerative Medicine, University of California, San Diego, La Jolla, CA, USA*

JUDITH M. RUNNELS • *Center for Systems Biology and Wellman Center for Photomedicine, Massachusetts General Hospital and Harvard Medical School, Boston, MA, USA*

HIMALEE SABNIS • *Division of Hem/Onc/BMT, Department of Pediatrics, Aflac Cancer and Blood Disorders Center, Emory University School of Medicine, Atlanta, GA, USA*

HESHAM A. SADEK • *Department of Internal Medicine, Division of Cardiology, University of Texas Southwestern Medical Center, Dallas, TX, USA*

HEIN SCHEPERS • *Department of Experimental Hematology, University Medical Center Groningen, University of Groningen, Groningen, The Netherlands*

JAN JACOB SCHURINGA • *Department of Experimental Hematology, University Medical Center Groningen, University of Groningen, Groningen, The Netherlands*

LEONARD D. SHULTZ • *The Jackson Laboratory, Bar Harbor, ME, USA*

PALLAVI SONTAKKE • *Department of Experimental Hematology, University Medical Center Groningen, University of Groningen, Groningen, The Netherlands*

EDWARD F. SROUR • *Indiana University School of Medicine, Indianapolis, IN, USA*

RIO SUGIMURA • *Stowers Institute for Medical Research, Kansas City, MO, USA*

TIFFANY A. TATE • *Department of Stem Cell and Regenerative Biology, Harvard University, Cambridge, MA, USA*

JOSÉ TELES • *Stem Cell Laboratory, University College London Cancer Institute, London, UK; Computational Biology and Biological Physics - Department of Astronomy and Theoretical Physics, Lund University, Lund, Sweden*

NIEK P VAN TIL • *Department of Hematology, Erasmus University Medical Center, Rotterdam, The Netherlands*

GRANT D. TROBRIDGE • *Washington State University, Pullman, WA, USA*

EVGENIA VEROVSKAYA • *Laboratory of Ageing Biology and Stem Cells, European Research Institute for the Biology of Ageing, University Medical Centre Groningen, University of Groningen, Groningen, The Netherlands*

GERARD WAGEMAKER • *Hospital Pharmacy, Erasmus University Medical Center, Rotterdam, The Netherlands*

ZHENGQI WANG • *Department of Pediatrics, Aflac Cancer and Blood Disorders Center, Emory University School of Medicine, Atlanta, GA, USA*

MITCHELL J. WEISS • *Department of Pediatrics, The Children's Hospital of Philadelphia, Philadelphia, PA, USA*

BRENDA WILLIAMS • *CMSE, Commonwealth Scientific and Industrial Research Organization (CSIRO), Clayton, Australia*

DAVID A. WILLIAMS • *Boston Children's Hospital and Dana Farber Cancer Institute, Harvard Medical School, Boston, MA, USA*

JUWELL W. WU • *Center for Systems Biology and Wellman Center for Photomedicine, Massachusetts General Hospital and Harvard Medical School, Boston, MA, USA*

RYO YAMAMOTO • *Division of Stem Cell Therapy, Center for Stem Cell Biology and Regenerative Medicine, The Institute of Medical Science, The University of Tokyo, Tokyo, Japan*

CHENGCHENG ZHANG • *Departments of Physiology and Developmental Biology, University of Texas Southwestern Medical Center, Dallas, TX, USA*

MENG ZHAO • *Stowers Institute for Medical Research, Kansas City, MO, USA*

JUNKE ZHENG • *Key Laboratory of Cell Differentiation and Apoptosis of Chinese Ministry of Education, Shanghai Jiao-Tong University School of Medicine, Shangha, China*

Part I

Overview

Chapter 1

The Hematopoietic Stem Cell Landscape

Kevin D. Bunting and Cheng-Kui Qu

Abstract

Hematopoietic stem cells (HSCs) play critical roles in regulating normal blood cell development. Although initially these cells were mysterious and difficult to study in isolation, those obstacles have progressively been rolled away in just a few decades to reveal a heterogeneity of repopulating activity, cell proliferation, and energy metabolism within defined stem cell populations based on drug transporter and cell surface marker expression. A wide range of new technologies have driven innovative discovery of the regulators of HSCs and continued to move the field forward toward a full view of the landscape of single HSCs at the gene and protein levels. It is the goal of this overview chapter to summarize the array of techniques included in the third edition of *Hematopoietic Stem Cell Protocols* which will aid investigators in the field.

Key words Hematopoietic stem cell, Flow cytometry, Embryonic stem cell, Retroviral vector, Hematology

1 Overview

The hematopoietic stem cell (HSC) has been of great interest for many decades due to the proven therapeutic potential. Limitations in the number of HSCs have always been the main drawback to a wider application of HSC-based therapies. Therefore, methods to better enrich these cells and to understand their biology have risen to the forefront of the field in an attempt to increase either the number in vitro or their repopulating capacity in vivo following transplantation. It is the goal of this chapter to briefly summarize the major techniques that are covered in this book. Rapid advances in our understanding of the biology of HSCs are driven by these techniques which extend well beyond the basic handling and manipulation of HSCs and progenitors.

Kevin D. Bunting and Cheng-Kui Qu (eds.), *Hematopoietic Stem Cell Protocols*, Methods in Molecular Biology, vol. 1185, DOI 10.1007/978-1-4939-1133-2_1, © Springer Science+Business Media New York 2014

2   Stem Cell Enrichment and Heterogeneity

There are multiple sources of HSCs for experimental study including fetal liver and adult bone marrow, and it is thus important to be able to accurately separate the HSCs from their microenvironment and isolate them for further characterization. This is especially difficult in embryonic development as the niche(s) are not defined to the degree that they are in adult bone marrow. Cao et al. (Chapter 2) summarize their technique for isolating fetal and newborn HSCs from various tissues. The location from which HSCs are isolated makes a big difference in the preservation of the self-renewal versus differentiation profile. Increased lymphoid lineage priming has been described in fetal liver versus bone marrow HSC using a technique called single-cell PCR. In the chapter by Teles et al. (Chapter 3), the single-cell PCR profiling technique is described in great detail in order to permit the investigator to assess the heterogeneity of the isolated HSC population.

Another measure of the quality of the isolated HSCs is measurement of the functional characteristics. These are more classically defined phenotypes including cell cycle status and survival. Furthermore, the ability to mobilize HSCs back out of the niche is clinically very important for isolation of normal HSCs for transplantation and may also be very important for stimulating dormant leukemic stem cells into the circulation where they are more vulnerable to chemotherapy. The chapter by Hoggatt et al. (Chapter 4) describes the techniques for HSC mobilization in mouse models and offers tips and advice on how to make these experiments more consistent and reproducible. Likewise, the chapters by Chitteti et al. (Chapter 5) and Dong et al. (Chapter 6) provide advanced flow cytometry-based assays for characterizing the response of HSCs to extracellular cues derived from the niche.

3   Molecular and Cellular Characterization of HSCs

Efforts to better understand HSC biology have been catalyzed by the advent of technologies capable of analyzing the whole genome. In the chapter by Raghavachari (Chapter 7) the technique of gene expression profiling is described, with emphasis on the coding region-specific mRNAs. In addition, regulation of gene expression secondary to changes in microRNA (miR) and DNA methylation changes is highlighted in the chapters by Park (Chapter 8) and Begtrup (Chapter 9). Once expressed, the critical biology occurs at the protein level, and thus proteomic based assays are required to discover changes in protein expression or posttranslational modification. Kocabas et al. (Chapter 10) describe how to characterize metabolic changes in HSCs using flow cytometry-based techniques.

The chapter by Bradley et al. (Chapter 11) describes how the NanoPro 1000 instrument can be applied to the study of small numbers of flow cytometry-sorted hematopoietic stem/progenitor cells in normal and leukemic hematopoiesis.

4 In Vitro Assays and Differentiation

Ultimately it is functional analyses that are critical for defining newly identified and isolated stem cell subpopulations. The chapter by Mills et al. (Chapter 12) starts back at the reprogramming stage in the development of induced pluripotent stem cells and teaches how to push these cells toward hematopoietic fates. Most commonly, studies of normal or leukemic stem cell populations start with cells that are already hematopoietic committed. Sontakke et al. (Chapter 13) describe how to study leukemic transformation of HSCs by using patient-derived tissues in long-term expansion culture and also incorporation of genetic modifications in order to develop useful animal models of human leukemia. Another major goal in the field is the ex vivo expansion of normal HSCs for the purpose of transplantation. This goal requires the HSCs to self-renew and maintain their ability to home and engraft into the appropriate niche. Endothelium constitutes one of the key niche components that have been characterized in recent years. Doan et al. (Chapter 14) describe methods to isolate and expand HSCs based on culture with an endothelial cell-derived factor called pleiotrophin. Likewise, Aha et al. (Chapter 15) demonstrate that HSCs can be cultured with endothelial cells using microfluidic engineering technology as a novel method for achieving HSC expansion in vitro.

5 Transplantation Assays and Imaging Engraftment

Transplantation is the gold standard assay for normal and leukemic stem cells. The chapter by Yamamoto et al. (Chapter 16) provides a new method for clonal analysis of transplanted HSCs in recipient mice. The localization of HSCs near bone and osteoblasts was one of the first descriptions of HSC/niche interactions. The chapter by Wu et al. (Chapter 17) provides key methods for in vivo intravital imaging of HSC engraftment in the mouse skull. Furthermore, human HSC engraftment requires an immunodeficient host in order to achieve engraftment. Aryee et al. (Chapter 18) provide the method for efficient immune system development following transplantation of human HSCs into mice. He et al. (Chapter 19) demonstrate methods for studying the homing and lodgement of HSCs using in vivo imaging of murine HSCs expressing a Scl-tTA:H2B-GFP reporter.

6   Genetic Modification

The final series of chapters in this book highlight the next generation of methods for genetic modifications of HSCs for further biology studies and also for therapeutic applications in gene therapy. Ciuculescu et al. (Chapter 20) highlight the key methods for introducing retroviral vectors into mouse and human HSCs. Van Til et al. (Chapter 21) provide a similar focus but rather using the lentiviral vector system which is more complicated but has a number of advantages. Beard et al. (Chapter 22) describe high-throughput genomic mapping of vector integration sites in gene therapy studies. Finally, barcoded vector libraries using retroviral and lentiviral systems are described by Bystrykh et al. (Chapter 23) as a novel tracking system for studies of hematopoiesis.

7   Conclusions

Altogether the techniques described in this book enable a full characterization of the molecular circuitry controlling HSCs. Continued understanding of the biology of HSCs is likely to lead to more rapid advances in therapeutics for a wide range of benign genetic disorders of hematology and for treatment of hematologic malignancies.

Acknowledgments

The authors thank the members of the Qu and Bunting labs for their helpful discussions during the preparation of this summary of the third edition of Hematopoietic Stem Cell Protocols.

Part II

Stem Cell Enrichment and Heterogeneity

Chapter 2

Investigating the Interaction Between Hematopoietic Stem Cells and Their Niche During Embryonic Development: Optimizing the Isolation of Fetal and Newborn Stem Cells From Liver, Spleen, and Bone Marrow

Huimin Cao, Brenda Williams, and Susan K. Nilsson

Abstract

Hematopoietic stem cells (HSCs) are maintained in a particular microenvironment termed a "niche." Within the niche, a number of critical molecules and supportive cell types have been identified to play key roles in modulating adult HSC quiescence, proliferation, differentiation, and reconstitution. However, unlike in the adult bone marrow (BM), the components of stem cell niches, as well as their interactions with fetal HSC during different stages of embryonic development, are poorly understood. During embryogenesis, hematopoietic development migrates through multiple organs, each with different cellular and molecular components and hence each with a potentially unique HSC niche. As a consequence, isolating fetal HSC from each organ at the time of hematopoietic colonization is fundamental for assessing and understanding both HSC function and their interactions with specific microenvironments. Herein, we describe methodologies for harvesting cells as well as the purification of stem and progenitors from fetal and newborn liver, spleen, and BM at various developmental stages following the expansion of hematopoiesis in the fetal liver at E14.5.

Key words Hematopoietic stem cells, Hematopoietic development, Liver, Spleen, Bone marrow

1 Introduction

In the past 30 years since the term "niche" was coined by Schofield [1] for the adult BM microenvironment in which hematopoietic stem cells (HSCs) reside, adult HSCs and their niche have been extensively investigated. To date, this has resulted in a huge body of literature identifying a number of cell types and extracellular matrix molecules as components of the niche that play a role in HSC regulation.

Hematopoiesis develops embryonically in an age-dependent and microenvironment-controlled process, with fetal HSC

Kevin D. Bunting and Cheng-Kui Qu (eds.), *Hematopoietic Stem Cell Protocols*, Methods in Molecular Biology, vol. 1185, DOI 10.1007/978-1-4939-1133-2_2, © Springer Science+Business Media New York 2014

migrating through multiple sites. Initially, in the mouse, primitive hematopoiesis starts in the yolk sac at E7.5 [2] and gives rise to erythrocytes. Hematopoiesis then becomes definitive, demonstrating multi-lineage reconstitution capacity, with evidence suggesting a contribution from the aorta-gonad-mesonephros (AGM) region, yolk sac, and placenta [3–5]. HSCs are first detected in the fetal liver at E11.5 [6], where they expand at E14.5 [7] prior to migrating to the BM at E16.5, the permanent location for hematopoiesis throughout adulthood [6]. In addition, the spleen is a temporary fetal hematopoietic organ from E12 shortly after birth [8]. Besides inherent intrinsic properties, fetal hematopoiesis receives extrinsic cues whilst residing within specific microenvironments [9]. However, unlike adult BM, the interaction between fetal HSCs and their niche during hematopoietic development has not been thoroughly investigated and hence remains poorly understood.

Herein, we describe methodologies for harvesting cells as well as the purification of stem and progenitors from fetal and newborn liver, spleen, and BM at various developmental stages following the expansion of hematopoiesis in the fetal liver at E14.5. These methodologies have been utilized to identify key interactions between HSC and specific developmental niches and assess their roles in HSC regulation.

2   Materials

2.1   Isolation of Liver, Spleen, and BM

1. Timed mated embryos or newborn pups.
2. Sterile #11 surgical blade and #3 handle.
3. Sterile straight surgical scissors (16 cm Kelly).
4. 30½G needle attached to a 1 ml syringe.
5. Sterile micro-dissecting knife (12 cm knife).
6. Sterile micro-dissecting scissors (100 mm).
7. Sterile fine tweezers.
8. Sterile beveled forceps.
9. Phosphate-buffered saline (PBS): pH 7.2, 310 mOsm (*see* **Note 1**) supplemented with 2 % serum.
10. Sterile 100 mm tissue culture petri dish.
11. Sterile 35 mm tissue culture petri dish.
12. Dissecting microscope.

2.2   Disaggregation of Liver, Spleen, and BM

1. PBS (pH 7.2), 310 mOsm.
2. PBS (pH 7.2), 310 mOsm supplemented with 2 % serum.
3. 50 ml Conical polypropylene centrifuge tube.
4. 3 ml Syringe plunger.

5. 40 μm Nylon cell strainer.

6. Sterile straight surgical scissors.

7. 3 mg/ml Collagenase I (we use *Clostridium histolyticum*) in PBS made fresh on the day of use.

8. 3 mg/ml Collagenase I/4 mg/ml Dispase II (*Bacillus polymyxa*, neutral protease) in PBS made fresh on the day of use.

9. 37 °C Orbital shaker (Eppendorf Thermomixer comfort model #5355 000.011).

10. 18- and 21-gauge needles attached to 1 ml syringes.

11. Sterile mortar and pestle.

12. Hemocytometer and microscope equipped with phase-contrast or an automated cell counter (Sysmex model KX-21N).

2.3 Density Gradient Separation

1. 300 ml Density media: Nycoprep Universal: 60 % (w/v) in water, 1.310 g/cm³, 580 mOsm mixed with 300 ml 20 mM tricine-NaOH (pH 7.2) and 676.6 ml of 0.65 % NaCl. Confirm that osmolarity is 265 mOsm and density is 1.077 g/ml at room temperature. 20 mM Tricine-NaOH is made by diluting 3.584 g of tricine in 900 ml milliQ water and adjusting the volume to 1 l and pH 7.2. 0.65 % of NaCl is made by adding 6.5 g of NaCl to 900 ml milliQ water and adjusting the volume to 1 l. Both the 20 mM tricine-NaOH and the 0.65 % NaCl should be sterile filtered before use.

2. Cannulas attached to 10 or 20 ml syringes.

2.4 Lineage Depletion

1. Lineage depletion antibody cocktail: A mixture of optimally titrated purified rat anti-mouse antibodies recognizing the cell surface antigens: B220, Gr-1, and Ter119 (*see* **Note 2**) in PBS (pH 7.2), 310 mOsm supplemented with 2 % serum (antibody concentrations are all ≤1 μg/ml) (*see* **Note 3**).

2. Polypropylene 5 ml round-bottom tubes.

3. Magnetic Dynal beads working buffer: PBS, 310 mOsm supplemented with 2 mM EDTA, and 0.1 % (w/v) fraction V bovine serum albumen (BSA; pH 7.4).

4. 4.5 μm Diameter sheep anti-rat IgG Dynabeads (Dynal Biotech ASA, Oslo, Norway).

5. Microtubes.

6. Dynal MPC-S magnet for 2–20 ml samples or MPC-L for 1–8 ml samples.

7. Tube rotator or similar suspension mixer, allowing both tilting and rotation at 4 °C (we use a MACSmix Tube Rotator placed in a fridge).

2.5 Hematopoietic Stem Cell Fluorescence-Activated Cell Sorting

1. HSC antibody cocktail: A mixture of optimally titrated allophycocyanin-cyanine 7 (APCCy7)-conjugated rat anti-mouse B220, Gr-1, CD3, and Ter119 (*see* **Note 4**); Pacific Blue™ (PB)-conjugated rat anti-mouse stem cell antigen 1 (Sca-1); AlexaFluor®647 (AF647)-conjugated rat anti-mouse c-Kit; fluorescein isothiocyanate (FITC)-conjugated rat anti-mouse CD48; and phycoerythrin (PE)-conjugated rat anti-mouse CD150 in PBS (pH 7.2) and 310 mOsm supplemented with 2 % serum (antibody concentrations are all ≤1 µg/ml).

2. Sterile polypropylene 5 ml round-bottom tubes.

3. Polystyrene, round-bottom tubes: 5 ml with 40 µm cell strainer cap. The cap is swapped to a 5 ml polypropylene tube for filtering cells.

4. Flow cytometer with sorting capability equipped with five solid-state lasers (355, 405, 488, 561, and 635 nm). Band-pass filter settings for the detection of fluorescence for FITC, PE, AF647, PB, and APCCy7 are 528 ± 19, 605 ± 20, 660 ± 10, 460 ± 25, and 780 ± 30, respectively. We use a 70 mm nozzle, sort at 30 psi, and drop delay frequency of 61 kHz for HSC sorting.

3 Methods

3.1 Timed Mating

1. House five sexually mature female mice in one cage (*see* **Note 5**) with mouse chow and acidified water ad libitum.

2. Tease females with bedding from the male for 3 days prior to mating (*see* **Note 6**).

3. Time mate by placing the females into the male's cage late in the afternoon for 12 h (*see* **Note 7**).

4. Separate the mice, and check each female for the presence of a vaginal plug. This is designated as 0.5 days (E0.5).

5. Confirm the pregnancy between E12.5 and E14.5.

6. Check pups' birth twice a day from E18.5. The day when pups are born is designated as day 0 (D0) and then sequentially as D1, D2, D3, etc.

3.2 Mice Harvesting

3.2.1 Embryos

1. Euthanize pregnant mouse by cervical dislocation. Remove uterus by opening the abdominal cavity and cutting away the connective tissue.

2. Isolate single embryos from the uterus by gently cutting and dissociating the outer muscular uterine layer and yolk sac. Remove the placenta, and euthanize each pup by decapitation (individual institutional animal ethics requirements may vary).

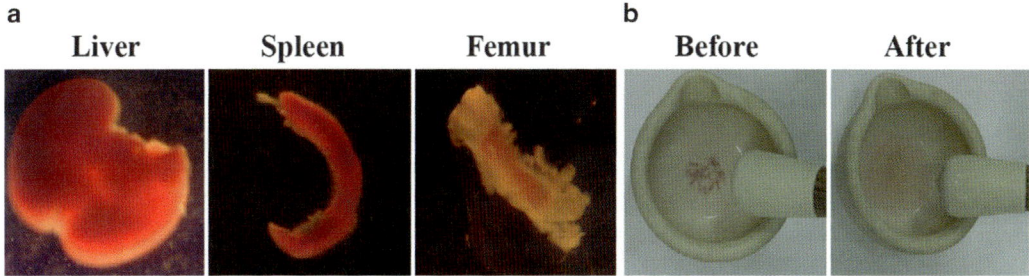

Fig. 1 Fetal organs and bone harvest. (**a**) Images show fetal liver, spleen, and femur. (**b**) Comparison of bones before and after grinding with a mortar and pestle

3.2.2 Newborn Pups

1. Euthanize pups by decapitation with a sharp pair of scissors.

3.3 Isolation of Liver, Spleen, and Long bones

3.3.1 Liver and Spleen

1. In a 100 mm petri dish separate the abdominal and hind portions using a surgical blade.

2. Using 30½G needles attached to 1 ml syringes (for embryos) or fine tweezers (for newborn pups) remove other organs from the posterior end of abdominal portion to expose the liver and spleen.

3. Under a dissecting microscope, use a pair of 30½G needles to remove any extra tissue attached to the liver and spleen (*see* **Note 8**) and place in a 35 mm petri dish containing PBS–2 % serum (Fig. 1a).

3.3.2 Long Bones

1. Lay the hind portion belly down in a 100 mm petri dish, and remove the skin with scissors to expose the transparently red long bones (femur and tibia).

2. Under a dissecting microscope, carefully remove muscles from the bones using a micro knife and/or a pair of micro scissors.

3. Dislocate the femur from the hip and knee, and cut the foot off the tibia at the ankle. Keep bones in individual groups (*see* **Note 8**) in separate 35 mm petri dishes containing PBS–2 % serum (*see* **Notes 9** and **10**) (Fig. 1a).

3.4 Single-Cell Suspensions of Liver, Spleen, and BM

3.4.1 Embryonic Livers

1. Gently push embryonic livers through a 40 μm strainer on top of a 50 ml centrifuge tube with the plunger of a 3 ml syringe.

2. Wash cells in PBS–2 % serum by centrifuging at $400 \times g$ for 5 min at 4 °C.

3. Decant supernatant, and resuspend the cell pellet in 10 ml PBS–2 % serum.

4. Refilter the cell suspension through a 40 μm strainer into a new 50 ml conical tube.

5. Perform a cell count.

3.4.2 Newborn Livers

1. Transfer newborn livers into a 50 ml conical tube and chop finely with a pair of long surgical scissors.

2. Add 1 ml Collagenase I for each minced liver, and agitate for 5 min at 37 °C in an orbital shaker, 750 rpm.

3. Repeatedly flush livers through an 18-gauge needle and then a 21-gauge needle until a single-cell suspension.

4. Add 40 ml PBS–2 % serum, and filter through a 40 µm strainer into a new 50 ml conical tube.

5. Wash cells twice in PBS–2 % serum by centrifuging cells at $400 \times g$ for 5 min at 4 °C.

6. Decant supernatant, and resuspend the cell pellet in 10 ml PBS–2 % serum.

7. Perform a cell count.

3.4.3 Spleens

Process embryonic and newborn spleens as for embryonic livers (*see* Subheading 3.4.1, **step 1**), except resuspend the cells in 5 ml PBS–2 % serum.

3.4.4 BM

1. Transfer bones into a sterile mortar.

2. Grind bones with a pestle (*see* **Note 11**).

3. Remove the supernatant and filter through a 40 µm strainer into a 50 ml conical tube.

4. Further grind bones if not dissociated into small fragments (Fig. 1b). Rinse the crushed bone fragments with PBS–2 % serum, and filter the supernatant as in **step 3**, until all bone fragments become white. Top up the 50 ml tube to 50 ml with PBS–2 % serum and set aside on ice until **step 9**.

5. Transfer the crushed bone fragments into a new 50 ml conical tube containing Collagenase I and Dispase II (1 ml per 12–18 bones) and place at 37 °C in an orbital shaker, 750 rpm, for 5 min (*see* **Note 12**).

6. Add 20 ml straight PBS to the digested bone fragments, and shake vigorously for 20 s.

7. Filter the cell suspension through a 40 µm strainer into a new 50 ml conical tube.

8. Repeat **steps 6** and **7**, and filter cells into the same 50 ml conical tube. Top the tube with 10 ml PBS–2 % serum.

9. Centrifuge the cell suspension tubes (from **steps 4** and **8**) at $400 \times g$ for 5 min at 4 °C.

10. Decant supernatant, resuspend pellets, pool cells in 10 ml PBS–2 % serum, and perform a cell count.

3.5 Density Separation for BM, Liver, and Spleen

1. Top up each cell suspension to 20 ml with PBS–2 % serum (*see* **Note 13**).

2. Underlay 10 ml Nycoprep using a cannula attached to a 10 or a 20 ml syringe.

3. Centrifuge the gradients at $600 \times g$ for 20 min at room temperature with no deceleration.

4. Using a 10 ml pipette, collect mononuclear cells from the interface between the PBS–2 % serum and the Nycoprep solution and place in a new 50 ml conical tube.

5. Fill the tube with PBS–2 % serum and centrifuge at $400 \times g$ for 5 min at 4 °C.

6. Decant supernatant, resuspend cells in 10 ml PBS–2 % serum, and perform a cell count.

3.6 Lineage Depletion

1. Pellet cells by centrifuging at $400 \times g$ for 5 min at 4 °C.

2. Stain cells at 1×10^7 cells/ml in the cocktail of lineage markers on ice for 20 min.

3. Wash cells with PBS–2 % serum by centrifuging at $400 \times g$ for 5 min at 4 °C to remove unbound antibodies.

4. Resuspend cells in 2 ml PBS supplemented with 2 mM EDTA and 0.1 % BSA and transfer into 5 ml sterile polypropylene tube. Perform a cell count (*see* **Note 14**) and set aside on ice until **step 10**.

5. Resuspend Dynabeads.

6. Place the required volume of Dynabead suspension, to give half a bead per cell, into two individual 1.7 ml microtubes. The optimal number of Dynabeads per cell has previously been established as half a bead per cell with the depletion repeated with a second half a bead per cell [10].

7. Remove the azide from the Dynabeads by adding 1 ml of 2 mM EDTA with 0.1 % BSA into each aliquot and mixing well. Place the tubes in the magnet for 1 min prior to decanting the supernatant and move from the magnet.

8. Repeat **step 7**.

9. Resuspend the Dynabeads in 500 μl 2 mM EDTA with 0.1 % BSA.

10. Add the first aliquot of washed Dynabeads to the cells, and mix well.

11. Incubate the mixture with gentle tilting and rotation for 5 min at 4 °C.

12. Place the cells in the magnet for 2 min.

13. Transfer the supernatant containing the unbound cells to a new 5 ml sterile polypropylene tube.

14. Rinse the rosetted bead–cell complexes with 1 ml 2 mM EDTA with 0.1 % BSA and place in the magnet for 1 min prior to collecting any residual unbound cells to the collection tube used in **step 13**.

15. Add the second aliquot of washed Dynabeads into the unbounded cell suspension.

16. Incubate the mixture with gentle tilting and rotation for 10 min at 4 °C.

17. Place the cells in the magnet for 2 min.

18. Transfer the supernatant containing the unbound cells to a new 5 ml sterile polypropylene tube.

19. Rinse the rosetted bead–cell complexes with 1 ml 2 mM EDTA with 0.1 % BSA and place in the magnet for 1 min prior to collecting any residual unbound cells to the collection tube used in **step 18**.

20. Measure the total volume of the unbound lineage-negative cell suspension, and perform a cell count.

3.7 HSC FACS

1. Pellet cells by centrifuging at $400 \times g$ for 5 min at 4 °C.

2. Stain cells at 1×10^7 cells/ml in the HSC antibody cocktail on ice for 20 min.

3. Add 3 ml PBS 2 % serum, and filter the cell suspension through a cell strainer into a new 5 ml sterile polypropylene tube (*see* **Note 15**). Centrifuge cells at $400 \times g$ for 5 min at 4 °C to remove unbound antibodies.

4. Resuspend cells at $25–30 \times 10^6$ cells/ml in PBS–2 % serum (*see* **Note 16**) and place on ice until sorted.

5. To set up HSC sorting by flow cytometry, the following samples are required for each tissue being sorted (*see* **Note 17**).

    (a) $0.5–1 \times 10^6$ Unstained cells to set voltage for forward scatter, side scatter, APCCy7, PB, AF647, PE, and FITC.

    (b) Individual tubes containing $0.5–1 \times 10^6$ cells stained with APCCy7, PB AF647, PE, and FITC for compensation controls.

    (c) $0.5–1 \times 10^6$ Adult BM cells stained with HSC antibody cocktail as a positive control.

6. Run the cell samples stained with HSC antibody cocktail and sequentially gate through FSC-H versus FSC-A, SSC-A versus FSC-A, SSC-A versus APCCy7, AF647 versus PB, and FITC versus PE (Fig. 2). Fetal HSCs phenotypically are defined as lineage$^+$Sca-1$^+$c-Kit$^+$CD150$^+$CD48$^-$.

7. Sort cells at predetermined optimal input speed and collect into culture medium or PBS–2 % serum depending on the functional assay requirement.

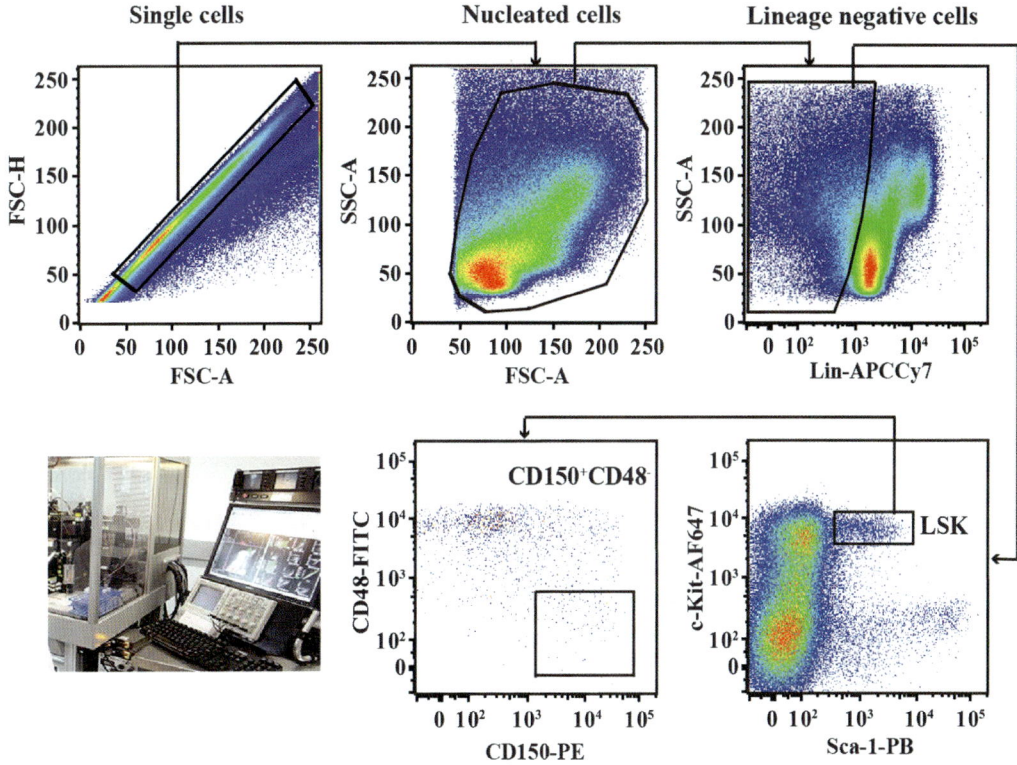

Fig. 2 HSC gating strategies for flow cytometry. HSCs are sequentially gated through single-cell gate, nucleated cell gate, and lineage-negative cell gate and then selected as Sca-1$^+$c-Kit$^+$CD150$^+$CD48$^-$

3.8 Summarized Optimizing Isolation of Fetal and Newborn Liver, Spleen, and BM at a Variety of Developmental Ages

Since fetal hematopoiesis is a developing process, maturation, expansion, and migration occur at different time points and in different organs. Hence, the strategies for isolating fetal HSC are not always the same. The optimized time points and methodologies for tissue harvesting, single-cell preparation, and HSC enrichment are summarized in Table 1.

4 Notes

1. This osmolarity is appropriate for murine cells and results in better cell recovery.

2. Mac-1 is excluded from the lineage depletion antibody cocktail due to its previously described presence on fetal and newborn HSC.

3. In order to obtain accurate and high-quality fluorescence-activated cell sorting (FACS) profiles, proper titration is required for all antibodies. The optimized working concentration allows the optimal separation of positive cells from negative

Table 1
Optimized methodologies for fetal and newborn HSC isolation

Organ		E14.5–16.5	E17.5	E18.5	D0–D4	D5–D8
Liver	Observation time points	HSC isolation				–
	Methods summary	Mash with plunger through a 40 µm strainer Density separation (*see* **Note 18**)			Mince with scissors Digest with Collagenase I Flush with 18G and 21G needles Filter through a 40 µm strainer	–
Spleen	Observation time points	–		HSC isolation		
	Methods summary	–		Mash with plunger through a 40 µm strainer Density separation Lineage depletion		
BM	Observation time points	–		HSC isolation		
	Methods summary	–		Grind with a mortar and pestle Digest with Collagenase I and Dispase II Density separation Lineage depletion		

cells without causing any shift for isotype control from unstained.

4. B220, CD3, Gr-1, and Ter119-APCCy7 are included to exclude residual lineage-committed cell contamination.

5. Co-housing females results in individual estrous cycles being synchronized.

6. The murine estrous cycle is 4–5 days, with ovulation on the third day. Therefore teasing females 3 days prior to mating will produce the maximum number of pregnancies.

7. Only using female mice that are confirmed to be actively in estrous can increase the success of the timed mating.

8. Due to the size of embryos and newborn pups, a number of organs need to be pooled to provide sufficient cells. Table 2 shows the number of organs required for harvesting 50×10^6 single cells from different organs at a variety of ages.

9. It is very difficult to clean embryonic and newborn bones as they are extremely soft and even gentle squeezing will result in the loss of marrow. As a consequence, holding the muscle around the bone whilst cleaning works best.

Table 2
The number of organs required for harvesting 50×10^6 cells

Mouse age	No. of livers	No. of femurs	No. of spleens
E14.5	6	–	–
E15.5	2	–	–
E16.5	2	–	–
E17.5	2	–	–
D0	3	220[a]	40
D1	5	160[a]	30
D2	5	120[a]	20
D3	4	80[a]	15
D4	4	60	7
D5	–	30	4
D6	–	30	3
D7	–	20	3
D8	–	15	2

[a]Calculated based on observed cellularity

10. As the cellularity of both the femur and tibia is low, the humerus is also often collected. We have experimentally shown that there is no significant difference in the cellularity or the HSC frequency between a femur, tibia, and humerus at E18.5 and D1–8 bones.

11. The processing of fetal and newborn BM is different from adult BM [11] in that due to their tiny size they are not separated into central and endosteal fractions.

12. 5 min is experimentally determined for ideally digesting bones, since after 5 min there is a progressive loss of c-Kit and Sca-1 [12], which are important surface markers for isolating HSC.

13. The maximum number of cells per tube for a density separation is 2×10^8.

14. The maximum number of cells per tube for a Dynal bead separation is 3×10^8.

15. As these cells tend to clump easily additional refiltering may be required immediately prior to sorting.

16. $25-30 \times 10^6$ Cells/ml is the ideal concentration for obtaining optimal yields and purity when sorting cells on our Influx1 cell sorter.

17. For convenience and saving our enriched cell samples, we use un-fractionated single cells for the compensation controls. Furthermore, CD45-conjugated antibodies are used for compensation as CD45 is highly expressed on un-fractionated cells.

18. Fetal liver HSCs are not enriched by lineage depletion, as E14.5 fetal liver cells do not express lineage antigens at high levels. Such a population only begins to appear after E16.5, but still comprises a very low proportion.

References

1. Schofield R (1978) The relationship between the spleen colony-forming cell and the haemopoietic stem cell. Blood Cells 4:7–25

2. Orkin SH, Zon LI (2008) Hematopoiesis: an evolving paradigm for stem cell biology. Cell 132:631–644

3. Kumaravelu P, Hook L, Morrison AM et al (2002) Quantitative developmental anatomy of definitive haematopoietic stem cells/long-term repopulating units (HSC/RUs): role of the aorta-gonad-mesonephros (AGM) region and the yolk sac in colonisation of the mouse embryonic liver. Development 129: 4891–4899

4. Gekas C, Dieterlen-Lievre F, Orkin SH, Mikkola HK (2005) The placenta is a niche for hematopoietic stem cells. Dev Cell 8:365–375

5. Ottersbach K, Dzierzak E (2005) The murine placenta contains hematopoietic stem cells within the vascular labyrinth region. Dev Cell 8:377–387

6. Christensen JL, Wright DE, Wagers AJ, Weissman IL (2004) Circulation and chemotaxis of fetal hematopoietic stem cells. PLoS Biol 2:E75

7. Lessard J, Faubert A, Sauvageau G (2004) Genetic programs regulating HSC specification, maintenance and expansion. Oncogene 23: 7199–7209

8. Wolber FM, Leonard E, Michael S, Orschell-Traycoff CM, Yoder MC, Srour EF (2002) Roles of spleen and liver in development of the murine hematopoietic system. Exp Hematol 30:1010–1019

9. Kiel MJ, Iwashita T, Yilmaz OH, Morrison SJ (2005) Spatial differences in hematopoiesis but not in stem cells indicate a lack of regional patterning in definitive hematopoietic stem cells. Dev Biol 283:29–39

10. Williams B, Nilsson SK (2009) Investigating the interactions between haemopoietic stem cells and their niche: methods for the analysis of stem cell homing and distribution within the marrow following transplantation. Methods Mol Biol 482:93–107

11. Grassinger J, Haylock DN, Williams B, Olsen GH, Nilsson SK (2010) Phenotypically identical hemopoietic stem cells isolated from different regions of bone marrow have different biologic potential. Blood 116:3185–3196

12. Haylock DN, Williams B, Johnston HM et al (2007) Hemopoietic stem cells with higher hemopoietic potential reside at the bone marrow endosteum. Stem Cells 25:1062–1069

Chapter 3

Single-Cell PCR Profiling of Gene Expression in Hematopoiesis

José Teles, Tariq Enver, and Cristina Pina

Abstract

Single-cell analysis of gene expression offers the possibility of exploring cellular and molecular heterogeneity in stem and developmental cell systems, including cancer, to infer routes of cellular specification and their respective gene regulatory modules. PCR-based technologies, although limited to the analysis of a predefined set of genes, afford a cost-effective balance of throughput and biological information and have become a method of choice in stem cell laboratories. Here we describe an experimental and analytical protocol based on the Fluidigm microfluidics platform for the simultaneous expression analysis of 48 or 96 genes in multiples of 48 or 96 cells. We detail wet laboratory procedures and describe clustering, principal component analysis, correlation, and classification tools for the inference of cellular pathways and gene networks.

Key words Single-cell quantitative RT-PCR, Microfluidics, Hierarchical clustering, Principal component analysis, Machine learning, Random forests, Logistic regression, Correlation-based gene networks

1 Introduction

Understanding the gene expression programs that regulate distinct cell states and the fate transitions between them is central to the successful delivery of regenerative medicine. An important part of this challenge lies in the prospective isolation of pure cell compartments [1–3] to reveal the transcriptional signatures of rare populations [4]. However, it is also increasingly clear that fluctuations in the molecular composition of individual cells, i.e., molecular 'noise', can contribute to cell potential and fate [5–9], and exploration of transcriptional heterogeneity requires single-cell methods of gene expression analysis.

Early attempts of single-cell transcriptional profiling relied on end-point multiplex RT-PCR using oligo-dT-based reverse transcription [10] or target-specific reverse transcription and amplification with sets of nested primers [11] for a small number of genes. While limited to qualitative analysis of transcriptional

Kevin D. Bunting and Cheng-Kui Qu (eds.), *Hematopoietic Stem Cell Protocols*, Methods in Molecular Biology, vol. 1185, DOI 10.1007/978-1-4939-1133-2_3, © Springer Science+Business Media New York 2014

composition, these approaches were instrumental in advancing the notion of multi-lineage primed states in hematopoietic stem and early progenitor cells [11, 12] as well as in supporting alternative hierarchies of hematopoietic lineage specification [1, 3]. However, quantitative approaches are required to capture the full spectrum of fluctuations in gene expression level and, significantly, to begin to reveal regulatory networks through pairwise correlation analysis of gene expression.

An early attempt at quantitative analysis of gene expression in defined hematopoietic populations, namely relatively pure hematopoietic stem cells, used global amplification and hybridization to microarrays [13]. While valuable in demonstrating the ability to capture biological variation of gene expression above technical variation, the analysis was limited by the small number of cells surveyed and did not afford a clear separation between cellular and molecular heterogeneity. Indeed, the cost associated with methodologies of global gene expression profiling limits the number of cells analyzed and compromises full appreciation of the data.

Quantitative RT-PCR is a more cost-effective alternative, particularly with the development of microfluidic devices that minimize the volume of reagents used and maximize the numbers of cells and genes inspected per run. While it does not allow transcriptome-wide analysis of gene expression, and the conclusions are necessarily biased by the identities of the genes studied, it is possible to analyze tens to hundreds of genes deemed relevant on the basis of prior knowledge or preliminary experiments to arrive at biologically meaningful conclusions. Identification of new cellular compartments [4, 14], inference of molecular mechanisms of lineage decisions [8, 15], and inference of small regulatory networks [14–16] are recent examples of applications in the hematopoietic field.

Several platforms—Fluidigm Dynamic Arrays, TaqMan Array Cards, and NanoString nCounter—are currently available for single-cell expression analysis, and they all rely on an initial step of target-specific amplification. Transcript detection and quantification are based on qPCR technology in the case of Fluidigm Dynamic Arrays and TaqMan Array Card methods and on hybridization of barcoded probes in the case of NanoString nCounter. In the absence of pre-amplification, nCounter technology allows direct quantification of the number of copies for each transcript, but this information is impacted by the efficiency of the pre-amplification reactions in single-cell applications. Application of the technology to single cells is very novel [14], and it has recently been used in hematopoiesis as confirmatory but not as an exploratory tool.

Fluidigm and Array Card technologies use quantification by PCR, which allows for relative quantification of transcripts. Absolute amounts of RNA and copy numbers can in principle be estimated against pre-quantified standards; however, in the case of single-cell analysis these can only be indicative rather than definitive, as they are not extracted through in-well cell lysis and have not been routinely used. Fluidigm Dynamic Arrays allow for the best high-throughput analysis (96 cells × 96 genes vs. a maximum of 8 cells in an Array Card) and constitute the most widely used method. As such, the experimental protocols in this chapter refer exclusively to the Fluidigm platform. The analytical tools are applicable to other methodologies.

2   Materials

2.1   Reagents

1. DEPC-treated or RNase-free water: Aliquot and store for a long term at –20 °C.

2. IGEPAL CA-630 (SIGMA): Prepare 10 % solution in DEPC-treated or RNase-free water; aliquot and store long-term at –20 °C.

3. RNase inhibitor (e.g., RNase OUT Recombinant Ribonuclease Inhibitor, Invitrogen).

4. CellsDirect One-Step RT-PCR kit (without ROX) (Invitrogen), *one-step protocol only.*

5. Platinum Taq polymerase (Invitrogen), *one-step protocol only.*

6. Superscript ViLO cDNA synthesis kit (Invitrogen), *two-step protocol only.*

7. T4 gene 32 protein, *two-step protocol only.*

8. TaqMan PreAmp Master Mix, *two-step protocol only.*

9. Exonuclease I, *two-step protocol only.*

10. TaqMan Universal PCR Master Mix.

11. 2× Assay Loading Reagent (Fluidigm).

12. 20× Sample Loading Reagent (Fluidigm).

13. 20× TaqMan assays for genes of interest—up to 96 (Applied Biosystems).

2.2   Microfluidics Chips and Plasticware

1. 0.2 ml Non-skirted 96-well PCR plates.

2. Optical PCR adhesive films.

3. Disposable 96-well plate plastic lids.

4. 1.5 ml Microfuge clear tubes, RNase and DNase free.

5. Light-touch ergonomic micropipettes (a good choice is the RAININ LTS system, including an 8-channel 1–10 μl micropipette) and corresponding tips.

6. 48.48 or 96.96 Dynamic Arrays for Gene Expression (Fluidigm): Note that a 192.24 Dynamic Array format (192

samples×24 assays) is also available, but requires specific instrumentation and is not discussed in this chapter.

7. Control Line Fluid (two syringes with 0.3 ml—48.48, or 0.15 ml of fluid—96.96 format (Fluidigm)).

8. Clear ("invisible") adhesive tape.

2.3 Instrumentation

1. Refrigerated benchtop centrifuge with plate adaptors.

2. Thermal cycler, preferably with multiple block capacity.

3. Integrated Fluid Circuit (IFC) Controller MX (for 48.48 arrays) or HX (for 96.96 arrays) (Fluidigm).

4. Biomark Reader (Fluidigm).

2.4 Software

1. Genesis   [17]   (http://genome.tugraz.at/genesisclient/genesisclient_download.shtml).

2. Matlab and Matlab Statistical toolbox (MathWorks).

3. R software for statistical computing (http://www.r-project.org/).

4. Rattle graphical user interface for R [18] (http://rattle.togaware.com/).

3   Experimental Procedures

Carefully read **Notes 1–6** at the end of this chapter, as they highlight simple procedures that should be adhered to in order to minimize amplicon contamination during pre-amplification and quantitative PCR setup.

3.1 Single-Cell Collection

3.1.1 Cell Sorting

Although the technicalities of fluorescence-automated cell sorting do not fall within the scope of this chapter, most experiments will require prospective isolation of defined hematopoietic cell compartments and single-cell deposition into lysis buffer or alternatively directly into the one-step RT/pre-amplification reaction mix (as detailed in Subheading 3.2.2). Given the low frequency of highly purified stem and progenitor hematopoietic compartments as well as the possibility of multiple-way sorting into tubes, but not plates, we recommend that a maximum number of cell types be bulk-sorted into tubes containing medium with serum or serum substitute, with the individual populations subsequently single-cell deposited onto 96-well PCR plates containing the appropriate buffer.

Each plate should include at least two wells into which no cells are deposited as well as three or more serial multiple-cell controls in duplicate wells for internal validation of the single-cell results. It is recommended that a maximum of 50 cells are used for any multiple-cell control, as the PCR can saturate above this cell number for the most abundant genes.

Prior to cell deposition, it is crucial to verify the alignment of the plates with the automatic cell deposition unit at two opposite corners and to ensure that the stream is directed at the center of the wells by test sorting 25–50 cells at different plate positions while keeping the plate covered with a clean lid. Sorters should be set on single-cell mode to prevent doublet deposition. In our experience, this sorting mode will use a five- to tenfold excess of cells, and a deposition efficiency of over 85 % is to be expected.

Plates should be covered with optical adhesive film and vortexed at maximum speed for 15–30 s to ensure efficient cell lysis and can be kept on ice for short periods (e.g., while depositing additional cells). The plates should then be centrifuged in a precooled centrifuge for 1 min at 2,000 ×g and either frozen at –80 °C for later use or processed immediately.

3.1.2 Cell Lysis

Although it is possible to sort cells directly into the reverse-transcription reaction buffer, we recommend that cells be initially sorted into a detergent-containing lysis buffer to maximize RNA extraction. This procedure also adds flexibility to work organization, as lysis plates can be frozen long-term at –80 °C prior to processing.

The lysis buffer can be prepared up to 24 h prior to cell deposition and stored at –20 °C. If prepared on the day of the experiment, lysis buffer-containing plates can be kept at 4 °C or on ice until use.

The reagents required for a 96-well plate (including a 10 % reagent excess) are indicated below:

RNase OUT 40 U/ml	5.3 μl
IGEPAL/NP40 10 %	21.1 μl
DEPC-treated or RNase-free H$_2$O	396.0 μl
Total	422.4 μl—USE 4 μl per well

When using a two-step reverse transcription and pre-amplification protocol (*see* Subheading 3.4) use this modified version of the buffer in order to keep reaction volumes as low as possible.

RNase OUT 40 U/ml	5.3 μl
IGEPAL/NP40 10 %	21.1 μl
Superscript VILO RT 5× reaction mix	105.6 μl
DEPC-treated or RNase-free H$_2$O	290.4 μl
Total	422.4 μl—USE 4 μl per well

The buffer mix should be prepared in a PCR hood, with reagents maintained on ice during preparation; it is possible to keep the tubes or the plates at room temperature for short periods for convenience. After aliquoting the lysis buffer, plates should be covered and centrifuged at $2,000 \times g$ for 1 min in a centrifuge pre-cooled to 4 °C.

3.2 Multiplex Reverse Transcription and Pre-amplification

One-step target-specific reverse transcription (RT) and pre-amplification are appropriate for murine primary hematopoietic cells as well as hematopoietic cell lines of human and mouse origin. It has also been used successfully in the analysis of non-hematopoietic cell types, including, among others, early stages of mouse embryonic development, mouse embryonic stem cells, and induced pluripotent (iPS) cells and normal and cancerous human epithelia [9, 19–21]. However, we and others have not been able to apply this method to single human primary hematopoietic cells [22] and thus describe an alternative two-step random primer-based RT and target-specific pre-amplification method that allows amplification of human primary hematopoietic material (Fig. 1). The two-step method is significantly more expensive, although the recent introduction by Fluidigm of small cell (5–10 μm)-capture microfluidics chips for execution of reverse transcription and pre-amplification steps on their C1 Single-cell Auto-prep System may allow for cost reduction and potentially for increased sensitivity and reduced variability of the reactions. However, a comprehensive comparison of one-step versus two-step reactions on PCR plate versus microfluidic chip formats is currently still lacking. Potential downsides of the microfluidic chip format are (1) the inability to analyze more than one cell type in the same run in the absence of strong fluorescent reporters than can be localized by microscopy on the chip and (2) the inability to store single cells in lysis buffer for later use.

Independently of the method used, we recommend that 'no-RT' reactions, where the reverse transcriptase is omitted from the reaction mix, are included for every gene set or in every plate. This allows quantification of genomic DNA detection in intron-less genes and, where primer design is not intron-spanning, attests to the specificity of the detection where exclusive detection of the mRNA species is expected. Note that 'no-RT' reactions can be run simultaneously with RT reactions on a plate, but not in the C1 chip format.

3.2.1 Assay Mix for Multiplex Analysis

We recommend the use of TaqMan assays (unlabelled primers + fluorescently-labelled probe with quencher moiety) for single-cell applications, as in our hands, single primer pairs with intercalating fluorescent dyes such as SYBR or Eva Green often result in nonspecific amplification of multiple products and impair data quantification at cell numbers lower than 10 to 30. However, specificity may be enhanced by the use of nested primers as recently published in

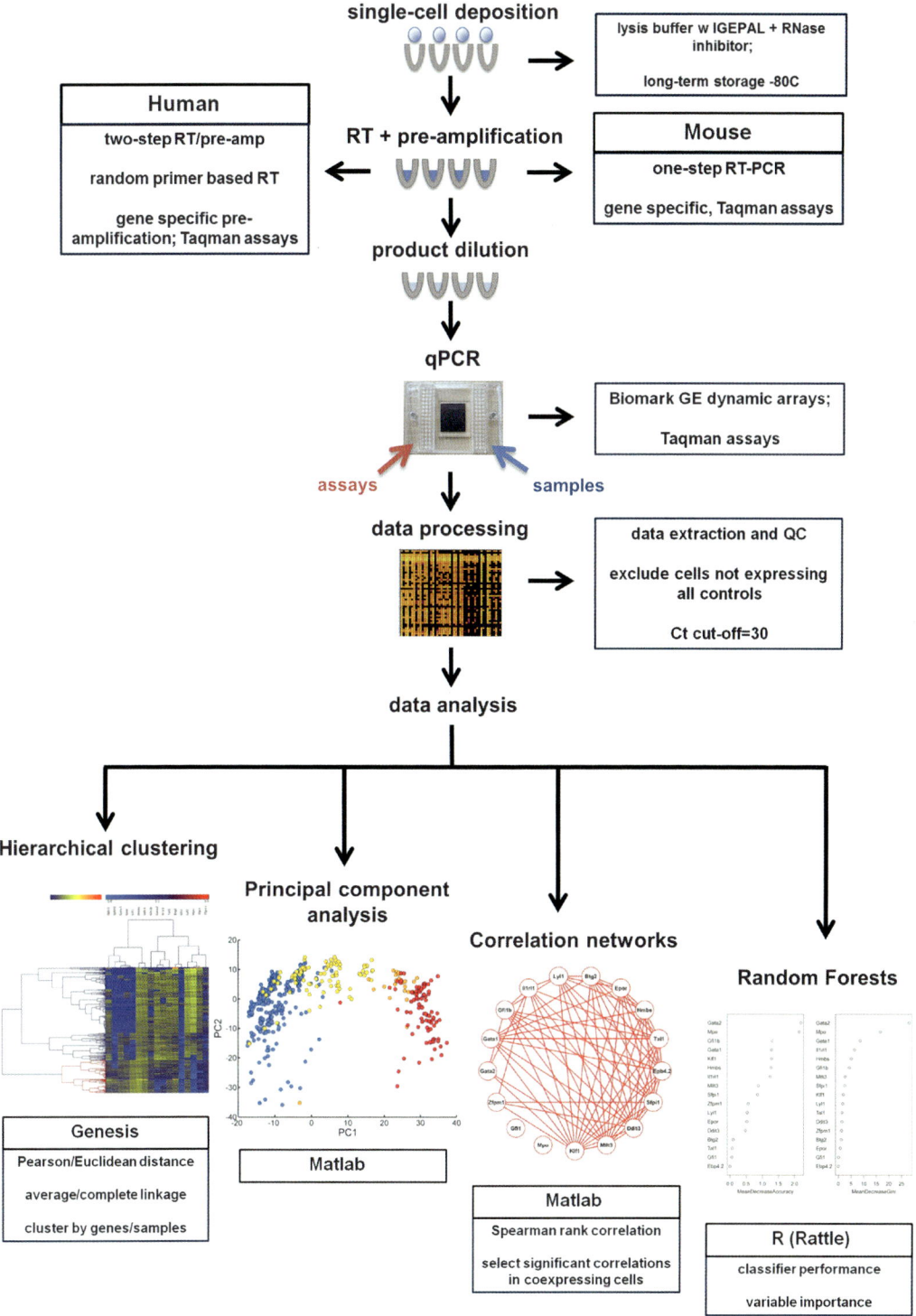

Fig. 1 Single-cell gene expression profiling by quantitative RT-PCR on a microfluidics platform: experimental and analytical workflow (see [15] for analysis panels)

a comprehensive single-cell RT-qPCR study of surface markers and regulators in mouse hematopoiesis [14]. Nested primer strategy was indeed central to early non-quantitative single-cell RT-PCR strategies and ensured specificity of end-stage products [11].

Prepare 1:96 dilutions of the TaqMan assays of interest by mixing equal volumes of the assays (commercially provided as 20× or 18 μM each primer and 3 μM probe) with either an equal total volume of RNase-free water (48.48 format) or no water added (96.96 format). Make up enough volume for all the chips planned with the same primer cocktail. Mix can be stored long-term at −20 °C.

3.2.2 One-Step Reverse Transcription and Pre-amplification

Prepare RT/pre-amplification and "no-RT" reaction mixes in a PCR hood while keeping reagents on ice.

RT + pre-amplification mix—per well:

Cell lysate	*4.0 μl*
CellsDirect 2× Master Mix	7.5 μl
TaqMan Assay Mix 187.5 nM	2.5 μl
Superscript III/Platinum Taq Mix	1.0 μl
Final volume	15 μl—aliquot 11 μl per well

No-RT control mix—per well:

Cell lysate	*4.0 μl*
CellsDirect 2× Master Mix	7.5 μl
TaqMan Assay Mix 187.5 nM	2.5 μl
Platinum Taq 5 U/μl	0.4 μl
DEPC-treated or RNase-free H_2O	0.6 μl
Final volume	15 μl—aliquot 11 μl per well

Thaw (if working from frozen lysates) and centrifuge plates to collect contents at the bottom of the wells prior to removing the film cover. Aliquot 11 μl of the reaction mix into each individual well, and mix gently by pipetting. Use a different tip in each well and align the box of pipette tips to mirror the plate layout in order to prevent cross-contamination of samples. Cover the plate with adhesive optical PCR film and spin down at 4 °C to collect contents prior to thermal cycler run.

Cycling conditions are as follows:

50 °C, 1 h (Superscript III reverse transcription).

95 °C, 2 min (inactivation of reverse transcriptase and activation of Platinum Taq polymerase).

22 or 25× (95 °C, 15 s; 60 °C, 4 min): Use a target number of 20–22 cycles for hematopoietic cell lines and 25 cycles for primary mouse bone marrow cells; refer to discussion in Subheading 4, Experimental Procedures.

25 °C, 10 s (end step).

Centrifuge plate to collect contents, and keep amplified material at –20 °C until the quantitative PCR run to prevent evaporation.

3.2.3 Two-Step Protocol: Random Primer-Based Reverse Transcription

Prepare RT and 'no-RT' reaction mixes in a PCR hood while keeping reagents on ice.

RT mix—per well:

Cell lysate	4.0 µl
T4 Gene 32 Protein 10 mg/ml	0.12 µl
Superscript VILO RT Enzyme Mix	0.15 µl
DEPC-treated or RNase-free H_2O	0.73 µl
Final volume	5 µl—aliquot 1 µl per well

'No-RT' mix—per well:

Cell lysate	4.0 µl
T4 Gene 32 Protein 10 mg/ml	0.12 µl
DEPC-treated or RNase-free H_2O	0.88 µl
Final volume	5 µl—aliquot 1 µl per well

Thaw (if starting from frozen lysates) and centrifuge plate and heat at 65 °C for 90 s to denature the RNA molecules; cool plate rapidly on ice, centrifuge in a precooled centrifuge to collect denatured lysates at the bottom of the wells, and keep on ice until addition of reaction mix. Only remove film cover at this stage.

Aliquot 1 µl of the reaction mix into each individual well and mix gently by pipetting. Use a different tip in each well and align the box of pipette tips to mirror the plate layout in order to prevent cross-contamination of samples. Cover the plate with adhesive optical PCR film and spin down at 4 °C to collect contents prior to thermal cycler run.

Thermal protocol conditions are as follows:

25 °C, 5 min.

50 °C, 30 min.

55 °C, 25 min.

60 °C, 5 min.

70 °C, 10 min.

25 °C, 10 s (end step).

Centrifuge plate in a precooled centrifuge to collect contents in the bottom of the well and keep on ice until addition of pre-amplification reaction mix.

3.2.4 Two-Step Protocol: Target-Specific Pre-amplification

Prepare pre-amplification reaction mix in PCR hood during the reverse transcription run; keep reagents and reaction mix on ice until use.

Pre-amplification mix—per well:

Reverse-transcribed material	*5.0 µl*
TaqMan Assay Mix 187.5 nM	5.0 µl
TaqMan PreAmp Master Mix 2×	10.0 µl
Final volume	20 µl—aliquot 15 µl per well

Remove film cover after centrifugation and immediately prior to addition of pre-amplification reaction mix. Aliquot 15 µl of the reaction mix into each individual well and mix gently by pipetting. Use a different tip in each well and align the box of pipette tips to mirror the plate layout in order to prevent cross-contamination of samples. Cover the plate with adhesive optical PCR film and spin down at 4 °C to collect contents prior to thermal cycler run.

Cycling conditions are as follows:

95 °C, 10 min (activation of AmpliTaq Gold polymerase).

25× (96 °C, 5 s; 60 °C, 4 min) (cDNA amplification; use 25 cycles as a reference number and refer to Subheading 4, Experimental Procedures, for additional discussion).

25 °C, 10 s (end step).

Centrifuge plate to collect contents, and keep amplified material at –20 °C until the quantitative PCR run to prevent evaporation. Optionally, the pre-amplified material can be treated by exonuclease I digestion to remove single-stranded DNA, i.e., non-annealed primers and probes. Addition of exonuclease I mix and all subsequent steps should be performed on the bench to avoid contamination of the PCR hood with amplified material.

3.2.5 Two-Step Protocol: Exonuclease I Treatment

Exonuclease I mix—per well:

Pre-amplified material	*20.0 µl*
Exonuclease I reaction buffer 10×	0.5 µl
DEPC-treated or RNase-free H$_2$O	3.25 µl
Exonuclease I 20 U/µl	1.25 µl
Final volume	25 µl—aliquot 5 µl per well

Remove film cover after centrifugation of pre-amplified material, taking care to do it away from working bench. Aliquot 5 µl of the exonuclease I reaction mix into each individual well and mix gently by pipetting. Use a different tip in each well and align the box of pipette tips to mirror the plate layout in order to prevent cross-contamination of samples. Cover the plate with adhesive optical PCR film and spin down to collect contents prior to thermal cycler run.

Thermal protocol conditions are as follows:

37 °C, 30 min.

80 °C, 10 min (inactivation step).

25 °C, 10 s (end step).

Centrifuge plate to collect contents, and keep amplified material at –20 °C until the quantitative PCR run to prevent evaporation.

3.3 Gene-Specific Quantitative PCR Analysis

The last step in the single-cell analysis of gene expression is a conventional quantitative PCR performed in a microfluidics chip that allows the partitioning of each pool of multiplexed pre-amplified cDNAs into 48 or 96 individual reactions.

The Fluidigm Gene Expression Dynamic Arrays give reproducible readings at Ct values between 5 and 27, and the number of pre-amplification cycles as well as the level of dilution of the pre-amplified material should be optimized to fit most or all of the reactions within this range. Fluidigm recommend a minimum 1:5 dilution of the pre-amplified material, and we have indicated the target numbers of pre-amplification cycles that in our experience meet these criteria. However, it should be noted that we have analyzed low-level expression events at the outset of lineage specification [8], and it is therefore likely that lower numbers of pre-amplification cycles be required for the detection and quantification of more abundantly expressed gene targets.

Prepare the pre-amplified material and the gene-specific assays in 96-well plate format that can be used for direct correspondence with chip locations as shown in Fig. 2. If using a 48.48 Dynamic Array format, arrange the samples on the right- and the assays on the left-hand side of the master plate to load directly onto the chip. In the case of a 96.96 chip format, two 96-well master plates are necessary, one for assays and the other for samples.

3.3.1 Preparation of Pre-amplified Material

Centrifuge the plate containing the pre-amplified material to collect contents at the bottom of the wells, and remove film cover away from the working bench. Keep plate covered with a clean disposable plastic lid, either on ice or at room temperature. Due to well-to-well inequalities in reaction volumes and the possibility of evaporation during the thermal cycler runs, we recommend

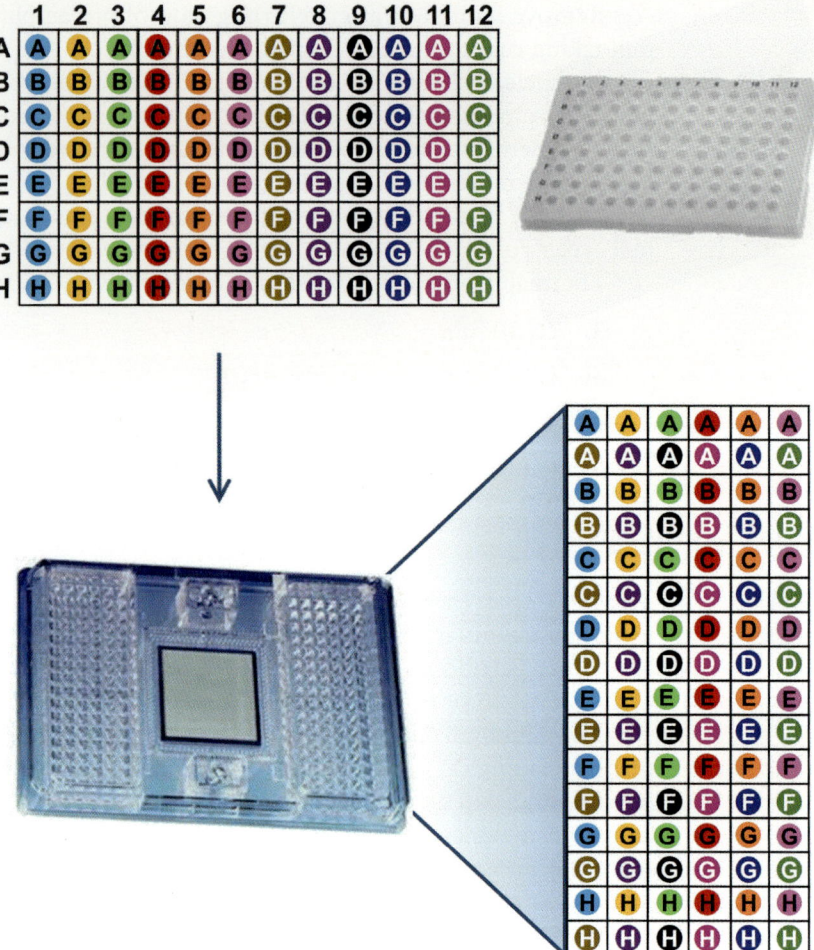

Fig. 2 Fluidigm Dynamic Arrays loading scheme depicting direct correspondences between cell locations on a 96-well plate and on the chip

performing the dilution into a new plate by adding 1 µl of cDNA to the appropriate volume of DEPC-treated or RNase-free water (e.g., 4 µl for a 1:5 dilution) while maintaining the original plate layout. Centrifuge at $2,000 \times g$ for 1 min to collect contents at the bottom of the wells.

Prepare the sample loading reaction mix (48/96 samples):

20× Sample loading reagent (yellow cap)	156/312 µl
2× TaqMan Universal PCR Master Mix	15.6/31.2 µl
Final volume	171.6/343.2 µl—aliquot 3.3 µl per well

Aliquot 3.3 µl of the sample reaction mix into each well of a master plate; use a multichannel pipette to add 2.7 µl of the diluted

sample into each individual well, and maintain the original plate layout. Mix gently by pipetting. The plate can be kept at room temperature all through the preparation.

3.3.2   Preparation of Gene-Specific Assays

Aliquot 3 µl of 2× assay loading reagent (blue cap) into each assay well of the 96-well master plate. Add each 3 µl of individual assay, and mix gently by pipetting. The plate can be kept at room temperature throughout the preparation. Centrifuge the master plate or plates after setup and load onto the chip as soon as possible.

3.3.3   Priming and Loading of Microfluidic Chip

Priming and loading of the chip are performed on a Fluidigm IFC Controller block (MX block for 48.48 chips, and HX block for the 96.96 format).

For priming, 0.3 ml/0.15 ml of control line fluid pre-aliquoted into a disposable syringe is loaded onto chambers of a 48.48 or a 96.96 chip, respectively. Gently pull back the plunger, tap the syringe to remove all the air bubbles, and then carefully release the plunger to fill the syringe tip with line fluid; clean the surface with a tissue. Load the chamber by pressing the syringe fully into the chamber valve and dispensing the fluid slowly while tilting the chip; note that failing to completely depress the valve or brisk loading will cause the fluid to overflow. Ensure that the surface of the chip is clean of fluid before loading onto the IFC controller block, and follow the touch screen instructions. Priming will take approximately 10 min for a 48.48 chip and 20 min for the larger format.

Load the assay and sample mixes onto the primed chip as soon as possible after priming. Follow the scheme in Fig. 2, and load 5 µl into each micro-well. Loading at a slight angle into the lower third of the micro-well and using the first stop position of the micropipette will minimize bubble formation. Check the chip carefully for bubbles, and remove these with the help of a large-bore (e.g., 1,000 µl) tip. Load onto the IFC controller, and follow the touch screen instructions. Loading will take approximately 1 h for the 48.48 chip and 90 min for the 96.96 format.

At this stage, switch on the Biomark Reader and computer, and launch the Biomark Data Acquisition software to switch on the lamp, which will require about 20 min to warm up.

At the end of the loading run, the chip should be moved as soon as possible to the Biomark Reader to execute the PCR step. Ensure that the surface of the chip is free of dust specks by using clear "invisible" tape, and remove the tape at the bottom of the chip; check for debris, and clean them with a tissue. Follow the instructions on the Data Acquisition software to start the run. We suggest inspecting the initial chip calibration photographs to ensure that no debris, visible as irregular bright spots, is present. Should debris be present, it is appropriate to abort the run, clean the chip again, and restart.

3.4 Data Analysis

3.4.1 Preliminary Data Processing and Quality Control

Gene expression data can be retrieved using Fluidigm PCR Analysis Software. Data are given as Ct values as well as ΔCt values to one or more chosen references, and the software allows for manual inspection of the amplification curves. Data are displayed as tables readily exportable into Excel as well as heatmaps of Ct or ΔCt values as per the chip layout. Threshold detections can be calculated automatically, either globally or on a per gene basis, or they can be customized. Similarly, default settings for amplification curve quality can be customized, and curves can be failed or passed manually upon inspection.

We recommend automatic calculation of Ct values and default settings for amplification curve quality, with customized decisions on individual reactions upon manual inspection of the curves. This is particularly important if using an intercalating dye (e.g., EvaGreen chemistry), as the software does not take into account the number or the identity of peaks in the melting curves.

Additional important quality controls are as follows:

1. Exclusion of reactions with Ct values close to background detection in no-template controls (no-cell wells); we typically exclude reactions less than two cycles different from background.

2. Exclusion of reactions with Ct values >30, as detection is spurious and unreliable; we have found that Ct values between 27 (Fluidigm's recommended cutoff) and 30 are reproducible for the large majority of genes, and include them for the benefit of low-level transcript detection. In practical terms, it is very rare to retrieve threshold amplifications later than cycle 30 that have passed all additional quality controls.

3. Exclusion of cells or wells where one or more of the control genes (we typically use three or four) are not detected.

Data can then be used as either Ct values or ΔCt values to one or more reference genes for downstream analysis.

3.4.2 Clustering Analysis

Clustering methods make use of the data to systematically group individual cells according to similarities in gene expression across all genes, allowing the identification of molecularly homogeneous subpopulations. Clustering can also be employed to group genes with similar expression patterns across cells, defining molecular signatures of a subpopulation.

We have used the Genesis software [17] to perform clustering analysis in our data. Step-by-step instructions are provided below, with background statistical context on the relevance of each step of the analysis.

1. Standardize gene expression values for each gene. For each raw expression value (x) subtract the mean (μ) and divide by the standard deviation (μ) calculated over all expression values measured for that particular gene:

Genesis:

– "Adjust" > "Mean Center Genes" followed by "Adjust" > "Adjust Genes by SD."

$$x \rightarrow \frac{x - \mu}{\sigma}$$

2. Calculate distance between individual cells based on their measured gene expression levels. Pearson correlation measures the level of linear dependency between expression levels of any two cells, while Euclidean distance calculates the geometric distance between two cells using their gene expressions as coordinates.

Genesis:

– "Distance" > "Pearson correlation" or "Distance" > "Euclidean distance"

3. Perform agglomerative hierarchical clustering analysis. The method builds clusters from bottom to top by systematically pairing the most similar genes in a stepwise manner according to the previously computed distances. In our experience, both distance metrics provided similar results.

Genesis:

– "Analysis" > "Calculate Hierarchical Clustering"

4. Define agglomeration rule as complete linkage or average linkage. Agglomeration rule sets the reference for how the distance between newly formed clusters is calculated: complete linkage uses their most distant points (in this case, genes or cells), while average linkage uses the mean distance between all pairs in each cluster.

Genesis:

– In the Hierarchical clustering window, "Agglomeration rule" > "Complete linkage clustering" or "Average linkage clustering" (*see* **Note 7**).

5. Cluster genes with similar expression patterns across all cells and/or cluster cells with similar expression patterns across all genes. The latter case is more useful in characterizing different cell populations.

Genesis:

– In the Hierarchical clustering window, "Calculation parameters" > "Cluster genes" or "Cluster experiments" (*see* **Note 8**).

Principal component analysis (PCA) allows the visualization of how individual cells position themselves in space in relation to each other on the basis of the similarity between their gene expression profiles. These representations have the potential to identify subpopulations of cells that are transcriptionally similar. Such subgroups usually have broad correspondence with clusters defined by hierarchical clustering as described in Subheading 3.4.2.

In practice, PCA reduces the high-dimensional space composed by the gene expression data (one dimension per gene) into a low-dimensional space, keeping most of the information of the original dataset and allowing more convenient data representations in two- or three-dimensional plots. The method consists of linearly transforming the gene expression data onto an alternative coordinate system, projecting it along the directions (or principal components) that retain the highest amount of variance from the original dataset: the first principal component is the projection that retains the most variance, the second component is the orthogonal projection that retains most of the remaining variance, and so forth. All principal components are linearly uncorrelated with each other, and typically the two or the three first principal components are used for data representation and analysis. We have used Matlab to perform PCA in our data and provide sample code below.

1. Prepare matrix *data*, with individual cells in rows and corresponding gene expression values in columns. Additionally prepare vector *geneNames* with the labels for all the gene names in the same order as they appear in *data*.

2. Perform PCA.

Matlab:

– Command *[coeff,score,latent]=princomp(data)* calculates the principal component coefficients (or loadings) ordered by decreasing component variance (coeff), principal component scores (score) corresponding to coordinates for representation in the principal component space, and eigenvalues of the covariance matrix of *data* (latent), which contain information on the percentage of variance explained by each principal component.

– Plot 1: Percentage of variance explained by each component:

explainedVariance=100*latent/sum(latent);

figure(1)

pareto(explainedVariance);

xlabel('Principal component');

ylabel('% Explained variance');

– Plot 2: Individual cells projected in principal component space (*see* **Note 9**):

figure(2)

plot(score(:,1), score(:,2), 'or');

xlabel('PC1');

ylabel('PC2');

– Plot 3: Gene contributions for position of cells along principal component space:

figure(3)

text(coeff(:,1),coeff(:,2),geneNames);

xlabel('PC1');

ylabel('PC2');

3.4.4 Correlation Analysis	PCR expression data from a large number of individual cells has the potential to suggest regulatory relationships between individual genes. The level of correlation between two variables provides a measure of dependency, which in the case of gene expression can indicate direct or indirect regulatory effects.

PCR expression data from a large number of individual cells has the potential to suggest regulatory relationships between individual genes. The level of correlation between two variables provides a measure of dependency, which in the case of gene expression can indicate direct or indirect regulatory effects.

A number of methods can be used to compute pairwise correlations, with Pearson product moment and Spearman rank correlations as two of the most common choices. Spearman correlation is a nonparametric method: data are ranked before correlation is computed, and as a consequence variable dependency is quantified as the strength of monotonic trends (increasing or decreasing). This is in contrast with Pearson correlation where expression values are used directly, making the method potentially more sensitive to PCR technical variation and the presence of outliers. We and others [15, 16, 23] have found the nonparametric nature of Spearman correlation to be a considerable advantage in analyzing our data. We have used Matlab to perform correlation analysis in our data and provide sample code below.

1. For each gene pair, select all cells for which expression was detected in both genes, and store the corresponding expression values in two columns, *data1* (expression for gene 1) and *data2* (expression for gene 2). Each row will correspond to a different cell.

2. Calculate the Spearman rank correlation using the gene expressions stored in *data1* and *data2*.

Matlab:

– Command *[rho,pval] = corr(data1, data2, 'type,' 'Spearman')* calculates the correlation coefficient (rho) and p-value (*p*val) for a given gene pair.

3. Select significantly correlated genes using the following parameters: Spearman correlation ≥0.4; *p*-value ≤0.01 (99 % significance level); and correlation calculated over a minimum of ten cells (*see* **Note 10**).

3.4.5 Classification Analysis

Machine learning methods can be used to extract information from the dataset that is not intuitive upon visual inspection. In particular, predictive classifier models can be trained with gene expressions from individual cells to discriminate between different populations. Identification of genes with high discriminant power can suggest potential functional roles in processes involving these populations.

Random forest classifiers are part of a family of nonlinear classifiers and can be considered an ensemble learning method which, simply put, consists of a collection of fully trained decision trees. Training of the model involves the random selection of a fraction of the dataset, with the remaining fraction being used for testing. We have used the Rattle package for R [18] to automatically train and validate random forest classifiers, providing measures for performance and variable importance. The example below consists in training a random forest classifier to discriminate between two populations using gene expressions of individual cells as input.

1. Prepare matrix *data*, with individual cells in rows and corresponding gene expression values in columns. The first column in the matrix should contain the cell identifiers, and the first row should contain all gene names. The last column of the matrix, which should have header "TYPE," specifies to which of the two populations individual cells belong to by attributing value 0 or 1.

Note: Data matrices can be prepared in Excel spreadsheets since Rattle accepts Excel files, among other formats, as input.

2. Train random forest classifier, and assess classifier performance as well as individual variable importance:

Rattle:

– "Data" tab > "Filename" (open file with matrix *data*) > "Execute"
– "Model" tab > Set "Type" to "Forest"
– Set "Number of Trees" to 1,000 and "Number of Variables" to 5 > Execute
– Select "OOB ROC" for classifier performance plot
– Select "Importance" for variable importance plot

Classifier performance is assessed through analysis of the out-of-the-bag (OOB) receiver operating characteristic (ROC) curve plot. The ROC curve depicts the false-positive rate (*x*-axis) versus the true-positive rate (y-axis). Performance is quantified by the area under the ROC curve (AUC): the higher the value, the better the classifier can discriminate between two populations based on the gene expression levels of individual cells. Discriminant power of individual genes can be assessed through the variable importance plot, where genes are shown in descending order of importance according to two different measures: permutation variable importance and Gini coefficient. These two methods compute importance in different ways and can therefore provide different rankings. We recommend a consensus approach, taking both measures into account.

Random forest classifiers encompass both linear and nonlinear relations between input (gene expressions) and output (cell populations) variables, but do not explicitly distinguish between the two types of effects. An alternative approach to achieve this distinction is to separately train a linear classifier (such as logistic regression) and a more complex nonlinear classifier (such as an artificial neural network with hidden nodes) using the same individual genes or gene combinations. Improved performance assessed by AUC with the nonlinear method is indicative of significant nonlinear relations which cannot be captured by linear methods alone. In this case, linearity represents a weighted sum of gene expressions, whereas nonlinearity involves more complex relations, such as combinations of gene expression products. Implementation of these methods involves a moderate level of programming, and presenting sample code would be beyond the scope of this protocol. There are nevertheless several packages and libraries that can facilitate implementation, both in Matlab and R environments.

4 Final Remarks

Single-cell PCR analysis of gene expression has afforded valuable insights into molecular configurations in cell fate transitions in the hematopoietic [8] and other stem cell systems [19] and can potentially be used for the inference of gene regulatory networks controlling individual cell states or routes of fate specification [14, 16] and as basis for computational modeling of gene expression regulation and cell fate decisions [15]. Despite its potential, the technologies are relatively new, and continued evaluation of their technical robustness in terms of sensitivity of detection, technical variability, and limits of quantification is critical for the correct employment of the methods and the soundness of biological conclusions. Technological optimization and evaluation of PCR-based technologies will be key to the standardization of global gene expression analysis of single cells by RNA sequencing, which is currently under intense development. Importantly, analytical tools described for PCR-based analysis of gene expression should be applicable to sequencing experiments.

5   Notes

1. All steps prior to completion of pre-amplification should be carried out in a PCR hood. All plate centrifugation steps prior to completion of pre-amplification should be performed at 4 °C (typically $2,000 \times g$, 1 min).

2. PCR hood, pipettes, tips, and plastics should be UV-treated for 20 min prior to use; after use, wipe surface with water, and wash all racks under running water. UV-treat again for 20 min.

3. All steps involving pre-amplified material should be carried out on the bench. Centrifugation steps can be performed at room temperature (typically $2,000 \times g$, 1 min).

4. Workbench should be cleaned with water and 70 % ethanol prior to use.

5. Always spin down plates immediately before use, and only remove film cover after centrifugation. Aim to remove film cover at a bench location distinct from that where the reactions are set up and preferably where materials from distinct cell types or different species are routinely handled. Cover the plate with a disposable plastic lid and keep on ice for the remainder of the procedure.

6. Change gloves after removing cover film from plates and whenever there has been risk of contamination during procedure.

7. A third alternative, single linkage, computes the distance by using the two closest points but in our experience tends to lead to very fragmented clusters.

8. For single-cell gene expression, the most frequent data formats have cells displayed in rows and genes in columns. Genesis assumes the converse order, so in order to cluster cells the option "Cluster genes" should be selected.

9. The Matlab code for Plot 2 uses the first two principal components only. In order to plot data using the three first principal components, a different Matlab function must be used: *plot3(score(:,1), score(:,2), score(:,3),'or');* instead of *plot(score(:,1), score(:,2),'or');*. Parameter *'or'* in the function specifies the shape and color of the symbols for individual cells (in this case red circles). Different shapes and colors can be attributed to individual cells by selecting their specific position in the *score* matrix. For instance, the command *plot(score(5,1), score(5,2),'sb');* will plot the cell in row 5 of the *score* matrix using the first two principal components and representing it with a blue square.

10. Due to the relative novelty of single-cell gene expression data, there is no established standard for significance. Upon extensive visual inspection of the data, we found that our parameters provided a good compromise in being inclusive while not risking a high number of false positives. While these thresholds can be experimented with, we do not recommend values lower than 0.3 for Spearman correlation and higher than 0.05 for the p-value.

Acknowledgements

We are grateful to Elizabeth Kruse and Mattias Ohlsson for critical reading of this chapter. We acknowledge Swedish Foundation for Strategic Research for funding to José Teles and Leukaemia and Lymphoma Research and Cancer Research, UK, for Programme Grants to Tariq Enver.

References

1. Adolfsson J, Mansson R, Buza-Vidas N et al (2005) Identification of Flt3+ lympho-myeloid stem cells lacking erythro-megakaryocytic potential a revised road map for adult blood lineage commitment. Cell 121:295–306

2. Kiel MJ, Yilmaz OH, Iwashita T et al (2005) SLAM family receptors distinguish hematopoietic stem and progenitor cells and reveal endothelial niches for stem cells. Cell 121:1109–1121

3. Pronk CJ, Rossi DJ, Mansson R et al (2007) Elucidation of the phenotypic, functional, and molecular topography of a myeloerythroid progenitor cell hierarchy. Cell Stem Cell 1:428–442

4. Karlsson G, Rorby E, Pina C et al (2013) The tetraspanin CD9 affords high-purity capture of all murine hematopoietic stem cells. Cell Rep 4:642–648

5. Chambers I, Silva J, Colby D et al (2007) Nanog safeguards pluripotency and mediates germline development. Nature 450:1230–1234

6. Hayashi K, Lopes SM, Tang F et al (2008) Dynamic equilibrium and heterogeneity of mouse pluripotent stem cells with distinct functional and epigenetic states. Cell Stem Cell 3:391–401

7. Kalmar T, Lim C, Hayward P et al (2009) Regulated fluctuations in nanog expression mediate cell fate decisions in embryonic stem cells. PLoS Biol 7:e1000149

8. Pina C, Fugazza C, Tipping AJ et al (2012) Inferring rules of lineage commitment in haematopoiesis. Nat Cell Biol 14:287–294

9. Macarthur BD, Sevilla A, Lenz M et al (2012) Nanog-dependent feedback loops regulate murine embryonic stem cell heterogeneity. Nat Cell Biol 14:1139–1147

10. Cheng T, Shen H, Giokas D et al (1996) Temporal mapping of gene expression levels during the differentiation of individual primary hematopoietic cells. Proc Natl Acad Sci U S A 93:13158–13163

11. Hu M, Krause D, Greaves M et al (1997) Multilineage gene expression precedes commitment in the hemopoietic system. Gene Dev 11:774–785

12. Miyamoto T, Iwasaki H, Reizis B et al (2002) Myeloid or lymphoid promiscuity as a critical step in hematopoietic lineage commitment. Dev Cell 3:137–147

13. Ramos CA, Bowman TA, Boles NC et al (2006) Evidence for diversity in transcriptional profiles of single hematopoietic stem cells. PLoS Genet 2:e159

14. Guo G, Luc S, Marco E et al (2013) Mapping cellular hierarchy by single-cell analysis of the cell surface repertoire. Cell Stem Cell 13(4):492–505

15. Teles J, Pina C, Eden P et al (2013) Transcriptional regulation of lineage commitment – a stochastic model of cell fate decisions. PLoS Comput Biol 9:e1003197

16. Moignard V, Macaulay IC, Swiers G et al (2013) Characterization of transcriptional networks in blood stem and progenitor cells using high-throughput single-cell gene expression analysis. Nat Cell Biol 15:363–372

17. Sturn A, Quackenbush J, Trajanoski Z (2002) Genesis: cluster analysis of microarray data. Bioinformatics 18:207–208

18. Williams G (2011) Data mining with rattle and R: the art of excavating data for knowledge discovery. Springer, New York

19. Buganim Y, Faddah DA, Cheng AW et al (2012) Single-cell expression analyses during cellular reprogramming reveal an early stochastic and a late hierarchic phase. Cell 150:1209–1222

20. Dalerba P, Kalisky T, Sahoo D et al (2011) Single-cell dissection of transcriptional heterogeneity in human colon tumors. Nat Biotechnol 29:1120–1127

21. Guo G, Huss M, Tong GQ et al (2010) Resolution of cell fate decisions revealed by single-cell gene expression analysis from zygote to blastocyst. Dev Cell 18:675–685

22. Goardon N, Marchi E, Atzberger A et al (2011) Coexistence of LMPP-like and GMP-like leukemia stem cells in acute myeloid leukemia. Cancer Cell 19: 138–152

23. Stahlberg A, Andersson D, Aurelius J et al (2011) Defining cell populations with single-cell gene expression profiling: correlations and identification of astrocyte subpopulations. Nucleic Acids Res 39:e24

Chapter 4

Hematopoietic Stem and Progenitor Cell Mobilization in Mice

Jonathan Hoggatt, Tiffany A. Tate, and Louis M. Pelus

Abstract

Hematopoietic stem cell transplantation (HSCT) can be performed with hematopoietic stem and progenitor cells (HSPC) acquired directly from bone marrow, from umbilical cord blood or placental tissue, or from the peripheral blood after treatment of the donor with agents that enhance egress of HSPC into the circulation, a process known as "mobilization." Mobilized peripheral blood stem cells (PBSC) have become the predominate hematopoietic graft for HSCT, particularly for autologous transplants. Despite the success of PBSC transplant, many patients and donors do not achieve optimal levels of mobilization. Thus, accurate animal models and basic laboratory investigations are needed to further investigate the mechanisms that lead to PBSC mobilization and define improved or new mobilizing agents and/or strategies to enhance PBSC mobilization and transplant. This chapter outlines assays and techniques for exploration of hematopoietic mobilization using mice as a model organism.

Key words Hematopoietic stem cell (HSC), Mobilization, Mouse models, Granulocyte-colony-stimulating factor (G-CSF)

1 Introduction

Mobilized autologous peripheral blood stem cells (PBSC) have replaced bone marrow as the primary source of hematopoietic stem and progenitor cells (HSPC) used for stem cell support following high-dose chemotherapy. PBSC provide more rapid neutrophil and platelet recovery [1], largely due to a higher yield of $CD34^+$ cells in PBSC grafts. Multiple studies and expert opinion have shown a significant correlation between $CD34^+$ cell dose and time to engraftment [2–9], including long-term platelet recovery [10]. It is generally agreed that 2×10^6 $CD34^+$ cells/kg is a minimum target threshold dose that enables complete hematopoietic recovery [11, 12]. $CD34^+$ cell doses below this threshold are utilized in cases where the minimum target threshold cannot be met, accepting that speed and completeness of blood cell recovery may be compromised. However, a $CD34^+$ cell dose of $\geq 5 \times 10^6$/kg

Kevin D. Bunting and Cheng-Kui Qu (eds.), *Hematopoietic Stem Cell Protocols*, Methods in Molecular Biology, vol. 1185, DOI 10.1007/978-1-4939-1133-2_4, © Springer Science+Business Media New York 2014

is associated with faster neutrophil and platelet recovery and less patient-to-patient variability as well as lower resource utilization and is considered by many as the "optimal" target for a single autologous stem cell transplant (ASCT) [2, 4, 13–15].

Various mobilization strategies using granulocyte, macrophage-colony-stimulating factor (GM-CSF) and more commonly granulocyte-colony-stimulating factor (G-CSF; filgrastim) have been used either alone or in combination with chemotherapy [16]. However, up to 40 % of patients fail to mobilize an "optimal" CD34+ cell dose [17–21], posing a greater problem in patients requiring tandem cycles of high-dose chemotherapy [22, 23]. Plerixafor (AMD3100), a small molecule CXCR4 antagonist, in combination with G-CSF has been shown to increase total CD34+ cells mobilized compared to G-CSF alone [24, 25] and is FDA approved for PBSC mobilization in patients with non-Hodgkin's lymphoma (NHL) and multiple myeloma (MM). However, a significant disadvantage of plerixafor is cost, adding $25,567 per patient compared to G-CSF alone in NHL patients in a recent economic analysis [26]. Furthermore, 14–24 % of MM and NHL patients receiving plerixafor plus G-CSF still failed to collect $\geq 2 \times 10^6$ CD34+ cells/kg in 4 days of apheresis in large trials [24, 25]. The search for novel, more effective mobilizing agents and strategies remains a need and an area of active investigation [12, 27].

Much of the advancements made in clinical HSPC mobilization have resulted from basic laboratory studies utilizing the mouse as a model organism. The purpose of this chapter is to provide general methods for exploring hematopoietic mobilization in mouse models and to provide a few "tips and tricks" that may be useful for investigators exploring novel mobilizing agents or working to discover new mechanistic insights into the regulation of hematopoietic trafficking.

2 Materials

2.1 Mice

At the core of basic mobilization studies is the laboratory mouse, which has proved to be a useful animal model in exploring the mechanisms of HSPC retention within the bone marrow and for the discovery and testing of novel mobilizing agents. With the expansion of commercial companies providing mice, and the rapid generation of genetically modified strains, there exists a large variety of mouse strains and models to choose from. We believe that there are several important factors to consider when choosing a mouse model. First, strains of mice vary widely in their mobilization response to various agents. For example, in a study comparing DBA/2, C57Bl/6J, and BALB/c mouse response to G-CSF, it was demonstrated that hematopoietic progenitor cell (HPC) mobilization in DBA/2 mice > BALB/c mice >> C57Bl/6J mice [28].

Strain differences were also reported for chemo-mobilization with cyclophosphamide, where FVB/nJ mice hardly mobilized, while SM/J mice robustly mobilized progenitor cells into the peripheral blood [29]. C57Bl/6J mice were also fairly low mobilizers in the chemo-mobilization study. However, the degree of mobilization is not the only criterion when picking a strain of mouse. In the field of murine hematopoietic transplantation, the gold standard for enumeration of long-term repopulating HSCs is the competitive repopulation assay, which commonly utilizes C57Bl/6J mice (Jackson Laboratories (JAX) stock number 000664) and their congenic, CD45.1 counterparts, B6.SJL-*Ptprc*^{a} *Pep3*^{b}/BoyJ (JAX stock number 002014), or similar strains. Therefore, if competitive hematopoietic transplantation is used to measure hematopoietic mobilization, the C57Bl/6J mouse would be preferred. Similarly, many mechanistic studies now employ knockout animals, all of which have different strain pedigrees. It is important to pick appropriate WT strain controls for these studies (i.e., non-knockout littermates) to acquire interpretable and accurate data. Secondly, age and sex have both been demonstrated to have effects on hematopoietic mobilization, with aged mice having increased mobilization [30] and male mice mobilizing more than their female counterparts [29].

In our laboratory we typically use 8–12-week-old, female, C57Bl/6J mice for mobilization studies. We allow these mice to acclimate in the animal facility for at least 1 week after shipping as stress from the shipping process causes baseline mobilization to increase. Similarly, acclimating the mice to handling prior to study initiation also reduces the stress responses, allowing for more consistent data (*see* **Note 1**). Unless age is an experimental variable, we do not recommend performing studies in mice younger than 8 weeks of age, since acquiring enough peripheral blood for subsequent analysis can be difficult. Given that strain differences do exist, and that these are likely due to specific genetic differences, we also recommend when studying new mobilizing agents to explore their effects in at least one other mouse strain (i.e., BALB/c, DBA/2).

2.2 Reagents and Supplies

Numerous reagents and supplies are used during the course of a hematopoietic mobilization study, and the specific reagents used will be determined by individual need and the analyses to be performed. However, we highlight a few key reagents and supplies that we have found helpful in our studies and will utilize these reagents in the methods described in the next section.

2.2.1 Blood Collection and Separation

1. Syringe wash: 0.1 M EDTA solution made with 1 part 0.5 M EDTA (UltraPure 0.5 M EDTA, Gibco #15575) and 4 parts Dulbecco's phosphate-buffered saline (DPBS without calcium and magnesium).

2. Syringes for cardiac puncture: Becton Dickinson 1 ml TB Syringe with 27 G × ½ needle (BD Ref #309623).

3. Drummond Scientific EDTA Capillary Tubes (#1-000-100-E).

4. K_2EDTA Collection Tubes (BD Ref #365974).

5. Red blood cell lysis: EasySep 10× RBC Lysis Solution (Stem Cell Technologies Cat# 20120) prepared by the manufacturer's instructions; Red Blood Cell Lysing Solution Hybri-Max (Sigma #R7757) or equivalent (*see* **Note 2**).

6. Lympholyte Mammal (Cedarlane Labs #CL5120).

7. Wash solution: DPBS/0.5 % BSA/2 mM EDTA.

2.2.2 Hematopoietic Colony Assays

Stem Cell Technologies (Stem Cell Technologies #M3434) markets a premixed methylcellulose medium that contains all of the growth factors necessary for enumeration of colony-forming unit (CFU)-granulocyte, macrophage (CFU-GM); CFU-granulocyte, erythrocyte, monocyte, megakaryocyte (CFU-GEMM); and blast-forming unit-erythrocyte (BFU-E). This media does not contain antibiotics however, and we recommend adding penicillin/streptomycin (Life Technologies #15070063 or equivalent) at ~50 U/ml (1:100 dilution in M3434) prior to use. While effective, the cost of this reagent may be prohibitive for many labs, particularly those conducting large-scale mobilization studies. Alternatively, methylcellulose medium for colony assays can be made in-house using the following procedure:

1. Prepare 500 ml of 2× Iscove's Modified Dulbecco's Medium (IMDM) using powdered IMDM mix (Life Technologies #12200-036) and following the manufacturer's instructions; however reduce the water content by 50 %.

2. Add 50,000 U of penicillin/streptomycin.

3. Sterile filter through a 0.2 μm vacuum filter flask, and place the resulting mixture in a 37 °C water bath.

4. Place 21 g of methylcellulose powder (4,000 centipoise; Sigma #M0512) in a sterile 2 l flask.

5. Place 550 ml of deionized water and a 3 in. magnetic stir bar in a separate 2 l flask and boil down to 500 ml.

6. Quickly take the boiled water flask to a biosafety hood along with the flask containing the powdered methylcellulose, pour the water and the stir bar into the powdered methylcellulose, and mix thoroughly making sure that all of the powder has been moistened. Cap the flask with a sterile piece of aluminum foil.

7. Place the flask on a magnetic stir plate, and stir the methylcellulose mixture until it cools to ~40 °C. Do not let it cool below this temperature.

8. Back in the hood, add the pre-warmed 2× IMDM mixture to the methylcellulose mixture and mix vigorously by hand until thoroughly mixed.

9. Recap with sterile aluminum foil, and continue to stir the mixture on a stir plate at 4 °C for about 24 h.

10. At this point, the product is now a 2× methylcellulose media that can be aliquoted and stored at –20 °C.

To make media appropriate for colony assays the following ingredients should be added aseptically to the 2× methylcellulose:

- 30 % by volume fetal bovine serum (FBS).
- 1 % by volume 100× L-Glutamine (Gibco GlutaMAX™ Supplement Recommended, Cat# 35050-061).
- 50 μM 2-Mercaptoethanol.
- 10 μg/ml Recombinant human insulin.
- 200 μg/ml Human transferrin.
- 50 ng/ml Recombinant mouse stem cell factor.
- 10 ng/ml Recombinant mouse GM-CSF (*see* **Note 3**).
- 3 U/ml Human erythropoietin (EPO) (we typically use prescription Procrit from the pharmacy, but recombinant protein from a reputable source is sufficient).
- QS with 1× IMDM.

Additional useful supplies for colony-forming assays include:

1. 14 ml Polystyrene round bottom tubes with snap caps.
2. 3 ml Syringes with 18 G × 1½ needle.
3. 35 mm × 10 mm Polystyrene tissue culture dish.
4. Gridded Scoring Dish (StemCell Technologies #27500).

3 Methods

3.1 A Note on G-CSF

It is beyond the scope of this particular chapter to delve into the many different agents that have been used for mobilization and outline specific protocols for each agent, though we have covered some of these agents in more depth elsewhere [31]. Most mouse studies used to study hematopoietic mobilization use G-CSF either to explore mechanisms specific to G-CSF or as a comparator to other mobilizing agents being studied. There is a wide range of dosing schedules used in mice to explore G-CSF, and there does not seem to be a consensus in the field regarding total dose, route of administration, or dosing schedule. We do have a few recommendations regarding G-CSF use in mice.

1. When possible, use Neupogen as the G-CSF of choice. We have seen varying degrees of reproducibility with other recombinant G-CSF products.

2. When using Neupogen, dilute it fresh prior to each injection in a sterile saline solution. We DO NOT recommend making a diluted

"stock" solution for an experiment, nor do we recommend adding any other proteins (i.e., BSA).

3. We recommend subcutaneous (SC) injection as the route of administration. Many publications use intraperitoneal (IP) injection; however we do not believe that this matches well to the kinetics used in human patients.

4. There is no current consensus on the dose or the frequency of G-CSF to use in mice. Many publications use twice-a-day dosing, and others use once a day. Some publications are also using upwards of 250 µg/kg daily G-CSF, and others less. Assuming a 20 g mouse (the approximate weight of a 10–12-week C57Bl/6J female mouse), and using a surface area conversion from man to mouse [32], the conversion of the recommended 10 µg/kg/day SC for human patients would be ~125 µg/kg/day SC in mice for at least 4 days. In our lab, we typically split this dose and give twice-daily subcutaneous injections 12 h apart for 4–5 days.

3.2 Routes of Administration

3.2.1 Subcutaneous Injection

Many of the mobilizing agents explored, including the FDA-approved filgrastim and plerixafor, use subcutaneous injection. There are numerous methods described for subcutaneous injection, and we encourage the reader to explore additional resources, including online videos or your institutional veterinary staff, to learn injection techniques. For subcutaneous injection our laboratory uses a method that utilizes a mouse restrainer (Braintree Scientific-TV-150). This method has the advantage of being extremely reproducible after practice, is good for those who are slightly timid with handling mice, and reduces the chances of needle sticks.

1. Using an ethanol wipe, wipe the hair on the flank of the mouse (Fig. 1a) opposite the direction the hair lies to clean the area and expose the skin.

2. Gently pull the mouse into the restrainer, and twist the mouse so that the wiped flank is exposed through the groove of the restrainer.

3. Using a 1 ml syringe with a 27 G needle, insert the needle through the open restrainer groove and into the skin of the mouse (Fig. 1b) and then proceed parallel with the skin surface. Pull up slightly to confirm that the needle is subcutaneous and not in a muscle.

4. Administer the mobilization agent in a volume of 100–200 µl. You should observe a slight bubbling of the skin, indicating correct SC injection. Do not use volumes larger than 200 µl, though too small volumes are difficult to administer accurately as well.

5. Withdraw the needle from the skin, and place the animal back in the cage. An advantage of this method is that the ethanol

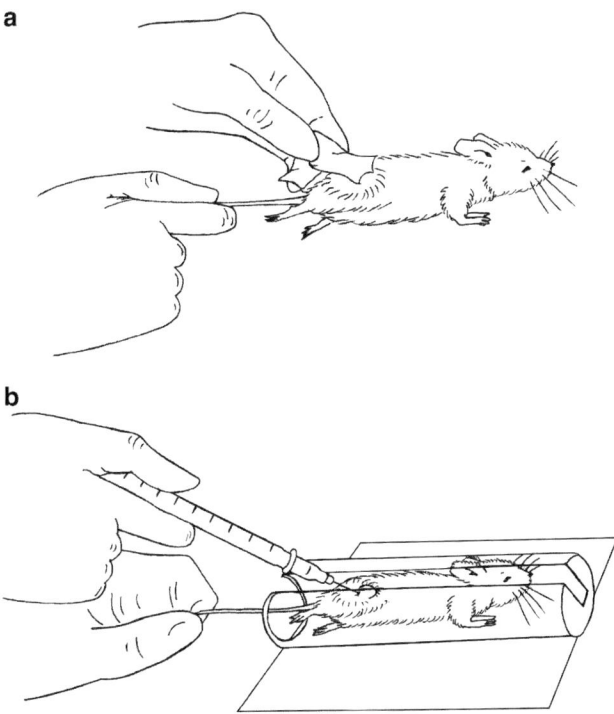

Fig. 1 Subcutaneous injections using a restrainer. (**a**) While gently holding the mouse by the tail on a firm surface (such as the top of the cage) use an ethanol wipe to expose the skin of the mouse on its flank as shown. (**b**) As you pull the mouse into the restrainer, gently rotate the mouse (~90°) so that the exposed skin is exposed with needle accessibility through the groove of the restrainer. Perform the subcutaneous injection through the groove at the exposed skin site as shown

wipe "marks" each mouse that has undergone an injection, allowing you to easily distinguish which remaining mice in the cage need to be treated. This allows the mice to maintain their home cage without constantly changing them to new environments (particularly in the settings of twice daily injections) and in our hands reduces experimental variability in mobilization.

6. If performing twice-daily injections, we recommend alternating sides of the mouse (left and right) to reduce irritation at a specific site.

3.2.2 Intraperitoneal Injection

1. Scruff the mouse with your non-injecting hand and restrain such that the abdomen is facing you and the head is slightly tilted down, shifting the viscera towards the head.

2. Using a 27 G needle insert the needle at about a 30° angle into the lower right abdominal quadrant of the mouse (with the mouse on its back, this is the left side of the midline). In female mice, you can use the upper right nipple as a landmark and be slightly cranial from there.

3. Inject the substance in a volume of 100–200 μl.

4. Gently withdraw the needle. As in the case of subcutaneous injection, you can mark the mouse with an ethanol swipe on its back and place it in the same cage.

3.2.3 *Intravenous Injection*

Although pharmacokinetics for some substances are similar between intraperitoneal and intravenous injection in the mouse, this similarity is not always the case and for many substances intravenous injection is preferred. Similarly, intravenous injection is the preferred method for HSPC transplant, an important assay for the evaluation of hematopoietic mobilization (Subheading 3.7). Out of all of the administration methods described here, intravenous injection tends to be the most demanding and difficult for new researchers to learn. Therefore, the technique should be extensively practiced prior to large-scale experiments. Several online videos demonstrating variances of the technique are available. In our lab, we use a custom-built restrainer that allows for entry of the mouse into the restrainer and will rotate the mouse 90° to make the lateral tail vein parallel with the working surface. This restrainer also elevates the mouse from the working surface at a height that allows the user to rest their holding hand, making them steadier. Braintree Scientific sells a commercial unit with similar capabilities called the "Rotating Tail Injector" (Cat# RTI STD). When performing tail vein injections, we recommend gently warming the mouse under a heat lamp to cause vasodilation prior to injection. We use 27 G needles and inject no more than 200 μl of volume per injection. Proper injection should require minimal force, and the syringe plunger should be pressed slowly. If properly inserted in the lateral tail vein, you will observe a blanching across the entire vein indicating that the vehicle is properly distributed. If the needle is not properly located in the vein, a local blanching and bulging will occur at the needle insertion site. If this happens, withdraw the needle, move slightly up the tail to a new injection site, and try again.

3.2.4 *Oral Gavage*

Some agents that may be explored during the course of a mobilization study will not be amenable to the above dosing methods, particularly if they are not soluble in aqueous solutions. Many agents that are insoluble in aqueous solutions can be administered in suspensions by oral gavage. Although oral gavage is used widely, the procedure can often result in increased animal stress, or even death if done improperly, which can be compounded by relative inexperience of the performing technician. The oral gavage technique requires stiff restraint of alert rodents to prevent technical complications (*see* **Note 4**) [33], and this type of restraint has been shown to increase plasma corticosterone levels in rats and mice [34, 35]. Recently, we developed a novel method of oral gavage that increases the ease of compound administration and reduces mouse stress responses, and we refer the reader for a detailed description of the method, including photos [36]. Briefly, we recommend using 1.5 in., curved, 20 G, stainless steel feeding needles with a 2.25 mm ball (Braintree Scientific, #N-010) attached to 1 ml tuberculin syringes. Immediately prior to gavage dip the gavage needle in a 100 % sucrose solution. This sweet substance will act to calm the mouse during the gavage procedure, facilitating easy passage into the esophagus and reducing stress responses.

3.3 Dosing and Kinetics

As discussed above, dosing of G-CSF varies considerably in the literature, although it typically requires 4–6 days of administration in mice and man to achieve optimal mobilization. Many other mobilizing agents have more rapid kinetics, with some reaching peak mobilization in as little as 15 min after a single treatment [31, 37–40]. For agents with rapid mobilizing properties it is important to plan how long your endpoints will take for each mouse, particularly blood and tissue harvesting, and stagger your injections appropriately to leave enough time to accurately harvest each mouse (*see* **Note 5**).

Kinetics of administration also has to be considered when exploring combinations of two or more mobilizing agents. In some cases, the peak mobilization kinetics of the agents on their own will not be the same as the agents in combination [31]. In other cases, creative staggering of two agents may be required to achieve peak mobilization. For example, we recently reported on the combinatorial use of nonsteroidal anti-inflammatory drugs (NSAIDs) along with G-CSF [41]. Initially, these agents were given together for 4 consecutive days, with blood harvesting the morning of the fifth day. When evaluating HSPC content by colony assays and flow cytometry, we saw significant synergy with the combination of NSAID and G-CSF. However, after transplantation, the grafts did not perform as predicted based on the HSPC content we had measured. Further analysis revealed that NSAIDs resulted in a reduction in HSPC expression of CXCR4, reducing the ability of the mobilized cells to home and engraft, despite the fact that HSPC number was increased. To compensate for this effect, we staggered the administration of NSAIDs and G-CSF by 48 h. In this scenario NSAIDs were given alone for 2 days, followed by a 2-day period where NSAIDs were co-administered with G-CSF, followed by a 2-day period with G-CSF on its own, with harvesting on the morning of day 7. This additional 2 days of protocol, with a staggered dosing regimen, led to the mobilization of a superior graft to that of G-CSF alone or the non-staggered G-CSF + NSAID regimen previously explored [41], even though the total amount of drug and dosing of the individual agents was the same as the non-staggered route. When performing mouse studies exploring combinatorial mobilization, several variations in dosing kinetics may be required to achieve optimal results.

**3.4 Peripheral Blood
Collection
and Processing**

3.4.1 Blood Collection

As discussed, peripheral blood should be collected at the peak of mobilization and efforts should be made to collect blood as close to the same time as possible amongst mice in the experiment. It has been reported that circadian rhythms can affect hematopoietic mobilization [42, 43]. In our laboratory, we typically time all of our mobilization experiments so that bleeding begins at about 10:00 a.m., which in our facility is ~4 h after initiation of light. Variations in bleeding time, if not controlled, can make interpretation of results from experiment to experiment difficult. This also means that in large-scale experiments blood needs to be harvested in the same time frame rather than extended throughout the day.

There are several different bleeding routes that can be used to collect blood from mice including cardiac puncture, retro-orbital bleed, tail bleed, facial bleeding using a lancet, and saphenous vein puncture. The choice of bleeding method is in part dependent on the requirements of the experiment and the agent being explored. For repeated bleeding, we recommend using the tail bleeding method, since you can control the volume of the bleed more precisely, reducing the variable effects of more or less blood loss on the mobilization process, and it is less harmful to the mouse than repeated bleeding via other methods (i.e., retro-orbital). A recently described method of bleeding using Goldenrod Lancets has been reported [44]; however we find that it is difficult to control the volume of blood released, and surprisingly, the procoagulant effects of G-CSF [45, 46] appear to specifically hinder the ability to reliably acquire blood by this method in settings of G-CSF mobilization.

For most endpoints exploring the peak of mobilization, multiple assays will need to be performed, requiring a terminal bleed to achieve the maximum amount of peripheral blood for subsequent analysis. We recommend cardiac puncture for this method as the blood values in our hands tend to be more consistent from mouse to mouse and is very rapid when performed correctly.

Prior to starting the experiment, rinse all syringes (1 ml syringe with a 27 G needle) with the EDTA wash solution, remove all air bubbles and expel the syringe wash into a waste container. This procedure fills the "headspace" of the syringe and needle with anticoagulant and removes air from the syringes.

1. Place the mouse in a CO_2 chamber and fill with 20 % CO_2 by volume for 2 min. Observe the mouse, and watch for no signs of movement or breath.

2. Remove the mouse and place on its back. If needed, a Styrofoam pin board can be used to secure the mouse. Spray the chest surface with 70 % ethanol.

3. Insert the needle into the abdominal wall just below the xiphoid process and slightly to the mouse's left at ~10–25° angle from the abdominal surface.

4. With just the tip of the needle inside the chest cavity, gently draw back on the syringe creating a negative pressure.

5. While maintaining the negative pressure, slowly push the needle cephalad towards the heart until a "flash" of blood is seen in the syringe.

6. Maintaining that position, slowly draw back on the syringe until all of the peripheral blood is collected (typically 800–1,000 µl in 8–12-week-old, female, C57Bl/6J mice).

7. Carefully remove the needle from the syringe, and distribute the blood into a 1 ml EDTA collection tube. Invert at least ten times, and then place on a roller.

Using this method we have achieved reliable blood amounts in mobilization experiments without any subsequent viability or clotting issues. Many published methods of mobilization studies describe the administration of heparin to mice prior to removal of blood by cardiac puncture. Intriguingly, heparin may regulate the mobilization response [47] perhaps acting as a confounding variable in certain studies. We have not seen the need for its use prior to blood collection and recommend avoiding heparin in mobilization studies.

3.4.2 Complete Blood Count

After collection of mobilized blood, analysis of complete blood count (CBC) is a critical parameter both for determination of the white blood cell (WBC) differentials, which indicate whether there is a differential effect on mature blood cell populations, and as a numerical count of the total number of lymphocytes per milliliter of blood in the animal, which can then be used for subsequent calculations in other assays (*see* **Note 6**). CBC can be performed using a veterinary cell counter (e.g., Hemavet 950FS, Drew Scientific; Vetscan HM5, Abaxis; or similar instrument) or WBC count can be determined manually with a hemacytometer and blood smears made for determination of CBC differential manually by microscopic analysis of Wright's stained smears. One advantage of automated veterinary cell counters is that they require very little blood for analysis (typically less than 25 µl) and can reliably perform counts in about 2 min per sample. For many of these machines the choice of anticoagulant is critical; EDTA is often the preferred anticoagulant, and some do not function optimally if the anticoagulant is heparin.

3.4.3 Low-Density Mononuclear Cell Isolation

Transplantation of a mobilized graft is typically done with peripheral blood mononuclear cells, the product obtained after apheresis of a mobilized patient or donor. For analysis of hematopoietic mobilization in mice obtaining a similar fraction using density centrifugation is useful prior to flow cytometry, hematopoietic transplantation, or in some cases hematopoietic colony assays. There are several density gradient solutions available on the market,

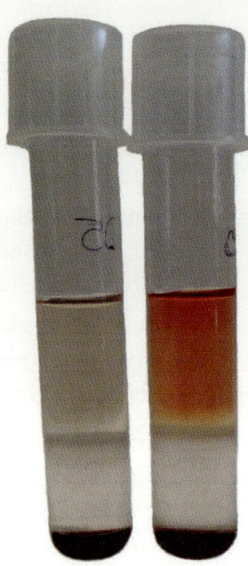

Fig. 2 Lymphocyte layer with DPBS or IMDM. Shown are lymphocyte, "buffy coat," layers after density centrifugation using Lympholyte Mammal with prior blood dilution using either DPBS (*left*) or IMDM (*right*). For some, the higher contrast between IMDM and the Lympholyte allows for easier viewing of the lymphocytes at the interface

each with subtle variances in the recommended protocol. In our lab, we have had good success with Lympholyte-Mammal from Cedarlane Labs (#CL5120). This solution is specifically designed for the isolation of viable lymphocytes and monocytes from the peripheral blood and consists of sodium diatrizoate combined with dextran to induce erythrocyte aggregation and reduce platelet aggregation.

1. Place 3 ml of Lympholyte Mammal into 14 ml round-bottom polystyrene tubes (BD #352051). These tubes are clear, allowing for easy viewing of the lympholyte layer after centrifugation.

2. Dilute peripheral blood from mice in a serum-free medium such as DPBS, McCoy's, and IMDM to a total volume of 4 ml. Using a medium like IMDM creates more of a contrast at the interface of the medium and the Lympholyte (Fig. 2), and for some people this is useful for observing the lymphocyte layer.

3. Carefully layer the diluted blood on top of the 3 ml of Lympholyte. Slightly angling the tube and allowing the diluted blood to slowly run down the side of the tube while pipetting help to create a good interface with little mixing.

4. Centrifuge at $800 \times g$ for 20 min at room temperature. Make sure that the brake of the centrifuge is turned off.

5. After centrifugation, there will be a well-defined lymphocyte layer or "buffy coat" sitting on top of the Lympholyte. Using a pipette, carefully collect this layer of cells and transfer to a new centrifuge tube.

6. Add wash solution to the lymphocytes to a total volume of 10 ml, and centrifuge at $800 \times g$ for 10 min to pellet.

7. Wash the lymphocytes 2–3 more times before further use.

3.4.4 Red Blood Cell Lysis

Our preferred method of assessing hematopoietic colonies from peripheral blood (Subheading 3.5) is by plating whole blood after RBC lysis.

1. Place blood (20–200 µl) into a 14 ml tube.

2. Add 2 ml of RBC lysis solution. Gently mix for 1–2 min.

3. Fill tube with wash solution, and centrifuge at $300 \times g$ for 10 min.

4. Repeat washes 2–3 times before further use.

3.5 Hematopoietic Colony Assays

Most agents being investigated for PBSC mobilization properties will mobilize mature blood cells as well as HSPC. While mature cells are easily discernible by automated cell counters or manual observation, hematopoietic cells are only definable by phenotype or function.

Colony-forming cell assays are available that identify lineage-restricted HPC based upon their ability to form colonies in semi-solid media, solid agar, methylcellulose, or plasma clot, when stimulated with appropriate hematopoietic growth factors. CFU-G, CFU-M, and CFU-GM progenitor cells can be characterized in soft agar or agarose when stimulated by G-CSF, M-CSF, GM-CSF with SCF, and other factors or conditioned medias [37, 40, 48, 49]. Combination cocktails can be utilized to stimulate earlier HPC, some having high proliferative potential or multilineage potential [50, 51]. Erythroid HPC identified as BFU-E or colony-forming unit-erythroid (CFU-E) are enumerated in 1 % methylcellulose containing EPO. In some experiments, multiple colony types, including CFU-G,M, GM, BFU-E, and multipotential CFU-GEMM, can be quantitated with 1 % methylcellulose containing EPO, GM-CSF, and SCF [52–56]. Additionally, megakaryocyte progenitor cells (CFU-Mk or CFU-Meg) can be selectively enumerated. Stem Cell Technologies markets specific reagents and culture media for quantitating this HPC (Megacult-C #04974). Our laboratory often utilizes the semisolid agar CFU-GM assay to define mobilization agent dose and regimens and later fully characterizes the repertoire of HPC mobilized using the methylcellulose assay.

The amount of cells plated in a colony assay is an important factor for accurate counting results. Too few cells results in plates with few or no colonies, and too many cells can result in plates that are overcrowded, prohibiting accurate counts. Trial runs with your particular assays and media with several dilutions may be required initially to optimize experiments. In methylcellulose assays in which we are distinguishing CFU-GM, CFU-GEMM, and BFU-E using M3434 or our in-house mix (Subheading 2.2.2) we typically target for 20–40 colonies per plate.

Prior to starting a colony assay, determine the total volume of methylcellulose media you will need and place in a 37 °C water bath. If factors were freshly added to the methylcellulose media (i.e., Pen/Strep) and there are bubbles from mixing, make sure that they have risen out before proceeding with use. We typically analyze the blood of every mouse independently, and perform the colony assays in triplicate per mouse.

1. In a 14 ml tube, place cells from the red blood cell lysis (typically 20–200 µl worth of whole blood) or low-density mononuclear cells from the Lympholyte layer (typically 200,000–800,000 lymphocytes) in a total volume of 400 µl in IMDM media.

2. Using a 5 ml syringe with an 18 G needle add 3.6 ml of pre-warmed methylcellulose media to the 400 µl of cells. Take note that this method is preparing a total of 4 ml, but for triplicate plating you will only use 3 ml. However, due to the viscosity and the bubbles on top, extra volume should always be made to allow for precise volume plating. Calculations of total cells per plate (i.e., per ml) should be made for accurate analysis of colonies per ml of blood for each animal accordingly.

3. Vortex all tubes vigorously, and place the tubes in a rack into a 37 °C incubator for at least 10 min to allow the bubbles to rise out of the media.

4. Using 3 ml syringes with 18 G needles, draw up 3 ml of the cells in methylcellulose and add 1 ml/35 mm dish. Gently tilt and rotate the dishes around to evenly distribute the mixture into a uniform layer covering the entire surface.

5. Place all of the plates in a 37 °C incubator with 5 % O_2 and 5 % CO_2 with >95 % relative humidity for 7–10 days.

6. After culture, place an individual dish into the gridded scoring dish and use an inverted microscope to count the total number of colonies per dish. Average the total of the triplicate dishes per mouse.

7. Calculations of CFUs per ml of blood are performed based on the source and number of cells plated. For example, if cells from an RBC lysis of 200 µl of whole blood were added to the tube, then each dish received the equivalent of 50 µl of blood. In this case, CFU/ml of blood would be calculated as average colonies per dish/0.05 ml.

3.6 Flow Cytometric Assays

Immunophenotypic analysis by flow cytometry is commonly used to determine the frequency of HSC and HPC in spleen, bone marrow, and peripheral blood. While phenotype and functional assays do not always directly correspond, the flow cytometry-based assays are rapid and are useful tools for screening for mobilizing effects. Repopulating stem cells were originally defined by the absence of lineage markers (Lin^neg) with expression of stem cell antigen-1 (Sca-1), low expression of Thy1.1 [57], and expression of SCF receptor (c-kit) [58–60]. These cells are commonly referred to as LSK cells. More recently, additional markers have been identified [61–63] that allow further enrichment of HSC and definition of HPC with different lineage capacity (reviewed in ref. 64). The signaling lymphocyte activation molecule (SLAM) family of receptors CD150, CD48, and CD244 define a highly enriched HSC population [65–67]. We have routinely utilized SLAM-LSK populations to evaluate a wide array of mobilization agents with success. However, interpretation of phenotypic analysis for HSC content in mobilized grafts should always be viewed with caution until results are validated by functional transplantation data. While the CD34 marker can be used in the mouse to distinguish HSC with short-term and long-term function at steady state, G-CSF-mobilized HSC express CD34 but lose CD34 when they return to steady state [68], perhaps reducing the reliability of CD34 when evaluating mobilization regimens. Detailed descriptions of flow cytometry techniques to identify HPC and HSC can be found in refs. 69–71.

For mobilization into peripheral blood, we recommend performing a Lympholyte separation as we describe (Subheading 3.4.3) prior to antibody staining. If at all possible, this method is preferred over RBC lysis for flow analysis of HSPC in peripheral blood. Sometimes, antigens such as c-kit receptor will also have reduced intensity compared to the bone marrow, and appropriate changes in gating may be necessary. As an example of the difference between bone marrow and mobilized peripheral blood, Fig. 3 shows flow cytometry plots gating on lineage-negative cells from bone marrow (left) or from mobilized peripheral blood (right) showing c-kit receptor and Sca-1 expression. Of note, the bone marrow plot was generated by collecting a total of 100,000 events, while the mobilized peripheral blood plot was generated by collecting 1,000,000 events (*see* **Note** 7).

3.7 Hematopoietic Transplantation Assays

True HSC can only be defined based upon their ability to fully repopulate a lethally irradiated host. If the mice live (>16 weeks) with reconstitution of all blood lineages, then by definition, the graft is considered to have contained HSC. However, monitoring "survival" does not permit enumeration or comparison of HSC number or function. Moreover, most institutional animal care and use committees are recommending alternative methods to survival studies. To address these problems, transplantation assays that

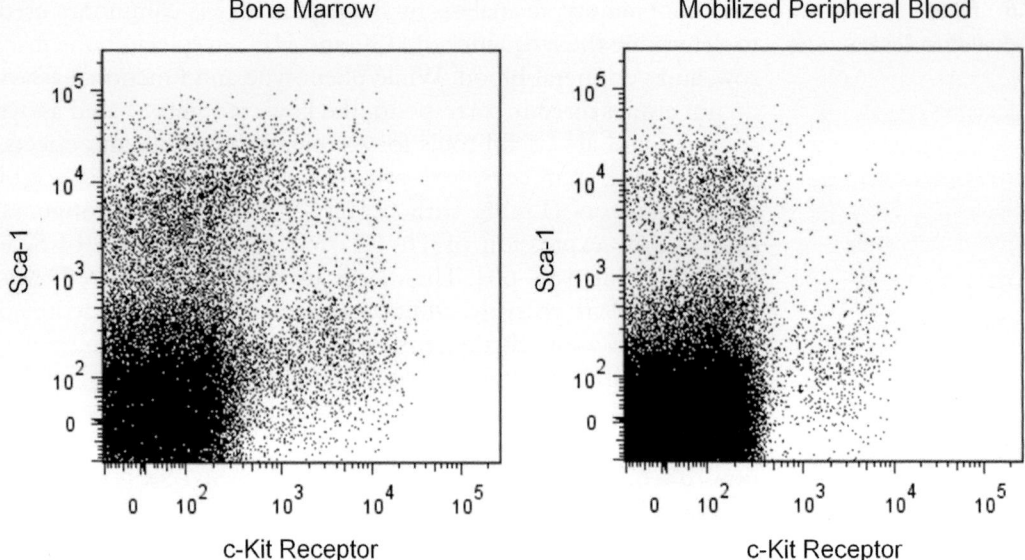

Fig. 3 Example LSK flow cytometry plots. Shown is a representative flow cytometry plot of gated lineage-negative cells from bone marrow (*left*) or mobilized peripheral blood (*right*). All staining conditions (number of cells, volume, antibody concentration, etc.) were equivalent between the two groups, and data acquisition was performed at the same time. Of note is that the bone marrow plot was generated by collecting a total of 100,000 viable cell events, while the mobilized peripheral blood took 1,000,000 events, illustrating the need to collect considerably more flow cytometry data for accurate analysis of rare HSPC in peripheral blood

assess long-term repopulating cells (LTRC), an HSC synonym, with comparison against a "competitor" graft were developed. Studies by Harrison first described a standard competitive HSC repopulation assay [72] and a mathematical method to calculate competing repopulating units (RU) [73]. In this assay, donor HSC are admixed with wild-type congenic bone marrow cells and transplanted into lethally irradiated recipients. Blood cell production from each source of cells is analyzed with fluorescently tagged antibodies that distinguish the two congenic strains, allowing for a comparison of the repopulating ability of each population of cells. The C57Bl/6 (CD45.2) and the B6.SJL-PtrcAPep3B/BoyJ (BOYJ) (CD45.1) mouse strains are most widely used for this assay differing only at the CD45 antigen [74]. Our animal facility maintains a colony of C57Bl/6× BoyJ F1 hybrid mice that offers a variation on this assay where two competing grafts, one CD45.1 and the other CD45.2, can be compared in the same animal using CD45.1/CD45.2 hybrid F1 bone marrow cells as competitors (*see* refs. [75, 76] for an example of head-to-head HSC population comparison).

Limiting-dilution competitive repopulation assay is a variation of the competitive repopulation assay that utilizes a series of dilutions of the donor, or "test" graft, compared to a standard number of competing cells (normally 2×10^5 whole bone marrow cells). A minimum threshold of peripheral blood cell reconstitution is set (~0.5–5 %) for the test graft, and the number of mice that fail to reconstitute with the test graft is used to determined CRU or HSC frequency by Poisson statistics [77–79]. A computer program, L-Calc™, which calculates the HSC frequency, can be obtained from Stem Cell Technologies.

3.7.1 Transplantation Assays Using Mobilized Peripheral Blood

Our laboratory routinely utilizes competitive transplantation to validate LT-HSC in a mobilized peripheral blood product and assess HSC quantity and function. For this analysis we use LDMC separated on Lympholyte® Mammal and compare these cells in ratios to 2×10^5 whole bone marrow competitor cells. Pilot studies to establish appropriate donor:competitor ratios for mobilized LDMC are often advantageous and may prevent performing a long and complicated transplant procedure that does not provide useful data (*see* **Note 8**). Typically, mobilized LDMC:competitor ratios of 1:1–5:1 are evaluated. We DO NOT recommend transplanting whole blood or RBC lysed blood into animals as a method to evaluate mobilization. These methods often lead to complications (such as platelet clotting) and do not as faithfully represent the graft used in clinical transplantation (apheresis product).

3.7.2 Serial Transplantation

Transplantation assays performed in competitive fashion are routinely analyzed at 16 weeks post-transplant, and if tri-lineage reconstitution is seen in peripheral blood one assumes that HSC were transplanted. However, HSC are heterogeneous, with varying capacities for self-renewal and capacity for extended repopulation. Recently, HSC with short-term (up to 16 weeks), intermediate-term (up to 32 weeks), and long-term (>32 weeks) capacity for multipotent differentiation into all blood lineages have been identified [80]. Thus, for some in the field, 16-week analysis of a primary graft is not sufficient to demonstrate LT-HSC function, often requiring the "gold standard" serial transplantation to be performed. It has been shown that in normal mice, the ability of HSC to self-renew is lost after four or five serial transplantations [81]. Provided a primary transplant is performed as described in Subheading 3.7, secondary transplantation methods are normally no different than those performed with more traditional bone marrow graft primary transplant.

3.8 Summary

With the discovery of chemo-mobilization and subsequently mobilization by G-CSF, the field of hematopoietic transplantation, particularly in the autologous setting, has been fundamentally changed. However, significant advancements can still be made in enhancing the yield of CD34⁺ cells acquired, reducing failure rates and the number of apheresis sessions, and improving the overall quality of the graft mobilized. With numerous therapies actively being explored using HSCs, including gene modification, easy, safe, and effective mobilization methods in a variety of donor and patient populations are likely to increase in need over the years to come. The mouse models described herein may prove to be useful tools for further laboratory investigation to identify and bring to practice the next generation of mobilization agents.

4   Notes

1. When mice first arrive in the facility, let them acclimate in their new environment for a few days. Then, have the laboratory technician or investigator handle the mice daily for ~5 days prior to initiating a mobilization study. You will find that the mice will be much calmer when handled during the experiment, and we have found that this reduces variability between animals in a given group.

2. Another product which can remove red blood cells is called HetaSep and is sold by StemCell Technologies (Catalog #07906). This is an erythrocyte aggregation agent that does not require lysis, but rather causes red blood cells to separate from nucleated cells. Higher levels of progenitor viability can sometimes be obtained using this agent rather than lysis agents.

3. Recombinant IL-3 and IL-6 (10–20 ng/ml) can be used in place of or in addition to GM-CSF. Other custom growth factor cocktails can be used depending on the progenitor types explored.

4. One slight alteration of the scruffing method has proven beneficial when performing oral gavage. If the mouse is scruffed with the non-gavaging hand, try to keep the index finger free, holding the skin of the neck and back with the thumb and three other fingers. With the free index finger, place on top of the head and gently pull back so that the nose and mouth are pointed upwards. This now gives a straight shot for the oral gavage procedure, and the index finger helps prevent additional head movement during compound administration.

5. Sometimes, something as simple as marking stripes on the tails of the mice with a Sharpie marker is a quick and convenient way to appropriately stagger treatments within a cage. For instance, sometimes we stagger injections by 3 min to allow for blood harvesting at precise times; the first mouse in the cage gets one stripe on the tail, followed by two stripes, etc. In this way, we can set a timer based on the first mouse and know which order to subsequently bleed the mice, so that dosing kinetics and analysis are at precise time points for each mouse.

6. It should also be noted that choice of blood collection site, e.g., cardiac, retro-orbital, facial, or tail, can affect WBC count and CBC. We recommend cardiac bleed for blood collection for mobilization studies.

7. HSPC in blood, even after mobilization, are still quite rare, and many more events than typically used for bone marrow analysis will need to be acquired for sufficient data.

8. Depending on the mobilizing agent or regimen, mobilized LDMC are often considerably less competitive than equivalent numbers of freshly isolated whole bone marrow cells, requiring higher ratios of cells to be used in limiting dilution analysis.

Acknowledgements

J. H. is supported by NIH Career Development Grant K99/R00 HL119559.

References

1. Korbling M, Anderlini P (2001) Peripheral blood stem cell versus bone marrow allotransplantation: does the source of hematopoietic stem cells matter? Blood 98:2900–2908

2. Weaver CH, Hazelton B, Birch R et al (1995) An analysis of engraftment kinetics as a function of the CD34 content of peripheral blood progenitor cell collections in 692 patients after the administration of myeloablative chemotherapy. Blood 86:3961–3969

3. Bender JG, To LB, Williams S et al (1992) Defining a therapeutic dose of peripheral blood stem cells. J Hematother 1:329–341

4. Bensinger W, Appelbaum F, Rowley S et al (1995) Factors that influence collection and engraftment of autologous peripheral-blood stem cells. J Clin Oncol 13:2547–2555

5. Ketterer N, Salles G, Raba M et al (1998) High CD34(+) cell counts decrease hematologic toxicity of autologous peripheral blood progenitor cell transplantation. Blood 91:3148–3155

6. Passos-Coelho JL, Braine HG, Davis JM et al (1995) Predictive factors for peripheral-blood progenitor-cell collections using a single large-volume leukapheresis after cyclophosphamide and granulocyte-macrophage colony-stimulating factor mobilization. J Clin Oncol 13: 705–714

7. Reiffers J, Faberes C, Boiron JM et al (1994) Peripheral blood progenitor cell transplantation in 118 patients with hematological malignancies: analysis of factors affecting the rate of engraftment. J Hematother 3:185–191

8. Yoon DH, Sohn BS, Jang G et al (2009) Higher infused CD34+ hematopoietic stem cell dose correlates with earlier lymphocyte recovery and better clinical outcome after autologous stem cell transplantation in non-Hodgkin's lymphoma. Transfusion 49:1890–1900

9. Siena S, Schiavo R, Pedrazzoli P et al (2000) Therapeutic relevance of CD34 cell dose in blood cell transplantation for cancer therapy. J Clin Oncol 18:1360–1377

10. Stiff PJ, Micallef I, Nademanee AP et al (2011) Transplanted CD34(+) cell dose is associated with long-term platelet count recovery following autologous peripheral blood stem cell transplant in patients with non-Hodgkin lymphoma or multiple myeloma. Biol Blood Marrow Transplant 17:1146–1153

11. DiPersio JF (2010) Can every patient be mobilized? Best Pract Res Clin Haematol 23: 519–523

12. Motabi IH, DiPersio JF (2012) Advances in stem cell mobilization. Blood Rev 26:267–278

13. Giralt S, Stadtmauer EA, Harousseau JL et al (2009) International myeloma working group (IMWG) consensus statement and guidelines regarding the current status of stem cell collection and high-dose therapy for multiple myeloma and the role of plerixafor (AMD 3100). Leukemia 23:1904–1912

14. Gertz MA, Kumar SK, Lacy MQ et al (2009) Comparison of high-dose CY and growth factor with growth factor alone for mobilization of stem cells for transplantation in patients with multiple myeloma. Bone Marrow Transplant 43:619–625

15. Schulman KA, Birch R, Zhen B et al (1999) Effect of CD34(+) cell dose on resource utilization in patients after high-dose chemotherapy with peripheral-blood stem-cell support. J Clin Oncol 17:1227

16. To LB, Haylock DN, Simmons PJ et al (1997) The biology and clinical uses of blood stem cells. Blood 89:2233–2258

17. Gordan LN, Sugrue MW, Lynch JW et al (2003) Poor mobilization of peripheral blood stem cells is a risk factor for worse outcome in lymphoma patients undergoing autologous stem cell transplantation. Leuk Lymphoma 44:815–820

18. Pavone V, Gaudio F, Console G et al (2006) Poor mobilization is an independent prognostic factor in patients with malignant lymphomas treated by peripheral blood stem cell transplantation. Bone Marrow Transplant 37:719–724

19. Akhtar S, Weshi AE, Rahal M et al (2008) Factors affecting autologous peripheral blood stem cell collection in patients with relapsed or refractory diffuse large cell lymphoma and Hodgkin lymphoma: a single institution result of 168 patients. Leuk Lymphoma 49:769–778

20. Pusic I, Jiang SY, Landua S et al (2008) Impact of mobilization and remobilization strategies on achieving sufficient stem cell yields for autologous transplantation. Biol Blood Marrow Transplant 14:1045–1056

21. Hosing C, Saliba RM, Ahlawat S et al (2009) Poor hematopoietic stem cell mobilizers: a single institution study of incidence and risk factors in patients with recurrent or relapsed lymphoma. Am J Hematol 84:335–337

22. Attal M, Harousseau JL, Facon T et al (2003) Single versus double autologous stem-cell transplantation for multiple myeloma. N Engl J Med 349:2495–2502

23. Cavo M, Tosi P, Zamagni E et al (2007) Prospective, randomized study of single compared with double autologous stem-cell transplantation for multiple myeloma: Bologna 96 clinical study. J Clin Oncol 25:2434–2441

24. DiPersio JF, Micallef IN, Stiff PJ et al (2009) Phase III prospective randomized double-blind placebo-controlled trial of plerixafor plus granulocyte colony-stimulating factor compared with placebo plus granulocyte colony-stimulating factor for autologous stem-cell mobilization and transplantation for patients with non-Hodgkin's lymphoma. J Clin Oncol 27:4767–4773

25. DiPersio JF, Stadtmauer EA, Nademanee A et al (2009) Plerixafor and G-CSF versus placebo and G-CSF to mobilize hematopoietic stem cells for autologous stem cell transplantation in patients with multiple myeloma. Blood 113:5720–5726

26. Kymes SM, Pusic I, Lambert DL et al (2012) Economic evaluation of plerixafor for stem cell mobilization. Am J Manag Care 18:33–41

27. Rettig MP, Ansstas G, DiPersio JF (2012) Mobilization of hematopoietic stem and progenitor cells using inhibitors of CXCR4 and VLA-4. Leukemia 26:34–53

28. Roberts AW, Foote S, Alexander WS et al (1997) Genetic influences determining progenitor cell mobilization and leukocytosis induced by granulocyte colony-stimulating factor. Blood 89:2736–2744

29. Watters JW, Kloss EF, Link DC et al (2003) A mouse-based strategy for cyclophosphamide pharmacogenomic discovery. J Appl Physiol 95:1352–1360

30. Xing Z, Ryan MA, Daria D et al (2006) Increased hematopoietic stem cell mobilization in aged mice. Blood 108:2190–2197

31. Hoggatt J, Pelus LM (2012) Hematopoietic stem cell mobilization with agents other than G-CSF. In: Kolonin MG, Simmons PJ (eds) Stem cell mobilization: methods and protocols. Springer, New York, pp 49–67

32. Reagan-Shaw S, Nihal M, Ahmad N (2008) Dose translation from animal to human studies revisited. FASEB J 22:659–661

33. Johnson MD, Gad SC (2007) The rat. In: Gad SC (ed) Animal models in toxicology, 2nd edn. CRC Press, Boca Raton, pp 150–173

34. Brown AP, Dinger N, Levine BS (2000) Stress produced by gavage administration in the rat. Contemp Top Lab Anim Sci 39:17–21

35. Dobrakovova M, Jurcovicova J (1984) Corticosterone and prolactin responses to repeated handling and transfer of male rats. Exp Clin Endocrinol 83:21–27

36. Hoggatt AF, Hoggatt J, Honerlaw M et al (2010) A spoonful of sugar helps the medicine go down: a novel technique to improve oral gavage in mice. J Am Assoc Lab Anim Sci 49:329–334

37. King AG, Horowitz D, Dillon SB et al (2001) Rapid mobilization of murine hematopoietic stem cells with enhanced engraftment properties and evaluation of hematopoietic progenitor cell mobilization in rhesus monkeys by a single injection of SB-251353, a specific truncated form of the human CXC chemokine GRObeta. Blood 97:1534–1542

38. Pelus LM, Bian H, King AG et al (2004) Neutrophil-derived MMP-9 mediates synergistic mobilization of hematopoietic stem and progenitor cells by the combination of G-CSF and the chemokines GRObeta/CXCL2 and GRObetaT/CXCL2delta4. Blood 103: 110–119

39. Pruijt JF, Verzaal P, Van Os R et al (2002) Neutrophils are indispensable for hematopoietic stem cell mobilization induced by interleukin-8 in mice. Proc Natl Acad Sci U S A 99: 6228–6233

40. Fukuda S, Bian H, King AG et al (2007) The chemokine GRObeta mobilizes early hematopoietic stem cells characterized by enhanced homing and engraftment. Blood 110:860–869

41. Hoggatt J, Mohammad KS, Singh P et al (2013) Differential stem- and progenitor-cell trafficking by prostaglandin E2. Nature 495:365–369

42. Mendez-Ferrer S, Lucas D, Battista M et al (2008) Haematopoietic stem cell release is regulated by circadian oscillations. Nature 452:442–447

43. Lucas D, Battista M, Shi PA et al (2008) Mobilized hematopoietic stem cell yield depends on species-specific circadian timing. Cell Stem Cell 3:364–366

44. Golde WT, Gollobin P, Rodriguez LL (2005) A rapid, simple, and humane method for submandibular bleeding of mice using a lancet. Lab Anim (NY) 34:39–43

45. Hill JM, Syed MA, Arai AE et al (2005) Outcomes and risks of granulocyte colony-stimulating factor in patients with coronary artery disease. J Am Coll Cardiol 46: 1643–1648

46. Lindemann A, Rumberger B (1993) Vascular complications in patients treated with granulocyte colony-stimulating factor (G-CSF). Eur J Cancer 29A:2338–2339

47. Saez B, Ferraro F, Yusuf RZ et al (2012) Hematopoietic stem/progenitor cell retention in the bone marrow depends on tissue specific heparan sulfate proteoglycans. Blood 112: Abstract 637

48. Broxmeyer HE, Mejia JA, Hangoc G et al (2007) SDF-1/CXCL12 enhances in vitro replating capacity of murine and human multipotential and macrophage progenitor cells. Stem Cells Dev 16:589–596

49. Pelus LM, Broxmeyer HE, Kurland JI et al (1979) Regulation of macrophage and granulocyte proliferation. Specificities of prostaglandin E and lactoferrin. J Exp Med 150: 277–292

50. Lowry PA, Zsebo KM, Deacon DH et al (1991) Effects of rrSCF on multiple cytokine responsive HPP-CFC generated from SCA + Lin- murine hematopoietic progenitors. Exp Hematol 19:994–996

51. Bradley TR, Hodgson GS (1979) Detection of primitive macrophage progenitor cells in mouse bone marrow. Blood 54:1446–1450

52. McLeod DL, Shreve MM, Axelrad AA (1976) Induction of megakaryocyte colonies with platelet formation in vitro. Nature 261:492–494

53. Johnson GR, Metcalf D (1977) Pure and mixed erythroid colony formation in vitro stimulated by spleen conditioned medium with no detectable erythropoietin. Proc Natl Acad Sci U S A 74:3879–3882

54. Hara H, Ogawa M (1978) Murine hemopoietic colonies in culture containing normoblasts, macrophages, and megakaryocytes. Am J Hematol 4:23–34

55. Fauser AA, Messner HA (1978) Granuloerythropoietic colonies in human bone marrow, peripheral blood, and cord blood. Blood 52:1243–1248

56. Fauser AA, Messner HA (1979) Identification of megakaryocytes, macrophages, and eosinophils in colonies of human bone marrow containing neutrophilic granulocytes and erythroblasts. Blood 53:1023–1027

57. Spangrude GJ, Heimfeld S, Weissman IL (1988) Purification and characterization of mouse hematopoietic stem cells. Science 241:58–62

58. Okada S, Nakauchi H, Nagayoshi K et al (1992) In vivo and in vitro stem cell function of c-kit- and Sca-1-positive murine hematopoietic cells. Blood 80:3044–3050

59. Ogawa M, Matsuzaki Y, Nishikawa S et al (1991) Expression and function of c-kit in hemopoietic progenitor cells. J Exp Med 174:63–71

60. Ikuta K, Weissman IL (1992) Evidence that hematopoietic stem cells express mouse c-kit but do not depend on steel factor for their generation. Proc Natl Acad Sci U S A 89: 1502–1506

61. Adolfsson J, Borge OJ, Bryder D et al (2001) Upregulation of Flt3 expression within the bone marrow Lin(-)Sca1(+)c-kit(+) stem cell compartment is accompanied by loss of self-renewal capacity. Immunity 15:659–669

62. Yang L, Bryder D, Adolfsson J et al (2005) Identification of Lin(–)Sca1(+)kit(+)CD34(+) Flt3– short-term hematopoietic stem cells capable of rapidly reconstituting and rescuing myeloablated transplant recipients. Blood 105: 2717–2723

63. Osawa M, Hanada K, Hamada H et al (1996) Long-term lymphohematopoietic reconstitution by a single CD34-low/negative hematopoietic stem cell. Science 273:242–245

64. Weissman IL, Shizuru JA (2008) The origins of the identification and isolation of hematopoietic stem cells, and their capability to induce donor-specific transplantation tolerance and treat autoimmune diseases. Blood 112: 3543–3553

65. Chen J, Ellison FM, Keyvanfar K et al (2008) Enrichment of hematopoietic stem cells with SLAM and LSK markers for the detection of hematopoietic stem cell function in normal and Trp53 null mice. Exp Hematol 36: 1236–1243

66. Yilmaz OH, Kiel MJ, Morrison SJ (2006) SLAM family markers are conserved among hematopoietic stem cells from old and reconstituted mice and markedly increase their purity. Blood 107:924–930

67. Kiel MJ, Yilmaz OH, Iwashita T et al (2005) SLAM family receptors distinguish hematopoietic stem and progenitor cells and reveal endothelial niches for stem cells. Cell 121: 1109–1121

68. Ogawa M, Tajima F, Ito T et al (2001) CD34 expression by murine hematopoietic stem cells. Developmental changes and kinetic alterations. Ann N Y Acad Sci 938:139–145

69. Challen GA, Boles N, Lin KK et al (2009) Mouse hematopoietic stem cell identification and analysis. Cytometry A 75:14–24

70. Johnnidis JB, Camargo FD (2008) Isolation and functional characterization of side population stem cells. Methods Mol Biol 430: 183–193

71. Srour EF, Yoder MC (2005) Flow cytometric analysis of hematopoietic development. Methods Mol Med 105:65–80

72. Harrison DE (1980) Competitive repopulation: a new assay for long-term stem cell functional capacity. Blood 55:77–81

73. Harrison DE, Jordan CT, Zhong RK et al (1993) Primitive hemopoietic stem cells: direct assay of most productive populations by competitive repopulation with simple binomial, correlation and covariance calculations. Exp Hematol 21:206–219

74. Shen FW, Tung JS, Boyse EA (1986) Further definition of the Ly-5 system. Immunogenetics 24:146–149

75. Hoggatt J, Singh P, Sampath J, Pelus LM (2009) Prostaglandin E2 enhances hematopoietic stem cell homing, survival and proliferation. Blood 113:5444–5555

76. Hoggatt J, Mohammad KS, Singh P, Pelus LM (2013) Prostaglandin E2 enhances long-term repopulation but does not permanently alter inherent stem cell competitiveness. Blood 122:2997–3000

77. Szilvassy SJ, Lansdorp PM, Humphries RK et al (1989) Isolation in a single step of a highly enriched murine hematopoietic stem cell population with competitive long-term repopulating ability. Blood 74:930–939

78. Szilvassy SJ, Humphries RK, Lansdorp PM et al (1990) Quantitative assay for totipotent reconstituting hematopoietic stem cells by a competitive repopulation strategy. Proc Natl Acad Sci U S A 87:8736–8740

79. Taswell C (1981) Limiting dilution assays for the determination of immunocompetent cell frequencies. I Data analysis. J Immunol 126: 1614–1619

80. Benveniste P, Frelin C, Janmohamed S et al (2010) Intermediate-term hematopoietic stem cells with extended but time-limited reconstitution potential. Cell Stem Cell 6:48–58

81. Ogden DA, Micklem HS (1976) The fate of serially transplanted bone marrow cell populations from young and old donors. Transplantation 22:287–293

Chapter 5

Cell Cycle Measurement of Mouse Hematopoietic Stem/Progenitor Cells

Brahmananda Reddy Chitteti and Edward F. Srour

Abstract

Lifelong production of blood cells is sustained by hematopoietic stem cells (HSC). HSC reside in a mitotically quiescent state within specialized areas of the bone marrow (BM) microenvironment known as the hematopoietic niche (HN). HSC enter into active phases of cell cycle in response to intrinsic and extrinsic biological cues thereby undergoing differentiation or self-renewal divisions. Quiescent and mitotically active HSC have different metabolic states and different functional abilities such as engraftment and BM repopulating potential following their transplantation into conditioned recipients. Recent studies reveal that various cancers also utilize the same mechanisms of quiescence as normal stem cells and preserve the root of malignancy thus contributing to relapse and metastasis. Therefore, exploring the stem cell behavior and function in conjunction with their cell cycle status has significant clinical implications in HSC transplantation and in treating cancers. In this chapter, we describe methodologies to isolate or analytically measure the frequencies of quiescent (G0) and active (G1, S, and G2–M) hematopoietic progenitor and stem cells among murine BM cells.

Key words Hematopoietic stem cells, Progenitor cells, Cell cycle, Hoechst 33342, Pyronin Y, CD71

1 Introduction

During the mammalian development, hematopoietic stem cells (HSC) first appear in the yolk sac, then migrate into fetal liver, and eventually migrate into the bone marrow (BM). At birth, hematopoiesis is almost entirely limited to the BM only. The lifelong production and maintenance of hematopoiesis is sustained by HSC within the BM [1–3]. These HSC reside in specialized areas called the "hematopoietic niche" (HN) within the BM microenvironment [4, 5]. At present, it is generally believed that the mitotic status of HSC in the BM microenvironment differ depending on the HN in which they reside. Quiescent HSC reside in the endosteal niche in close proximity to osteoblasts where they receive signals that maintain their quiescence [6–8]. On the other hand, more active HSC reside in the vascular niche and are primed to

Kevin D. Bunting and Cheng-Kui Qu (eds.), *Hematopoietic Stem Cell Protocols*, Methods in Molecular Biology, vol. 1185, DOI 10.1007/978-1-4939-1133-2_5, © Springer Science+Business Media New York 2014

respond quickly to hematopoietic stress and entry into circulation [9, 10]. Mitotic quiescence enables HSC to remain in deep dormancy for long periods of time thus avoiding exhaustion of the stem cell pool [4, 5]. Experimental evidence suggests that quiescent HSC possess superior functional characteristics such as long-term repopulation relative to more active or cycling stem cells [11]. Therefore, transplantation of HSC isolated based on their cell cycle status may have clinical implications as far as their ability to sustain BM repopulation, maintain the stem cell pool, and support long-term hematopoiesis [12].

Recent studies demonstrate that most leukemias and tumors possess tumorigenic cells with characteristics like self-renewal, extensive proliferation, and quiescence similar to those of normal stem cells. These cells are called cancer stem cells (CSC). It is speculated that mitotic quiescence of CSC helps preserve the root of malignancy and contributes to the relapse and metastasis of cancer [13, 14]. More importantly quiescent CSC exhibit drug resistance and escape chemotherapy and other cytotoxic interventions. Therefore, understanding the molecular mechanisms of quiescence has tremendous clinical significance in targeting quiescent and drug resistant CSC directly.

Simple DNA staining either by Hoechst 33342 (Hst) or 4′,6-diamidino-2-phenylindole (DAPI) or Propidium iodide enables us to distinguish cells in either G0/G1 ($2n$ DNA) phases of cell cycle from those in S or G2–M phases ($4n$ DNA). This simple DNA staining cannot identify and thus separate G0 cells from G1 cells. Since quiescent (G0) HSC have different physiological and molecular properties as well functional capacities compared to primed (G1, S, G2–M) HSC [15–17], it is critical to distinctly separate and isolate viable HSC in all phases of cell cycle. In addition to DNA staining, several procedures were introduced to stain other cellular components such as mitochondria [18–21], nuclear proteins [22], and RNA [23, 24] to help distinguish between quiescent and cycling or metabolically active cells. However, most of these procedures were detrimental for viable cell recovery as the techniques involve permeabilization or fixation or due to toxicity of the dyes. In 1981, Shapiro [23] described a double DNA–RNA staining procedure to identify cells in different phases of cell cycle and eventually to sort viable and functional cells by flow cytometry. Later on, with a readjusted Pyronin Y (PY) concentration, the double DNA–RNA staining was applied to analyze human BM-derived CD34+ cells [25]. To do this, Hst was used to stain DNA, and PY was used to stain RNA. Hst is a relatively nontoxic, water-soluble, and cell-permeable bisbenzimide dye that stains the minor groove of the double stranded DNA helix. PY is a nucleic acid dye that stains DNA and RNA as well as mitochondrial membranes [24, 26]. However, when cells are first stained with Hst, PY is blocked from staining DNA, and at moderate concentrations it predominantly stains RNA rather than mitochondrial membranes. To distinguish cells in G0 or G1, which have the same DNA content but different amounts of RNA content, cells in G0 phase are identified by their $2n$ DNA and minimal RNA content; cells in G1 phase are identified by their $2n$ DNA and high RNA content. Cells in G2–M are identified by their $4n$ DNA and high RNA content. The S phase cells are identified by their varying levels of DNA and RNA content distributed between $2n$ and $4n$ DNA peaks. Though Hst and PY staining is well suited for the analysis and viable cell sorting of human BM cells, PY concentrations that are normally used for the staining of human cells are toxic to murine cells. Therefore, viable cell sorting of murine BM cells with PY is less favorable (BRC and EFS, unpublished observations). However, for analytical studies of murine cells, protocols employing Hst and PY staining combined with a fixation step produce consistent reliable results that are easy to interpret. Obviously, this approach precludes the isolation of viable cells for use in functional in vitro and in vivo assays. We have successfully replaced PY with the Transferrin receptor protein 1 (CD71) to identify metabolically active cells. CD71 is a 95 kDa, type II membrane glycoprotein. It transports iron into proliferating cells by binding to Fe(Apo)-transferrin [27]. CD71 expression is low on quiescent cells, but its expression elevates on activated lymphocytes, monocytes, macrophages, and most dividing cells [28, 29]. Combining CD71 and Hst staining allows for the identification and isolation of cells in G0 versus cells in G1 as will be discussed below.

In this chapter, we describe a detailed protocol (Subheading 3.1) to concurrently stain mouse BM cells with various cell surface markers along with HST and PY staining. This method allows us to hierarchically identify and quantify mature BM cells and immature stem and progenitor cells while simultaneously assessing the cell cycle status of each of these groups or classes of hematopoietic cells. We also describe (Subheading 3.2) Hst and CD71 staining along with cell surface markers where viable cell isolation is required.

2 Materials

1. Ficoll purchased as a ready-to-go solution. Make sure that the bottle of Ficoll is warmed up to room temperature before using.

2. Heparin medium consisting of Hank's Balanced Salt Solution (HBSS) medium supplemented with 1 % Penicillin–Streptomycin and 20 U/mL heparin.

3. Stain wash: 1× Phosphate buffered saline (PBS) supplemented with 1 % bovine calf serum and 1 % Penicillin–Streptomycin.

4. Fluorochrome-labeled primary antibodies for immunostaining. These antibodies may vary in number and specificities depending on the phenotypic definition of the cells that will be identified and analyzed by flow cytometry.

5. Cell fixative reagent such as BD Cytofix/Cytoperm.

6. Hst buffer: HBSS, 20 mM (4-(2-hydroxyethyl)-1-piperazineethanesulfonic acid (HEPES), 1 g/L glucose, 10 % fetal calf serum (FCS), pH 7.2.

7. Hst: Hst is purchased from Molecular Probes at 10 mg/mL solution in water. Both the intermediate and working solutions must be prepared from the stock on the same day for an application, light-protected, and used within a few hours of preparation.

   (a) Hst intermediate solution: Intermediate solution is prepared in Hst buffer. To do this, add 20 μL of the 10 mg/mL stock to 1.98 mL of Hst buffer.

   (b) Hst working solution: Add 50 μL of the intermediate solution to 4.95 mL Hst buffer. These dilutions will result in a final concentration of 1 μg Hst per mL, which is equivalent to 1.5 μM.

8. PY: PY is purchased from Polysciences as powder and dissolved at 100 mg of PY per mL of water containing 10 % acetic acid. From this solution prepare a stock solution of 10 mg/mL in Hst buffer.

   PY working solution: Dilute 10 mg/mL PY stock 1:50 in Hst buffer (10 μL stock in 490 μL Hst buffer). The final PY concentration, after the PY working solution is added to the Hst working solution with which cells are stained (as described in Subheading 3.1), will be 3.3 μM.

9. Verapamil-containing Hst buffer: Verapamil can be added to the Hst buffer to stop the Hst efflux at 50 or 100 μM. Dissolve verapamil in DMSO at 100 mg/mL as stock solution. From this stock, in order to make 50 μM final concentration of verapamil, add 50 μL of stock to 200 mL of Hst buffer; to make 100 μM final concentration of verapamil, add 100 μL of stock to 200 mL of Hst buffer. Importantly, verapamil-containing Hst buffer should be prepared fresh.

3   Methods

3.1 Cell Cycle Analysis of Murine Progenitor Cells Using Hst and PY (Fixation Involved)

This method describes the simultaneous staining of murine bone marrow low-density cells with cell surface markers such as lineage (CD3, CD4, CD45R, Ter119, and Gr1), Sca-1, and cKit to gate hematopoietic progenitor cells along with Hst and PY for cell cycle analysis. In order to measure the cell cycle status of progenitor cells across the samples, and if further live cell condition is not required, our experience is that fixing the cells after cell surface staining followed by Hst/PY staining yields better results. If isolation of viable cells in different phases of cell cycle is required, see Subheading 3.2.

1. In a sterile environment (e.g., biosafety cabinet) euthanize a 6–10-week-old mouse (*see* **Note 1**).

2. Collect all four limbs into heparin medium.

3. Strip the bones of muscle and soft tissue using sterile gauze.

4. Flush the bones thoroughly using 27 G, 1/2 in. needle with 8–10 mL of heparin medium. To accomplish this, cut long bones in half keeping the epiphyses intact at the end of each half of the long bones. With sterile forceps, grab a piece of the cut bone and introduce the needle into the shaft of the bone aiming toward the epiphysis. Inject 1–2 mL of heparin medium into the bone while retracting and pushing the needle into the bone shaft. Make sure that the tip of the needle reaches the epiphysis to ensure that the epiphysis is thoroughly washed with the injected medium. If necessary, inject more medium to release the majority of the BM cells and eject them out of the bone shaft. A well-flushed bone will normally lose the red color of the BM and become almost translucent.

5. Using Ficoll, separate the low-density BM cells from RBC and other cells contained in the BM (*see* **Note 2**). This is accomplished by layering flushed BM cells on top of Ficoll and centrifuging the mix for 30 min at room temperature. Collect low-density (LD) cells at the interface of medium and Ficoll. Use repeated slow aspiration of the cells at interface to ensure the collection of all of the cells (for details on how to Ficoll BM cells please *see* **Note 3**).

6. Place the cells into a 50 mL tube and fill the tube with heparin medium to dilute the Ficoll as much as possible.

7. Centrifuge at $500 \times g$ for 10 min at 4 °C.

8. Decant supernatant and resuspend the cells in stain wash buffer at a concentration of 1.0–2.0 million cells per mL.

9. Count the BM LD cells by differential trypan blue staining.

10. As an example, we explain here what is required to measure the cell cycle status of mouse BM derived progenitor cells defined as Lineage-Sca1+ cKit+(LSK) cells (*see* **Note 4**). These cells are identified by three sets of markers. Lineage markers (a cocktail of lineage-specific markers that identifies differentiated hematopoietic cells within specified cell lineages such as T cells (CD3, CD4, and CD8), B cells (B220), and myeloid cells (GR-1, Mac1, CD11b, CD14)), Sca-1, and cKit (CD117). The lineage marker cocktail can vary in composition, but it should at least cover the three lineages listed above.

11. For this combination, a total of seven tubes (six control tubes and one sample tube) are required.

    (a) Isotype control tube.

    (b) Lineage only tube.

    (c) Sca-1 only tube.

    (d) cKit only tube.

    (e) Hst only tube.

    (f) PY only tube.

    (g) Sample tube with Lineage, Sca-1, cKit, Hst, and PY.

12. Since cells become photosensitive while cell surface and Hst/PY staining, wrap all tubes containing cells in aluminum foil and work with the lights in the hood turned off.

13. Using tubes compatible with the available flow cytometric equipment, prepare tubes with 5×10^5–1×10^6 cells in each tube for one isotype control, five single-color positive controls, and a sample tube. The number of cells to be used in the sample tube depends on the degree of characterization of these cells and the final size of the group of cells to be collected. For regular phenotypic analyses 2–3 million cells should be enough.

14. If desired, fluorescence minus one (FMO) tubes can be added to this schema as appropriate. However, FMO tubes for Hst and PY are in general not very useful and do not necessarily aid in the identification of positive events for each dye or in the analysis of cells in different phases of cell cycle.

15. Fill these tubes with 3–4 mL of stain wash and centrifuge at $500 \times g$ for 10 min at 4 °C.

16. Decant the supernatant and vortex the pellet gently.

17. Cells can be stained in 50 μL volume of stain wash.

18. Add all the isotype control antibodies to tube a; add all the lineage antibodies to tube b (this cocktail can be designed to cover any combination required and all antibodies used should be conjugated to the same fluorochrome to allow for the selection of negative cells based on the collective positive signal from all markers combined); add Sca-1 antibody to tube c; add CD117 antibody to tube d; add all the lineage markers plus Sca-1 and cKit to the sample tube (g). At this point, do not stain the cells in tubes e and f (*see* **Note 5**).

19. Stain cells on ice for 15 min.

20. Wash cells once with 3–4 mL of stain wash by centrifuging at $500 \times g$ for 10 min at 4 °C.

21. Decant supernatant and resuspend cells in 250 μL of BD Cytofix/Cytoperm fixative reagent.

22. Incubate the cells at 4 °C for 30 min.

23. Centrifuge at $500 \times g$ for 10 min at 4 °C.

24. Decant the supernatant (fixative reagent) and vortex the pellet gently.

25. Resuspend the cells in tubes a through d (Isotype and single color positive control tubes) in 300 μL of stain wash and keep them on ice until their analysis on flow cytometer.

26. Proceed with Hst/PY staining on sample tube (tube #g), Hst only tube (tube #e), and PY only tube (tube #f).

27. Resuspend the cells in these tubes in 3–4 mL of Hst buffer.

28. Centrifuge at $500 \times g$ for 10 min at 4 °C.

29. Decant the supernatant and vortex the pellet gently.

30. Resuspend the cells by adding 0.5 mL of the Hst working solution to the sample tube and to the Hst only tube. Add 0.5 mL of Hst buffer to the PY tube. If more than 1×10^6 cells are used in the sample tube, the amount of Hst has to be increased appropriately (*see* **Note 6**).

31. Incubate in water bath at 37 °C for 45 min. Vortex the tubes gently every 15 min. during incubation.

32. At the end of the 45 min incubation add 2.5 μL of PY working solution to the sample tube and to the PY only tube. Vortex the tubes gently every 15 min. during incubation. If more than 1×10^6 cells are used in the sample tube, the amount of PY has to be increased appropriately. A good formula is to add 2.5 μL of PY working solution to every 0.5 mL of Hst working solution used.

33. Incubate the cells for an additional 45 min at 37 °C in the water bath.

34. Wash the cells with 3–4 mL of verapamil-containing Hst buffer by centrifuging at $500 \times g$ for 10 min at 4 °C.

35. Decant the supernatant and resuspend the cells in 300 μL of verapamil-containing Hst buffer. Keep the tubes on ice until their analysis on a flow cytometer.

36. Data acquisition: All tubes, except sample tube/s, are used as single color controls to adjust the auto-fluorescence and compensation that is required for any type of standard flow cytometry. However, it is important to note that the Hst and PY signal must be collected with the PMT in linear mode. A flow cytometer providing 50 mW of UV light at a wave length of approximately 350 nm is required for the excitation of the Hst. PY can be excited by 100 mW of 488 nm light emitted from an argon laser. Other fluorochromes that are used to gate on the progenitor cells such as PE-Cy7 can be excited by 488 nm laser

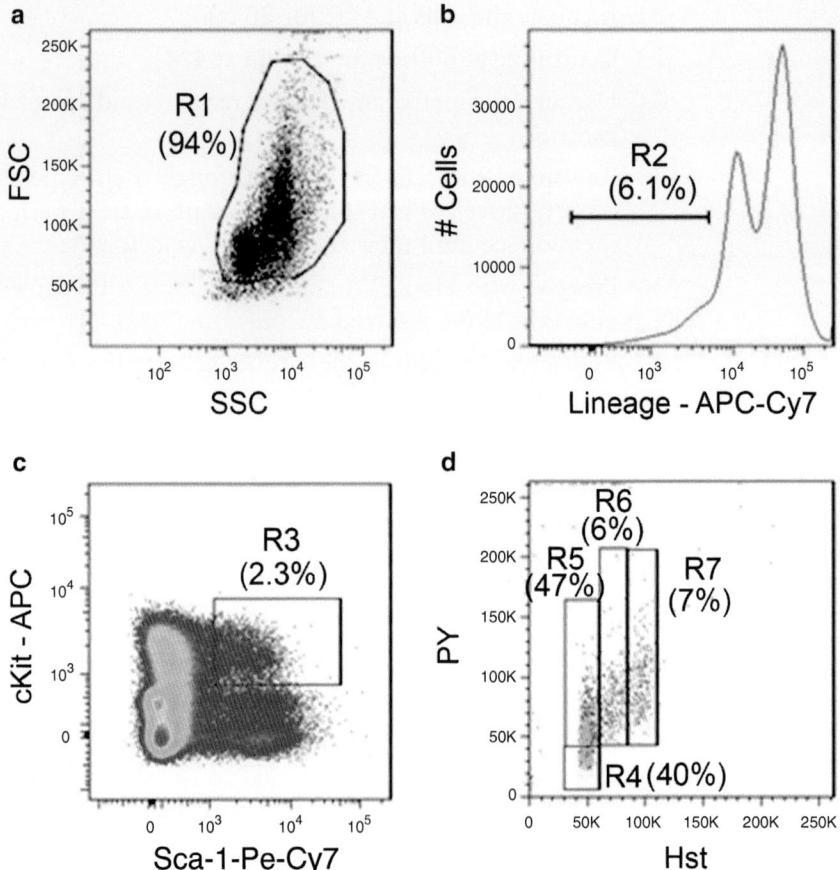

Fig. 1 Analysis of mouse bone marrow low-density cells with Hst and PY. Low-density cells were collected on Ficoll-Paque, and then stained with APC-Cy7 conjugated Lineage markers (CD3, CD4, CD45R, Ter119, and Gr1), PE-Cy7 conjugated Sca-1, and APC conjugated cKit monoclonal antibodies. Stained cells were fixed with BD Cytofix/Cytoperm followed by Hst and PY staining as described in Subheading 3.1. Dot plot (**a**) represents light-scatter distribution of stained low-density BM cells. Gate R1 in (**a**) was established to contain live cells. Histogram (**b**) shows only live cells, and gate R2 represents the lineage negative cells. Dot plot (**c**) shows Sca-1 and cKit fluorescence distribution of R1 and R2 gated lineage negative cells. Gate R3 represents the double positive cells for Sca1 and cKit. Events contained within R3 were considered Lin-Sca1+cKit+(LSK) cells. Events satisfying the selection criterion of R1, R2, and R3 were then analyzed for Hst and PY fluorescence distribution simultaneously as shown in (**d**). Dot plot (**d**) depicts a typical two-dimensional distribution of DNA (*X*-axis) and RNA (*Y*-axis) content. Gates R4, R5, R6, and R7 represent G0, G1, S, and G2–M cells, respectively

whereas APC and APC-Cy7 can be excited by 633 nm excitation laser. PY, which has a λ max 552 nm, can be detected through a 575 ± 13 nm dichroic filter, and Hst can be detected through a 424 ± 22 nm dichroic filter. A typical dot plot of LSK cells concurrently stained with Hst and PY is shown in Fig. 1.

3.2 Cell Cycle Analysis of Murine Progenitor Cells Using Hst and CD71 (for Live Cells)

This method describes the simultaneous staining of mouse BM cells with cell surface markers to gate on progenitor cells along with Hst for DNA staining, and CD71 to gate proliferating cells. This method is appropriate where isolation of live cells is required for further biological assays. Although the method described in Subheading 3.1 is recommended for analytical purposes, our experience is that PY is toxic to the cells. The majority of murine BM cells exposed to PY die within 96 h of staining making it very difficult to accurately assess the number of cells used in a particular assay in vitro or transplanted in vivo for repopulating assays (BRC and EFS, personal observations). Therefore, replacing PY with CD71 can overcome this problem. However, we still employ Hst and PY method with human cells for both analytical and functional studies (*see* **Note 7**).

1. Follow **steps 1–10** as described in Subheading 3.1.

2. Since cells become photosensitive during Hst and cell surface staining, wrap all tubes containing cells in aluminum foil and work with the lights in the hood turned off.

3. Using tubes compatible with the available flow cytometric cell sorting equipment, prepare one tube with 5×10^5 to 1×10^6 cells as an Hst only control tube. Fill this tube with 3–4 mL of Hst buffer.

4. Prepare the second tube as sample tube with a sufficient number of low-density BM cells to yield the required number of isolated populations. The number of cells to be used depends on the degree of characterization of these cells and the final size of the group of cells to be selected and sorted. If more than 10×10^6 cells are used, they can be stained in 50 mL conical tube. Fill the sample tube with appropriate volume of Hst buffer.

5. Centrifuge both tubes at $500 \times g$ for 10 min at 4 °C.

6. Decant the supernatant and vortex the pellet gently.

7. Resuspend the cells by adding 0.5 mL of the Hst working solution to the Hst only tube and to the sample tube. If more than 1×10^6 cells are used in sample tube, the amount of Hst has to be increased appropriately (*see* **Note 6**).

8. Incubate in a water bath at 37 °C for 1.5 h. Vortex the tubes gently every 15 min. during incubation.

9. Wash the cells with an appropriate volume of verapamil-containing Hst buffer by centrifuging at $500 \times g$ for 10 min at 4 °C.

10. Decant the supernatant and resuspend the cells in Hst only tube with 300 µL of verapamil-containing Hst buffer and leave it on ice until analysis. Continue staining the sample tube with cell surface markers.

11. Prepare another 5 tubes with 5×10^5 to 1×10^6 BM LD cells each as an isotype control tube, and four single-color positive controls. Fill these tubes with 3–4 mL of stain wash and centrifuge at $500 \times g$ for 10 min at 4 °C. Decant the supernatant and vortex the pellet gently.

12. Cells can be stained in 50 μL volume of stain wash in case of control tubes, or Hst buffer in case of sample tube.

13. Add all the isotype control antibodies to tube 1. Add all the lineage antibodies to tube 2 (this cocktail can be designed to cover any combination required and all antibodies used should be conjugated to the same fluorochrome to allow for the selection of negative cells based on the collective positive signal from all markers combined). Add the Sca-1 antibody to tube 3. Add the CD117 antibody to tube 4. Add CD71 antibody to tube 5. Add all the lineage marker, Sca-1, cKit, and CD71 to sample tube.

14. Stain cells on ice for 15 min.

15. Wash cells in control tubes with stain wash and resuspend in 300 μL of stain wash. Wash cells in sample tube with verapamil-containing Hst buffer and resuspend in appropriate volume of verapamil-containing Hst buffer.

16. Analyze or sort the cells on the flow cytometer. All single color control tubes are used to adjust the auto-fluorescence and compensation. It is important to note that the Hst signal must be collected with the PMT in linear mode. CD71 signal can be collected on a log scale. A flow cytometer providing 50 mW of UV light at a wave length of approximately 350 nm is required for the excitation of the Hst. Other fluorochromes such as PE or PE-Cy7 can be excited by 488 nm laser where as APC or APC-Cy7 can be excited by 633 nm excitation laser. A typical dot plot of LSK cells stained with Hst and CD71 is shown in Fig. 2.

4   Notes

1. Although adult mouse BM cells are used to describe the procedure, the procedure is same for BM cells of any age.

2. Ficolling the BM cells gives superior results to lysing the RBC with RBC lysis buffer. In the course of lysing, more cells may die and in general, dead cells interfere with cell surface and cell cycle staining.

3. Ficolling: Suspend BM cells in 30 mL medium or PBS/tube in a 50 mL conical tube. Carefully underlay 13.5 mL Ficoll beneath the cell suspension by inserting a 10 mL pipette at the bottom of the tube and releasing the Ficoll slowly allowing it to settle below the cell suspension. Alternatively, the cell suspension can be overlaid on top of the Ficoll. As mentioned

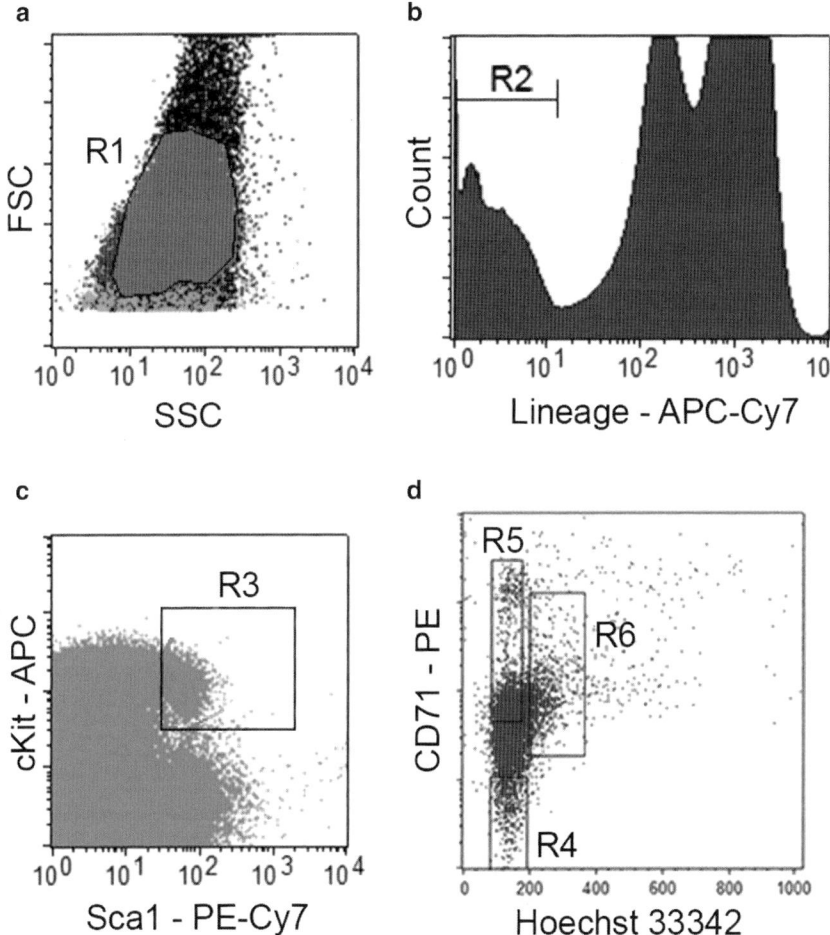

Fig. 2 Analysis of mouse BM low-density cells with Hst and CD71. Low-density cells were stained with Hst as described in Subheading 3.2 followed by cell surface staining. Hst stained cells were stained with APC-Cy7 conjugated Lineage markers (CD3, CD4, CD45R, Ter119, and Gr1), PE-Cy7 conjugated Sca-1, APC conjugated cKit, and PE conjugated CD71 monoclonal antibodies. Dot plot (**a**) represents light-scatter distribution of stained low-density BM cells. Gate R1 in (**a**) was established to contain live cells. Histogram (**b**) shows lineage expression of R1 gated cells. Gate R2 defines lineage negative cells. Dot plot (**c**) shows Sca-1 and cKit fluorescence distribution of R1 and R2 gated lineage negative cells. Gate R3 represents the double positive cells for Sca1 and cKit. Dot plot (**d**) depicts a typical two-dimensional fluorescence distribution of R3 gated cells (LSK cells) depicting Hst and CD71-PE signals on the *X*- and *Y*-axis, respectively. Gate R4 contains CD71 negative or low cells which represent quiescent cells; whereas gate R5 contains CD71 high cells which represent proliferating or cycling cells. Cells in gate R6 represent S/G2+M cells based on their DNA content as determined by their Hst fluorescence signal

above, the volume of Ficoll required for a 50 mL conical tube is 13.5 mL and if the cells are layered in a 15 mL conical tube, then only 4 mL of Ficoll are used (delivered in a 5 mL pipette to avoid spilling). In general it is easier to underlay Ficoll in a 50 mL tube and to overlay cells (on top of 4 mL Ficoll) in a 15 mL tube.

4. The phenotype can be extended with several other markers with non-overlapping conjugated fluorochromes (i.e., five or six color analysis) depending on the configuration of the flow cytometer.

5. In order to avoid major compensation issues, if permitted, it is recommended to use the fluorochromes with emission spectra far from Hst and PY emission. In our studies, we use Lineage markers in Allophycocyanin-Cy7 (APC-Cy7), Sca1 in Phycoerythrin-Cy7 (PE-Cy7), and cKit in Allophycocyanin (APC).

6. If the number of cells to be stained is more than 1.0×10^6 cells, the volume of Hst working solution has to be increased. A good formula to follow is to use 0.5 mL of Hst working solution for every 1.0×10^6 cells. Bigger tubes such as conical 50 mL tubes can be used to accommodate larger number of cells. In the following steps, the volume of PY working solution has to be adjusted too if a larger volume of Hst is used in order to maintain the same final concentration.

7. Selected human CD34+ cells derived either from BM or cord blood or mobilized peripheral blood can be stained with Hst and PY along with other cell surface markers. Human cells are more resistant to PY toxicity compared to murine cells. Therefore, sorted CD34+ cells based on Hst/PY staining can be used for further functional studies including in vivo transplantation into NSG mice. In order to do this, selected CD34+ cells along with appropriate single color control tubes as mentioned in Subheading 3.1 are washed once with Hst buffer, then stained with Hst followed by PY at 37 °C in the dark for a total of 1.5 h as explained in **steps 30–34** in Subheading 3.1. The stained cells are washed once with vera-pamil-containing Hst buffer and resuspend in appropriate cell surface antibody cocktail mix. Stain for another 15 min on ice. Wash once with verapamil-containing Hst buffer and resuspend in appropriate volume of verapamil-containing Hst buffer for their analysis or sorting.

Acknowledgements

The authors thank the operators of the Indiana University Melvin and Bren Simon Cancer Center Flow Cytometry Resource Facility for their outstanding technical help and support. Indiana University is an NIDDK designated Center of Excellence in Molecular Hematology (NIDDK P01 DK090948). The Flow Cytometry Research Facility is partially funded by NCI P30CA082709.

References

1. Till JE, McCulloch EA (1961) A direct measurement of the radiation sensitivity of normal mouse bone marrow cells. Radiat Res 14: 213–222

2. Abramson S, Miller RG, Phillips RA (1977) The identification in adult bone marrow of pluripotent and restricted stem cells of the myeloid and lymphoid systems. J Exp Med 145(6):1567–1579

3. Snodgrass R, Keller G (1987) Clonal fluctuation within the haematopoietic system of mice reconstituted with retrovirus-infected stem cells. EMBO J 6:3955–3960

4. Van Zant G, Scott-Micus K, Thompson BP, Fleischman RA, Perkins S (1992) Stem cell quiescence/activation is reversible by serial transplantation and is independent of stromal cell genotype in mouse aggregation chimeras. Exp Hematol 20:470–475

5. Ogawa M (1993) Differentiation and proliferation of hematopoietic stem cells. Blood 81(11):2844–2853

6. Zhang J, Niu C, Ye L, Huang H, He X, Tong WG, Ross J, Haug J, Johnson T, Feng JQ, Harris S, Wiedemann LM, Mishina Y, Li L (2003) Identification of the haematopoietic stem cell niche and control of the niche size. Nature 425(6960):836–841

7. Arai F, Hirao A, Ohmura M, Sato H, Matsuoka S, Takubo K, Ito K, Koh GY, Suda T (2004) Tie2/angiopoietin-1 signaling regulates hematopoietic stem cell quiescence in the bone marrow niche. Cell 118(2):149–161

8. Calvi LM, Adams GB, Weibrecht KW, Weber JM, Olson DP, Knight MC, Martin RP, Schipani E, Divieti P, Bringhurst FR, Milner LA, Kronenberg HM, Scadden DT (2003) Osteoblastic cells regulate the haematopoietic stem cell niche. Nature 425(6960):841–846

9. Heissig B, Hattori K, Dias S, Friedrich M, Ferris B, Hackett NR, Crystal RG, Besmer P, Lyden D, Moore MAS, Werb Z, Rafii S (2002) Recruitment of stem and progenitor cells from the bone marrow niche requires MMP-9 mediated release of kit-ligand. Cell 109(5): 625–637

10. Wilson A, Oser GM, Jaworski M, Blanco-Bose WE, Laurenti E, Adolphe C, Essers MA, Macdonald HR, Trumpp A (2007) Dormant and self-renewing hematopoietic stem cells and their niches. Ann N Y Acad Sci 1106:64–75, Hematopoietic Stem Cells VI

11. Gothot A, van der Loo JCM, Clapp DW, Srour EF (1998) Cell cycle-related changes in repopulating capacity of human mobilized peripheral blood CD34+ cells in non-obese diabetic/ severe combined immune-deficient mice. Blood 92(8):2641–2649

12. Jones RJ, Celano P, Sharkis SJ, Sensenbrenner LL (1989) Two phases of engraftment established by serial bone marrow transplantation in mice. Blood 73(2):397–401

13. Jordan CT, Guzman ML, Noble M (2006) Cancer stem cells. N Engl J Med 355(12): 1253–1261

14. Lacorazza HD, Yamada T, Liu Y, Miyata Y, Sivina M, Nunes J, Nimer SD (2006) The transcription factor MEF/ELF4 regulates the quiescence of primitive hematopoietic cells. Cancer Cell 9(3):175–187

15. Chitteti BR, Liu Y, Srour EF (2011) Genomic and proteomic analysis of the impact of mitotic quiescence on the engraftment of human CD34+ cells. PLoS One 6(3):e17498

16. Gothot A, Pyatt R, McMahel J, Rice S, Srour EF (1998) Assessment of proliferative and colony-forming capacity after successive in vitro divisions of single human CD34+ cells initially isolated in G0. Exp Hematol 26:562–570

17. Gothot A, Pyatt R, McMahel J, Rice S, Srour EF (1997) Functional heterogeneity of human CD34+ cells isolated in subcompartments of the G0/G1 phase of the cell cycle. Blood 90(11):4384–4393

18. Johnson LV, Walsh ML, Chen LB (1980) Localization of mitochondria in living cells with rhodamine 123. Proc Natl Acad Sci U S A 77:990–994

19. Bertoncello I, Hodgson GS, Bradley TR (1985) Multiparameter analysis of transplantable hematopoietic stem cells. I. The separation and enrichment of stem cells homing to marrow and spleen on the basis of rhodamine 123 fluorescence. Exp Hematol 13:999–1006

20. Visser JWM, deVries P (1988) Isolation of spleen-colony forming cells (CFU-S) using wheat germ agglutinin and rhodamine 123 labeling. Blood Cells 14:369–384

21. Darzynkiewicz Z, Traganos F, Staiano-Coico L, Kapuscinski J, Melamed MR (1982) Interactions of rhodamine 123 with living cells studied by flow cytometry. Cancer Res 42: 799–806

22. Jordan CT, Yamasaki G, Minamoto D (1996) High-resolution cell cycle analysis of defined phenotypic subsets within primitive human hematopoietic cell populations. Exp Hematol 24:1347–1352

23. Shapiro HM (1981) Flow cytometric estimation of DNA and RNA content in intact cells stained with Hoechst 33342 and Pyronin Y. Cytometry 2:143–151

24. Darzynkiewicz Z, Kapuscinski J, Traganos F, Crissman HA (1987) Application of pyronin Y (G) in cytochemistry of nucleic acids. Cytometry 8:138–145

25. Ladd AC, Pyatt R, Gothot A, Rice S, McMahel J, Traycoff CM, Srour EF (1997) Orderly process of sequential cytokine stimulation is required for activation and maximal proliferation of primitive human bone marrow CD34+ hematopoietic progenitor cells residing in G0. Blood 90(2):658–668

26. Cowden RR, Curtis SK (1983) Supravital experiments with pyronin Y, a fluorochrome of mitochondria and nucleic acids. Histochemistry 77:535–542

27. Judd W, Poodry CA, Strominger JL (1980) Novel surface antigen expressed on dividing cells but absent from nondividing cells. J Exp Med 152(5):1430–1435

28. Jefferies WA, Brandon MR, Hunt SV, Williams AF, Gatter KC, Mason DY (1984) Transferrin receptor on endothelium of brain capillaries. Nature 312(5990):162–163

29. Kemp JD, Thorson JA, McAlmont TH, Horowitz M, Cowdery JS, Ballas ZK (1987) Role of the transferrin receptor in lymphocyte growth: a rat IgG monoclonal antibody against the murine transferrin receptor produces highly selective inhibition of T and B cell activation protocols. J Immunol 138(8):2422–2426

Chapter 6

Flow Cytometric Analysis of Signaling and Apoptosis in Hematopoietic Stem Cells

Lei Dong and Cheng-Kui Qu

Abstract

The analysis of protein phosphorylation in hematopoietic stem cells (HSCs) provides a powerful tool for studying the cell signaling activities that mediate HSC fate decisions, such as self-renewal, differentiation, and apoptosis. The first part of this chapter describes a method of intracellular staining for phosphorylated proteins in conjunction with membrane staining for multiple hematopoietic cell-surface markers, and subsequent flow cytometric analysis of protein phosphorylation levels [indicated by mean fluorescence intensity (MFI) of specific fluorochromed phospho-antibodies] in primitive hematopoietic cells. The second part describes a method for assessing the frequency of apoptosis in HSCs using extracellular staining with recombinant Annexin V and 7-Amino-Actinomycin (7-AAD). Both parts involve an initial magnetic enrichment of hematopoietic stem/progenitor cells from bone marrow. Because of the intracellular detection required for the HSC signaling assay, this assay also includes cell fixation and permeabilization. Gating strategies for assessing MFI and the frequency of Annexin V+ apoptotic cells in a complex population are also described along with representative examples.

Key words Hematopoietic stem cell, Cell signaling, Apoptosis, Flow cytometry

1 Introduction

Hematopoietic stem cells (HSCs) are a subset of rare precursor cells which are at the apex of a hierarchy of lineage-specific progenitors that develop in a committed manner to fully mature blood cells. HSCs possess both self-renewal and differentiation capabilities and the balance between self-renewal and differentiation must be finely tuned in order to maintain hematopoietic homeostasis [1]. In order to accomplish this homeostasis, HSCs receive constant input from growth factors/cytokines, extracellular matrix components, and hormones from extracellular milieu. The intracellular signaling pathways initiated from the receptors of these

Kevin D. Bunting and Cheng-Kui Qu (eds.), *Hematopoietic Stem Cell Protocols*, Methods in Molecular Biology, vol. 1185, DOI 10.1007/978-1-4939-1133-2_6, © Springer Science+Business Media New York 2014

factors enable the cells to integrate diverse signals to determine the final outcome. Several evolutionarily conserved pathways, such as Erk, Akt, Jak/Stat, Wnt, Notch, TGFβ/SMAD, etc. regulate self-renewal, differentiation, proliferation, apoptosis, and senescence of HSCs [2]. Apoptosis is one of the physiological processes that also control HSC homeostasis [3]. The HSC compartment maintains sufficient HSC numbers to sustain hematopoiesis, while avoiding inappropriate expansion.

Studies of HSC signaling determining the cell fate in physiology and diseases often require quantitative assays to assess cell signaling activities (e.g., phosphorylation levels of key signaling proteins) and the frequency of apoptosis in the rare HSC compartment. Flow cytometry is an extremely powerful multiparameter method for analyzing phospho-specific epitopes that has many advantages over western blotting or ELISA [4]. A critical advantage of phospho-epitope analysis by flow cytometry is the capability to measure signaling events in a single cell. Another advantage of flow cytometry analysis is its ability to analyze specific cell subsets, such as HSCs, within complex populations. It may also permit simultaneous analyses of intracellular parameters in multiple cell subsets, such as HSCs, common myeloid progenitors, common lymphoid progenitors, granulocyte and macrophage progenitors, and megakaryocyte erythroid progenitors. In addition, flow cytometric data of phosphorylation levels of signaling proteins are easy to visualize and compare among samples. They are more informative than those obtained by western blotting, although some considerations need to be addressed for this analysis (*see* **Note 1**). These assays will facilitate the studies of how genetic mutations, disease, or therapeutic manipulations impact HSC activities.

In this protocol, we present approaches that enable quantitation of phosphorylation levels of signaling proteins and apoptotic cells in the HSC population. In brief, we will enrich HSCs from bone marrow cells by removing differentiated cells with lineage cocktail antibodies, combined with positive selection for markers known to be expressed on HSCs (Sca-1$^+$ and c-Kit$^+$) [5, 6]. This strategy selects a population of LSK (lineage$^-$Sca-1$^+$c-Kit$^+$) cells that contains only ~10 % long-term HSCs. To obtain highly pure HSCs, several additional selection strategies have been developed in different laboratories [7]. In this protocol, we use the LSKCD150$^+$CD48$^-$ strategy from Morrison's lab to identify HSCs [8]. Phosphorylation levels of signaling proteins and apoptotic cells in HSCs are then determined following phospho-epitope antibody intracellular staining and extracellular staining with recombinant Annexin V and 7-Amino-Actinomycin (7-AAD), respectively.

2   Materials

2.1   Buffers and Reagents Required for Both the Signaling and Apoptosis Assays

1. Tissue culture medium: Iscove's modified Eagle's medium, 2 % fetal bovine serum (FBS), 10 mM HEPES-HCL, pH 7.2–7.5. Stored at 4 °C.

2. Red blood cell (RBC) lysis buffer: Purchased from e-Bioscience, Inc. (San Diego, CA, USA).

3. Phosphate-buffered saline (PBS): Purchased from HyClone laboratories, Inc. (Logan, UT, USA).

4. Staining buffer: PBS supplemented with 2 % FBS. Stored at 4 °C.

5. 70 µm cell strainer: Purchased from BD Biosciences, Inc. (Bedford, MA, USA).

6. Trypan blue: Purchased from GIBCO, Inc. (Carlsbad, CA, USA).

7. MACS separation columns: Purchased from Miltenyi Biotec, Inc. (Auburn, CA, USA).

8. All fluorochrome-conjugated antibodies used in the signaling and apoptosis assays can be purchased from BD Biosciences, Inc. or e-Bioscience, Inc.

2.2   Conjugated Antibodies Required for Both the Signaling and Apoptosis Assays

1. Lineage cell depletion kit: Purchased from Miltenyi Biotec, Inc. (Cat. No. 130-090-858).

    (a) Cocktail of biotin-conjugated monoclonal antibodies against CD5, CD45R (B220), CD11b, anti-Gr-1 (Ly-6G/C), 7-4, and Ter-119.

    (b) Anti-biotin microbeads.

2. Lineage antibodies conjugated to biotin, such as anti-B220 (Cat. No. 13-0452-75; Clone: RA3-6B2), anti-CD3 (Lot. No. E02344-1631; Clone: 145-2C11), anti-Gr1 (Cat. No. 13-5931-75; Clone: RB6-8C5), anti-Mac1 (Cat. No. 13-0112-75; Clone: M1/70), and anti-Ter119 (Cat. No. 13-5921-75; Clone: TER-119): Purchased from e-Bioscience, Inc. (San Diego, CA, USA).

3. eFluor 450-conjugated streptavidin: Purchased from e-Bioscience, Inc. (Cat. No. 48-4317-82).

4. PE-cy7-conjugated anti-Sca-1: Purchased from BD Biosciences, Inc. (Cat. No. 558162; Clone: D7).

5. APC-cy7-conjugated anti-c-Kit: Purchased from BD Biosciences, Inc. (Lot. No. E08461-1633; Clonc: 2B8).

6. APC-conjugated anti-CD150: Purchased from BD Biosciences, Inc. (Cat. No. 562373; Clone: TC15-12F12.2).

7. PE-conjugated anti-CD48: Purchased from BD Biosciences, Inc. (Cat. No. 12-0481-82; Clone: HM48-1).

2.3 Reagents for Intracellular Staining	1. FITC-conjugated anti-mouse IgG κ isotype control: Purchased from BD Biosciences, Inc. (San Diego, CA, USA) (Cat. No. 556028).
	2. BD Cytofix/Cytoperm: Purchased from BD Biosciences, Inc. (Cat. No. 51-2090KZ). Stored at 4 °C.
	3. BD perm/wash: Purchased from BD Biosciences, Inc. (Cat. No. 554723). Stored at 4 °C.
	4. FITC-conjugated anti-phospho-protein antibodies: In this protocol, we use Alexa-488-conjugated anti-phosphorylated Erk (p-Erk) as an example. p-p44/42 MAPK (T202/Y204) was purchased from Cell Signaling Technology, Inc. (Danvers, MA, USA) (Lot. No. 26; Clone: E10).
2.4 Reagents for Apoptosis Analysis	1. Annexin V-FITC. Purchased from BD Biosciences, Inc. (Cat. No. 5165875X).
	2. 1× Annexin V-binding buffer: Diluted from 10× Annexin V-binding buffer. Purchased from BD Biosciences, Inc. (Cat. No. 51-66121E).
	3. 7-AAD: Purchased from BD Biosciences, Inc. (Cat. No. 51-68981E).
2.5 Flow Cytometer	1. Flow cytometer: BD LSR II flow cytometer (BD Biosciences, Inc.).
	2. FlowJo software: Purchased from Treestar, Inc. (Ashland, OR, USA).

3 Methods

3.1 HSC Signaling Analysis by Flow Cytometry	1. Isolate bone marrow cells from mouse femurs and tibias by flushing with 3 ml of tissue culture media (*see* **Note 2**).
	2. Prepare single cell suspension by passing through a 70 μm cell strainer and pellet cells at $300 \times g$ for 5 min at 4 °C.
	3. Aspirate supernatant and resuspend cell pellet in 2 ml of RBC lysis buffer.
	4. Mix well and leave the tube at room temperature (RT) for 2 min and then dilute with 2 ml of cold PBS.
	5. Centrifuge at $300 \times g$ for 5 min at 4 °C.
	6. Decant supernatant and resuspend cell pellet in 5 ml of cold staining buffer.
	7. Count viable cells on a hemacytometer:
	(a) Take 10 μl aliquot and add 30 μl of PBS in a microcentrifuge tube.

(b) Add 40 µl of trypan blue and mix well.

(c) Take 10 µl of the mixture and load onto a hemacytometer, count cells, and determine cell concentration.

8. Spin and resuspend cells in staining buffer at 1×10^7 cells/40 µl for lineage depletion.

9. Add 10 µl of cocktail of biotin-conjugated monoclonal antibodies [CD5, CD45R (B220), CD11b, anti-Gr-1 (Ly-6G/C), 7-4, and Ter-119] per 10^7 cells.

10. Mix well and incubate at 4 °C for 10–15 min.

11. Add 30 µl of staining buffer per 10^7 cells.

12. Add 20 µl of anti-biotin microbeads per 10^7 cells, mix well and incubate at 4 °C for additional 15 min.

13. Wash cells by adding 1 ml of PBS and centrifuge at $300 \times g$ for 5 min at 4 °C.

14. Resuspend up to 10^8 cells in 500 µl of staining buffer for magnetic separation.

15. For the MACS depletion step, follow the manufacturer's instruction. In brief, run samples through a fresh PBS Pre-run MACS separation column, collect pass-through cells which are negatively labeled Lin⁻ rich fraction, and wash the column twice with staining buffer and collect pass-through cells. Count cell numbers for subsequent HSC and intracellular staining (*see* **Note 3**).

16. Pellet Lin⁻ rich cell fraction at $300 \times g$ for 5 min at 4 °C and count cell number as described in **step 7**.

17. Resuspend cells in 40 µl of staining buffer. Proceed with cell surface staining for the HSC phenotypes of choice. For HSCs, perform the following surface staining (*see* **Note 4**):

(a) Repeat the lineage staining step by using lineage antibodies conjugated to biotin (anti-B220, anti-CD3, anti-Gr1, anti-Mac1, and anti-Ter119) (0.25 µg antibody per 10^6 cells). Stain at 4 °C for 15–20 min and then wash with 1 ml of cold PBS. Pellet cells at $300 \times g$ for 5 min at 4 °C and remove supernatant.

(b) Resuspend the cells in 40 µl of staining buffer containing anti-biotin-eFlour 450, Sca-1-PE-cy7, c-Kit-APC-cy7, CD48-PE, and CD150-APC (0.5 µg antibody per 10^6 cells). Stain at 4 °C for 15–20 min in the dark (*see* **Note 5**).

18. Add 1 ml of cold PBS, mix, and pellet cells at $300 \times g$ for 5 min at 4 °C.

19. Add 200 µl of BD Cytofix/Cytoperm to cell pellets and resuspend cells. Incubate at RT for 30 min in the dark on an orbital shaker (*see* **Note 6**).

20. Wash cells with 1 ml 1× BD perm/wash and centrifuge at $300 \times g$ for 5 min.

21. Resuspend cells in 50 μl of BD perm/wash containing 0.5 μg of FITC labeled phospho-protein antibody, for example, p-Erk-Alex-488 (0.5 μg per 10^6 cells). Mix and incubate at RT for 1 h in the dark (*see* **Note 7**).

22. Wash cells with 1 ml of BD perm/wash, pellet cells at $300 \times g$ for 5 min at 4 °C.

23. Resuspend cells in 300 μl PBS/tube. Analyze samples on BD LSR II. Collect 100–1,000 HSC events per staining.

3.2 Annexin V Staining to Measure Apoptosis in HSCs by Flow Cytometry

1. Collect bone marrow cells and perform lineage depletion as described in Subheading 3.1.1–3.1.17 (*see* **Note 8**).

2. Add 1 ml of PBS, mix, and pellet cells at $300 \times g$ for 5 min at 4 °C.

3. Resuspend the cell pellet in 1× binding buffer at the concentration of 1×10^6 cells/50 μl.

4. Add 5 μl of Annexin V-FITC and 5 μl of 7-AAD, gently vortex cells, and incubate at RT for 15 min in the dark (*see* **Note 9**).

5. Add additional 400 μl of 1× binding buffer to each tube. Analyze samples on BD LSR II within 1 h. Collect 100–1,000 HSC events per staining.

3.3 Gating Strategies for Flow Cytometric Analyses of HSC Signaling and Apoptosis

1. After collecting raw data on a flow cytometer, apply non-rectangular gate (*see* Fig.1a, Gate 1) based on forward and side scatter parameters. This gate excludes cell debris and unusually large cells.

2. Apply a rectangular Lin⁻ gate analogous to that shown in Fig. 1a (Gate 2). This gate excludes the fraction of Lin⁺ cells which passed through the lineage depletion column.

3. Apply a rectangular c-Kit⁺Sca-1⁺ gate analogous to that shown in Fig. 1a (Gate 3). This gate captures the fraction of LSK cells.

4. Apply a quad gate analogous to that shown in Fig. 1a (Gate 4). This gate captures the HSC (Lin⁻c-Kit⁺Sca-1⁺CD150⁺CD48⁻) population [9].

5. Select the HSC population and use histogram to show mean fluorescence intensity (MFI) of phospho antibody (for example, anti-phospho-Erk-Alex-488) staining. Combine this and isotype control as shown in Fig. 1b (*see* **Note 6**).

6. Select the HSC population and apply a quad gate analogous to that shown in Fig. 1c to estimate the proportion of Annexin V⁺/7-AAD⁻ early apoptotic and Annexin V⁺/7-AAD⁺ late apoptotic cells.

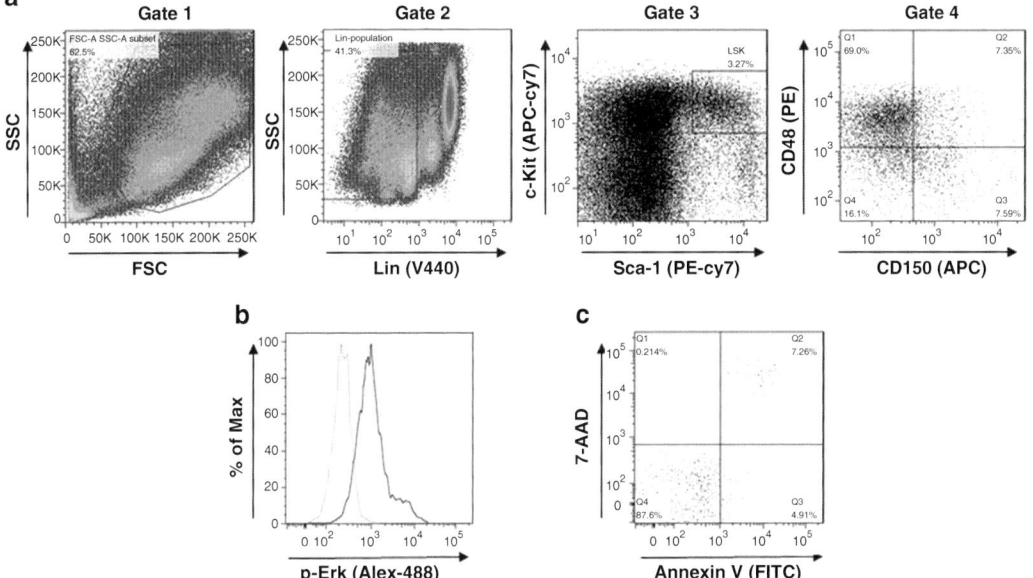

Fig. 1 Gating scheme for assessing Erk phosphorylation levels and apoptosis in the hematopoietic stem cell (HSC) compartment. (**a**) Sequence of electronic gates applied to the flow data collected from wild-type bone marrow cells following magnetic depletion of lineage positive (Lin+) cells and HSC surface marker staining as described in Subheading 3.1. The percentage shown in each plot represents the fraction of events that fall within the indicated gate. FlowJo7.6.1 software was used to analyze the collected events. This example shows gating strategies for assessing Erk activities and apoptotic cells in phenotypic (LSKCD150+CD48−) HSCs. (**b**) Mean fluorescence intensity of phosphorylated Erk in the HSC population. *Gray line* indicates the isotype control. (**c**) *Dot plots* showing the frequency of apoptosis (Annexin V+7-AAD− early apoptotic cells and Annexin V+7-AAD+ late apoptotic cells) in HSCs

4 Notes

1. Some considerations need to be addressed for protein phosphorylation analysis by flow cytometry.

   (a) Antigen accessibility: Unlike surface proteins, phospho-epitopes are often buried by protein–protein interactions. Experimental evidence suggests that with proper permeabilization reagents, most antigens can be measured efficiently. For staining certain phosphoproteins, stronger permeabilization reagents, such as methanol, may be required.

   (b) Stability of phospho-epitopes: Because of the transient nature of intracellular signaling events, fixation techniques used for protein phosphorylation analyses must be rapid and efficient in freezing signaling to prevent dephosphorylation or further phosphorylation.

(c) Antibody selection: Phospho-specific flow cytometry requires careful selection and screening of antibodies to stain the antigen of interest. Successful analyses of protein phosphorylation may also need you preceding each step carefully to reduce loss of proportion of starting cells post fixation/permeabilization and washing steps.

2. Harvest of tissues: To obtain bone marrow cells from leg bones, we flush the marrow out of bones with a 3-cc needle and syringe. You can also obtain the marrow from arm bones by flushing or from vertebral columns by crushing with a mortar and pestle.

3. After lineage cell depletion, we typically obtain $1-3 \times 10^6$ Lin$^-$ rich cells per mouse from tibias and femurs. Each sample uses the same number of cells for the following HSC staining and intracellular staining in order to get better results.

4. For multicolor flow cytometric analyses, calibration of a flow cytometer and compensation for spectral overlap are always necessary.

(a) Calibration of instrument's response to fluorescence signal: Calibrate the flow cytometer by using a series of beads with at least four different predefined levels of fluorescence intensity values of fluorescent microbeads to distinguish positive from negative cells in immunophenotyping assays.

(b) Setting electronic compensation for spectral overlap (color compensation): accurate compensation can be achieved by staining a control sample containing mutually exclusive populations of the same fluorochrome as used in experimental samples, and by processing the control sample in the same way to experimental samples.

5. It is important to use an antibody of the same clone for each sample and in all experiments to produce reproducible results. After staining, it is important to keep samples in the dark in order to protect fluorescence.

6. The speed of the orbital shaker can set at 400–600 rpm. Gentle agitation is advised to avoid cell clumping during fixation.

7. It is important to include a sample for isotype control staining. When compared with the isotype control, protein phosphorylation levels defined by MFI of the samples should exceed MFI of the isotype control.

8. The RBC lysis step can be skipped in order to get better results in measuring apoptosis in HSCs.

9. Annexin V is used to quantitatively determine the percentage of the cells that are actively undergoing apoptosis. In apoptotic cells, the membrane phospholipid phosphatidylserine is translocated from the inner leaflet to the outer leaflet of the plasma

membrane. 7-AAD is a standard viability probe for distinguishing viable from nonviable cells. Viable cells with intact membranes exclude 7-AAD, whereas damaged membranes are permeable to 7-AAD.

Acknowledgements

This work was supported by the National Institutes of Health grant DK092722 (to C. K. Q.).

References

1. Orford KW, Scadden DT (2008) Deconstructing stem cell self-renewal: genetic insights into cell-cycle regulation. Nat Rev Genet 9:115–128

2. Clements WK, Traver D (2013) Signalling pathways that control vertebrate haematopoietic stem cell specification. Nat Rev Immunol 13:336–348

3. Chao MP, Majeti R, Weissman IL (2012) Programmed cell removal: a new obstacle in the road to developing cancer. Nat Rev Cancer 12:58–67

4. Kalaitzidis D, Neel BG (2008) Flow-cytometric phosphoprotein analysis reveals agonist and temporal differences in responses of murine hematopoietic stem/progenitor cells. PLoS One 3:e3776

5. Okada S, Nakauchi H, Nagayoshi K et al (1992) In vivo and in vitro stem cell function of c-kit- and Sca-1-positive murine hematopoietic cells. Blood 80:3044–3050

6. Morrison SJ, Lagasse E, Weissman IL (1994) Demonstration that Thy(lo) subsets of mouse bone marrow that express high levels of lineage markers are not significant hematopoietic progenitors. Blood 83:3480–3490

7. Mayle A, Luo M, Jeong M et al (2013) Flow cytometry analysis of murine hematopoietic stem cells. Cytometry A 83:27–37

8. Kiel MJ, Radice GL, Morrison SJ (2007) Lack of evidence that hematopoietic stem cells depend on N-cadherin-mediated adhesion to osteoblasts for their maintenance. Cell Stem Cell 1:204–217

9. Kiel MJ, Yilmaz OH, Iwashita T et al (2005) SLAM family receptors distinguish hematopoietic stem and progenitor cells and reveal endothelial niches for stem cells. Cell 121:1109–1121

Part III

Molecular and Cellular Characterization

Chapter 7

Gene Expression Profiling of Hematopoietic Stem Cells (HSCs)

Nalini Raghavachari

Abstract

Transcriptomic analysis to decipher the molecular phenotype of hematopoietic stem cells, regulatory mechanisms directing their life cycle, and the molecular signals mediating proliferation, mobilization, migration, and differentiation is believed to unravel disease-specific disturbances in hematological diseases and assist in the development of novel cell-based clinical therapies in this era of genomic medicine. The recent advent in genomic tools and technologies is now enabling the study of such comprehensive transcriptional characterization of cell types in a robust and successful manner. This chapter describes detailed protocols for isolating RNA from purified population of hematopoietic cells and gene expression profiling of those purified cells using both microarrays (Affymetrix) and RNA-Seq technology (Illumina Platform).

Key words HSC, mRNA, Gene expression profiling, Transcriptome, Microarrays, Next-generation sequencing, RNA-seq

1 Introduction

Recent developments in stem cell biology have generated much excitement about the potential for regenerative medicine and cell-based therapies in a variety of clinical applications, such as treating Parkinson's disease, leukemia, lymphoma, and spinal cord injuries [1–8]. Crucial to the success of these applications is the detailed understanding of how the cells remain stem cells and the cues that they require to differentiate and commit themselves to specific cell fates. Given that hematopoietic stem cells are a particularly interesting class of stem cells with a well-characterized cellular differentiation system, a number of studies have recently been undertaken to decipher their genetic program both in culture and in vivo [9–12].

Hematopoiesis is the term applied to the myriad processes wherein all the different cell lineages that form the blood and immune system are generated from a common pluripotent stem cell [12, 13]. The definitive hematopoietic system produces all adult blood cell types including erythrocytes and cells of the

Kevin D. Bunting and Cheng-Kui Qu (eds.), *Hematopoietic Stem Cell Protocols*, Methods in Molecular Biology, vol. 1185, DOI 10.1007/978-1-4939-1133-2_7, © Springer Science+Business Media New York 2014

myeloid and lymphoid lineages. A complex interplay between the intrinsic genetic processes of hematopoietic cells and their environment including the effects of specific cytokines such as interleukins and granulocyte/monocyte-stimulating factors determines whether stem cells, lineage-specified progenitors, and mature blood cells self-renew, remain quiescent, proliferate, differentiate, or undergo apoptosis [13, 14]. Catastrophic consequences to aberrant hematopoiesis have been described in diseases such as leukemia and lymphoma [13–16]. Hence, understanding the nature of the hematopoietic stem cells, as well as the molecular process by which these cells acquire their specific cell fate, is crucial for understanding disease pathogenesis and for the success of cell-based therapies. To sustain lifelong hematopoiesis, HSCs must self-renew to maintain or expand the HSC pool and must differentiate to form committed hematopoietic progenitor cells (HPCs) that progressively lose self-renewal potential and become increasingly restricted in their lineage potential. A combination of extrinsic and intrinsic signals is thought to converge to regulate HSC differentiation versus self-renewal decisions. Hematopoietic differentiation is known to be strictly regulated by complex network of transcription factors that are controlled by ligand binding to cell surface receptors [13–20].

HSCs can be isolated from bone marrow or peripheral blood using enrichment (magnetic cell separation—MACS) and/or single-cell sorting (fluorescence-activated cell sorting—FACS) based on cell surface markers and/or vital dye staining [21, 22]. The HSC has served as the paradigm for adult stem cell populations by virtue of a well-defined differentiation cascade with distinct intermediaries connecting the differentiation of LT-HSCs into mature, functional hematopoietic cells. Each of the cell stages of HSC differentiation can be purified from the bone marrow or the peripheral blood using characteristic cell surface markers to facilitate the study of hematopoietic biology [23–28].

The transcriptome, the entire repertoire of transcripts in a species, represents a key link between information encoded in DNA and phenotype of a cell/organ. The molecular phenotype of stem cells, the regulatory mechanisms directing their mode of living, and the molecular signals mediating mobilization, migration, and differentiation are only partially understood [29–31]. From a clinical standpoint, deciphering the pattern of gene expression during hematopoiesis may help unravel disease-specific mechanisms in hematopoietic malignancies. As the phenotype of any given cell is ultimately the product of the genes, it is critical to identify gene expression patterns during lineage-specific differentiation of stem cells. It is also believed that an important source of diversity in the transcriptome of differentiated cells is due to the splicing process in multi-exon genes [32]. Alternative splicing is thought to regulate differentiation through coordination of gene networks where each network coordinates a different cell function [12, 33, 34].

In the recent past, gene expression profiling has developed into a robust and standardized profiling technology that allows a direct translation from comprehensive gene expression analysis to biological interpretation of a system. The tools for gene expression profiling have been available for years, as Northern blots, reverse-transcription PCR (RT-PCR), expressed sequence tags (ESTs), and serial analysis of gene expression (SAGE). But the rapid and high-throughput quantification of the expressed genes became a possibility only with the development of gene expression microarrays. In this regard, high-density human Exon 1.0 ST arrays with 1.2 million probe sets that aim to target every known and predicted exon in the entire genome have been successfully employed in many clinical investigations to obtain gene expression profiles and associated alternative splicing events in disease processes [35]. Despite the success of most of the microarray studies in biomarker discovery, inherent limitations on the dynamic range of arrays and the lack of complete coverage for detecting alternative splicing events have constrained the application of the technology [35].

The recent advent of next-generation sequencing (NGS) technologies for direct sequencing of the transcriptional output of the genome is now enabling the complete transcriptional characterization of all the cells of an organism. RNA sequencing (RNA-seq) leverages the capacity of NGS technology to representatively sample a population of RNA templates with a large number of "reads" or parallel reactions on discrete template [36–38]. RNA-seq involves direct sequencing of complementary DNAs (cDNAs) using high-throughput NGS technologies, followed by mapping of the sequencing reads to the reference genome or the gene sets for gene expression analysis and polymorphism detection. The advantages of RNA-seq include generation of digital information of individual annotated genes with literally unlimited dynamic range and ability to comprehensively detect novel transcripts and mRNA variants resulting from alternative promoter usages, splice sites, and polyadenylation [39, 40]. However, the technology also brings with it new issues to resolve such as the requirement of large amounts of starting material, cumbersome library preparation, and novel systematic biases during sample preparation and sequencing which must be accounted for when analyzing the data [41, 42].

The choice of using either of the platforms really depends on multiple factors at this point including both strategy and costs. A major advantage of RNA-seq relative to microarray technology is that sequencing generates data free from any bias of what may or may not be represented with specific probes on any given commercial microarray. Nonetheless, these two technologies are in such constant evolution that it is not possible yet to conclude which of these two approaches to expression profiling is the best. In addition, selection of an appropriate tool for genomic studies is mostly driven by the biological question underlying the study: whether a hypothesis is being tested or the study is conducted purely to obtain deeper coverage of the transcriptome and to discover novel transcripts. There is an emerging approach to apply both RNA-seq and arrays in combination where RNA-seq is first conducted to uncover all the transcriptome elements associated with the disease in question followed by high throughput and reliable screening of these elements on a large sample size using the arrays. Integrating data from both microarray and RNA-seq experiments may open up new possibilities for creating meaningful informational networks which will aid our understanding of disease pathology and development of novel therapeutics.

This chapter describes detailed protocols for isolating RNA from cells purified by FACS or MACS, gene expression profiling using both microarrays (Affymetrix), and RNA-Seq technology (Illumina Platform).

2   Materials

2.1   RNA Isolation

1. RNeasy Mini Kit (Qiagen).

2. RNAqueous Micro Kit (Ambion).

3. Buffer RLT (lysis buffer) contains guanidine isothiocyanate (Qiagen)—use appropriate safety measures when handling.

4. Buffer RW1 (wash buffer) contains guanidine isothiocyanate and alcohol (Qiagen)—use appropriate safety measures when handling.

5. Buffer RPE (wash buffer) supplied as a concentrate (Qiagen): Before using for the first time, add ethanol (90–100 %) as indicated on the bottle to obtain a working solution.

6. RNeasy mini columns (Qiagen).

7. Collection tubes for elution.

8. 14.3 M β-Mercaptoethanol (β-ME).

9. Ethanol (96–100 %).

10. Ethanol (70 % in water).

11. RNase-Free DNase Set (Qiagen).

12. Agilent 2100 Bioanalyzer.

13. Nanodrop ND-1000 Spectrophotometer.

2.2   Microarrays

1. Affymetrix Gene Chips—Exon 1.0 ST arrays or Human Transcriptome Array 2.0 or comparable arrays according to the cells from species other than humans.

2. GeneChip Eukaryotic Hybridization Control Kit.

3. Affymetrix WT Labeling and Controls Kit (Affymetrix).

4. Affymetrix HT HWS Kit for GeneTitan and WT Array Plates.

5. 2× Hybridization buffer: 200 mM MES, 2 M [Na+], 40 mM EDTA, 0.02 % Tween-20.

6. Hybridization Oven 640.

7. Affymetrix Fluidics Station 400 or 450/250.

8. Wash buffer A: Non-stringent wash buffer: 6× SSPE, 0.01 % Tween-20.

9. Wash buffer B: Stringent wash buffer: 100 mM MES, 0.1 M [Na+], 0.01 % Tween-20.

10. 2× Stain buffer: 200 mM MES, 2 M [Na+], 0.1 % Tween-20.

11. Affymetrix GeneChip® Scanner 3000 or Agilent GeneArray® Scanner.

2.3 RNA-Seq

2.3.1 Preparation of Sequencing Libraries from Total RNA (100 ng to 3 μg)

1. TruSeq Stranded Sample Preparation Kit—provides all reagents, enzymes, barcoded adapters, and beads for library preparation.

2. Agencourt AMPure XP Beads (Beckman Coulter Genomics).

3. Ethanol for bead purifications.

4. Magnetic separation stand (Invitrogen™, Life Technologies).

5. Qubit Fluorometer 2.0 (Invitrogen™, Life Technologies).

6. 96-well thermal cycler with heated lid.

7. Agilent 2100 Bioanalyzer (Agilent Technologies).

8. MicroSeal 96 well PCR plates and adhesive seals (Bio-Rad).

9. Certified low-range agarose.

10. Illumina Cluster Kit (Illumina, Inc.).

11. Illumina Cbot.

12. Illumina HiSeq 2000/2500.

3 Methods

3.1 Preparation of Total RNA

Many RNA purification methods have been developed, but not all yield RNA that is suitable for gene expression analysis. It is highly critical to prepare a good-quality RNA as starting material for reproducible results from both microarrays and RNA-seq (*see* **Note 1**). Figure 1 illustrates good-quality RNA. Moreover, specific types of biological specimens may require specific treatments and purification protocols to yield top-quality RNA. This section describes methods that have been optimized in our laboratory for RNA purification from cells.

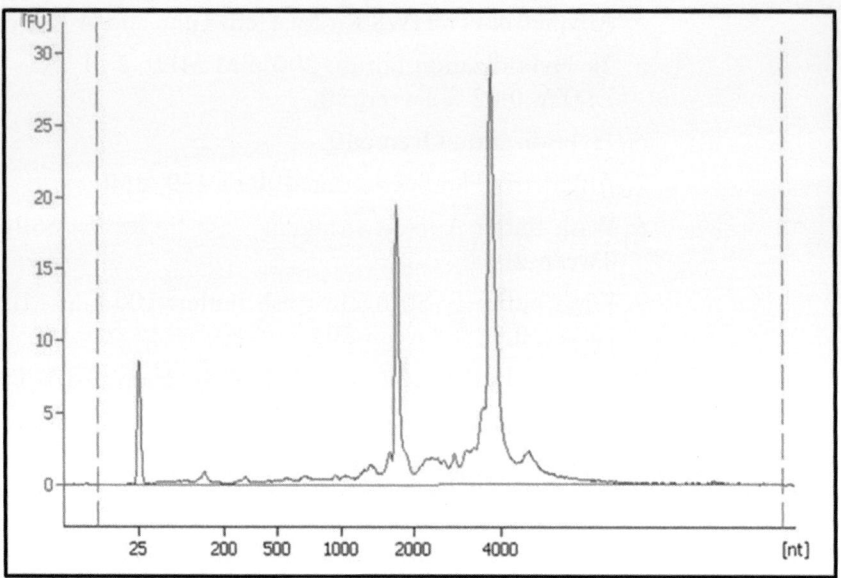

Fig. 1 Example of a good-quality total RNA. *X*-axis shows in bases (*nt* nucleotides). *Y*-axis shows relative fluorescent units. The peak at approximately 25 nt is a size marker. The larger peaks near 1,700 and 4,000 nt show the two major ribosomal RNA transcripts

3.1.1 Preparation of Lysate from Cell Pellets from Cultured Cells

1. Loosen cell pellet thoroughly by flicking the tube. Add 700 μl of buffer RLT plus β-ME per column to lyse the cells (5×10^6 to $3 \times 5 \times 10^7$/column).

2. Homogenize cells by passing the cell suspension in QIAshredder column at max speed for 2 min. The lysate can be stored at −80 °C until further use if the extraction is not going to be performed immediately.

3.1.2 Purification of Total RNA with DNase Digestion from Purified Cells (>100,000)

All centrifugations need to be done in microcentrifuge.

1. Add 44 μl of 100 % EtOH to buffer RPE.

2. Make up 70 % EtOH in DEPC water.

3. If lysate was frozen in RLT, thaw in 37 °C water bath for 20 min (RLT precipitates when frozen and thawed).

4. If there are precipitates, spin at $5,000 \times g$ for 5 min and collect supernatant.

5. Add an equal volume of 70 % EtOH to the homogenized lysate. Mix well by pipetting.

6. Apply up to 700 μl of sample to RNeasy mini spin column.

7. Centrifuge for 30 s at $>8,000 \times g$. Discard flow-through.

8. Repeat with the remainder of the sample. For higher yields try passing flow-thru a second time over the column.

9. Add 350 µl of buffer RW1 onto the mini spin column.

10. Centrifuge for 30 s at >8,000×g. Discard flow-thru.

11. The kit protocol may say that DNAse treatment is optional, but it is important to remove even traces of DNA (*see* **Note 2**). Add 10 µl DNAse I stock solution to 70 µl buffer RDD.

12. Mix gently by inverting. Do not vortex.

13. Pipette this 80 µl mix onto the spin column and leave at room temperature for 15 min.

14. Add 350 µl RW1 onto spin column.

15. Centrifuge for 30 s at 8,000×g. Discard flow-through.

16. Add 500 µl buffer RPE onto the mini spin column.

17. Centrifuge for 30 s at 8,000×g. Discard flow-through.

18. Add 500 µl buffer RPE on the mini spin column.

19. Centrifuge for 2 min at 12,000×g. Discard flow-through.

20. Dry spin the mini column.

21. Centrifuge at 12,000×g for 1 min.

22. Transfer RNeasy column into a new 1.5 ml collection tube.

23. To elute, transfer RNeasy spin column to a new collection tube. Pipette 20 µl of RNase-free water directly onto the column membrane. Close the tube lightly, let it stand for 1 min, and spin for 1 min.

24. Add another 20 µl of RNase-free water, and spin for 1 min. To obtain a higher total RNA concentration, this second elution step may be performed by using the first eluate. The yield might be 15–30 % less than the yield obtained using a second volume of RNase-free water, but the final concentration will be higher.

25. Mix and transfer the eluted RNA to a sterile 1.5 ml microcentrifuge tube.

26. Quantitate and store RNA at –80 °C.

27. RNA should be snap frozen and stored at –80 °C or over liquid nitrogen in a freezer. Keep on ice when pulled out to use.

3.1.3 Isolation of Total RNA from Flow-Sorted Cells (<100,000)

To isolate total RNA from micro-sized samples, such as flow-sorted cells or MACS or other procedures that yield only a few hundred cells, Ambion's RNAqueous-Micro Kit can be used. This kit employs a guanidinium-based lysis/denaturant as well as glass fiber filter separation technology. DNA-free™ DNA removal reagents are included in this kit, making the isolated RNA suitable for most downstream applications.

1. Prepare a cell lysate with 100 µl cell lysis buffer provided in the kit. For a standard prep of 100 µl of lysate, add 50 µl of 100 % ethanol, and vortex briefly but thoroughly. If desired, centrifuge the tube briefly to collect the sample at the bottom of the tube.

2. Pass the lysate/ethanol mixture through a micro filter cartridge assembly using 150 μl at a time, and close the cap. Centrifuge for ~10 s at maximum speed or until all of the mixture has passed through the filter. For lysate/ethanol mixtures >150 μl, load and filter the first 150 μl, and then repeat with additional aliquots until the entire sample has passed through the filter. The collection tube has a capacity of ~700 μl when assembled with a micro filter cartridge. The RNA is now bound to the filter in the micro filter cartridge.

   Note: All centrifugation in the following steps should be done in a microcentrifuge at 10,000–12,000×*g*.

3. Wash the filter with 180 μl wash solution 1. Open the micro filter cartridge, add 180 μl of wash solution 1 (working solution mixed with ethanol) to the filter, and close the cap. Centrifuge for ~10 s to pass the solution through the filter.

4. Wash the filter with 2×180 μl wash solution 2/3. Open the micro filter cartridge, add 180 μl of wash solution 2/3 (working solution mixed with ethanol) to the filter, and close the cap. Centrifuge for ~10 s to pass the solution through the filter. Repeat with a second 180 μl aliquot of wash solution 2/3.

5. Discard the flow-through, and centrifuge the filter for 1 min at max speed. Open the micro filter cartridge assembly, remove the filter cartridge from the collection tube, and pour out the flow-through. Replace the micro filter cartridge into the same collection tube, close the cap, and centrifuge at maximum speed for 1 min to remove residual fluid and dry the filter.

6. Elute the RNA into a micro elution tube with 2×5–10 μl preheated elution solution. Label a micro elution tube (1.5 ml tubes provided with the kit), and transfer the micro filter cartridge into it. Apply 5–10 μl of elution solution, preheated to 75 °C, to the center of the filter. Close the cap, and store the assembly for 1 min at room temperature. Centrifuge the assembly for ~30 s to elute the RNA from the filter. Repeat with a second 5–10 μl aliquot of preheated elution solution, collecting the eluate in the same micro elution tube.

   Note: The exact volume of elution solution used is not critical and may be increased if desired. In general, ~75–85 % of the RNA will be recovered from samples derived from up to 2×5 μl of elution solution, and ≥85 % will be recovered using 2×10 μl of elution solution.

7. Add 1/10 volume of 10× DNase I buffer and 1 μl of DNase I. Add 1/10 volume 10× DNase I buffer (e.g., 2 μl for RNA eluted in 20 μl) and 1 μl of DNase I (provided with the kit) to the sample, and mix gently but thoroughly.

8. Incubate for 20 min at 37 °C. Incubate the DNase reaction for 20 min at 37 °C. Remove the DNase inactivation reagent from

–20 °C, and allow it to thaw at room temperature during this incubation.

9. Add 2 μl or 1/10 volume DNase inactivation reagent, mix well, and leave at room temp for 2 min. Vortex the DNase inactivation reagent vigorously to completely resuspend the slurry. Use 2 μl or 1/10 volume DNase inactivation reagent, whichever is greater. For example, if the RNA volume is 50 μl, and 1 μl DNase I and 5 μl of DNase buffer were used in **step 1**, add 5.6 μl of DNase inactivation reagent.

10. Store the reaction at room temperature for 2 min, vortexing once during this interval to disperse the DNase inactivation reagent.

11. Pellet the DNase inactivation reagent, and transfer the RNA to a fresh tube. Centrifuge the reaction for 1.5 min at maximum speed to pellet the DNase inactivation reagent. Transfer the RNA to a fresh RNase-free tube (not supplied with the kit) and store at –80 °C.

3.2 Gene Expression Profiling Using Microarrays

3.2.1 Preparation of Labeled Targets from Total RNA for Hybridization onto Gene Chips

The WT Expression Kit is designed to generate amplified sense-strand cDNA ready for fragmentation and labeling using the Affymetrix GeneChip WT Terminal Labeling Kit. The WT Expression Kit uses a reverse transcription priming method that specifically primes non-ribosomal RNA from your sample. This is important because otherwise ribosomal RNA is so abundant in all cells that it would interfere with detection of mRNAs in the sample. The cDNA is then in vitro transcribed (IVT) to amplify the target, and finally the RNA generated in the IVT reaction is copied back into cDNA with dUTP incorporated for the downstream fragmentation and hybridization steps. It is essential to get updates from the manufacturer on the latest methods that can be used for improvement in the labeling process (*see* **Notes 3** and **4**).

1. First-strand synthesis—Prepare first-strand master mix as follows for a single reaction on ice. Adjust as needed for multiple reactions.

First-strand buffer mix	4 μl
First-strand enzyme mix	1 μl
Total	5 μl

2. Add 5 μl of total RNA/poly-A RNA control mix (20–50 ng T.RNA) to the appropriate tube for a total volume of 10 μl. Mix well by pipetting.

3. Incubate in a thermal cycler using the following program: *Note*: Use heated lid:
25 °C 1 h; 42 °C 1 h; and 4 °C hold (hold for at least 2 min).

4. Second-strand cDNA synthesis: Prepare second-strand master mix on ice as follows for a single reaction. Adjust as needed for multiple reactions.

Nuclease-free H$_2$O	32.5 µl
Second-strand buffer mix	12.5 µl
Second-strand enzyme mix	5 µl
Total	50 µl

5. Transfer 50 µl of second-strand master mix to each sample for a total volume of 60 µl. Mix by pipetting.

6. Incubate in a thermal cycler using the following program: *Note*: Do not use the heated lid:
16 °C 1 h; 65 °C 10 min; and 4 °C hold (hold for at least 2 min).

7. Transfer sample to ice, and immediately proceed to cRNA synthesis.

8. In vitro transcription—synthesis of antisense cRNA.
Prepare IVT master mix at RT as follows for a single reaction. Adjust as needed for multiple reactions:

IVT buffer mix	24 µl
IVT enzyme mix	6 µl
Total	30 µl

9. Transfer 30 µl of IVT master mix to each cDNA sample for a total of 90 µl. Mix well by pipetting.

10. Incubate in a thermal cycler using the following program: *Note*: Use heated lid:
40 °C 16 h and 4 °C hold overnight.
Note: Samples may be stored at –20 °C overnight before proceeding to cRNA purification.
Prepare the following reagents before starting the purification process.

11. Preheat elution solution to between 50 and 58 °C for at least 10 min.

12. Add 100 % EtOH to the nucleic acid wash solution concentrate.
Make sure that the nucleic acid-binding buffer concentrate is completely dissolved. Warm solution to <50 °C until solubilized if necessary. Vortex the nucleic acid-binding beads vigorously before use ensuring full dispersion.

13. Prepare the cRNA-binding mix as follows for a single reaction. Adjust volume as needed for multiple reactions. A 10 % excess may be added to compensate for pipetting losses.

Nucleic acid-binding beads	10 µl
Nucleic acid-binding buffer concentrate	50 µl
Total	60 µl

14. Add 60 µl of cRNA-binding mix to each sample. Pipette up and down three times to mix.

15. Transfer each sample to the well of a U-bottom plate.

16. Add 60 µl of 100 % isopropanol. Pipette up and down three times to mix.

17. Gently shake the plate for 2 min to bind cRNA to the beads.

18. Transfer the plate to the magnetic stand for 5 min. The samples should be clear when the beads have been completely captured.

19. Pipette the supernatant without disturbing the beads. Discard the supernatant. Remove plate from magnetic stand.

20. Add 100 µl of nucleic acid wash solution to each sample, and moderately shake the plate for 1 min.

21. Transfer plate to the magnetic stand for 5 min.

22. Carefully pipette the supernatant without disturbing the beads and discard. Remove plate from magnetic stand.

23. Repeat **steps 20–22** to wash beads again with 100 µl nucleic acid wash solution.

    After wash buffer has been aspirated, transfer plate to shaker and shake vigorously for 1 min to evaporate residual ethanol.

24. Elute the cRNA by adding 40 µl of preheated elution solution. Incubate for 2 min at RT.

25. Transfer plate to magnetic stand for 5 min.

26. Transfer supernatant to an appropriately labeled 1.5 ml tube and place on ice.

27. Quantitate the cRNA using the Nanodrop. Samples can be stored at –20 °C overnight.

28. Purification of cRNA—Prepare 10 µg of cRNA in a volume of 22 µl (455 ng/µl). Adjust with H_2O or vacuum concentrate as needed.

29. Combine cRNA and random primers in the supplied PCR tube on ice as follows:

cRNA 10 µg	22 µl
Random primers	2 µl
Total	24 µl

30. Incubate in a thermal cycler using the following program: Use heated lid:

70 °C	5 min
25 °C	5 min
4 °C	Hold (hold for at least 2 min)

31. Prepare the second-cycle master mix on ice as follows for a single reaction. Adjust as needed for multiple reactions:

Second-cycle buffer mix	8 µl
Second-cycle enzyme mix	8 µl
Total	16 µl

32. Transfer 16 µl of the second-cycle master mix to the 24 µl cRNA/random primer for a 40 µl total volume.

33. Incubate in a thermal cycler using the following program:

    25 °C—10 min; 42 °C—90 min; 70 °C—10 min; and 4 °C—hold (hold for at least 2 min).

34. RNaseH hydrolysis: Add 2 µl of RNaseH on ice to the second-cycle DNA for a total volume of 42 µl. Mix by pipetting.

35. Incubate in a thermal cycler using the following program:

    37 °C—45 min; 95 °C—5 min; and 4 °C—hold (hold for at least 2 min). Samples can be stored at –20 °C overnight.

36. Purification of cDNA: Prepare the cDNA-binding mix as follows at RT. Adjust volume as needed for multiple reactions:

Nucleic acid-binding beads	10 µl
Nucleic acid-binding buffer concentrate	50 µl
Total	60 µl

37. Add 18 µl nuclease-free H_2O to each hydrolyzed sample for a total volume of 60 µl.

38. Add 60 µl of cDNA-binding mix to each sample. Pipette up/down three times to mix.

39. Transfer each sample to a well of a U-bottom plate.

40. Add 120 µl of 100 % ethanol to each sample. Pipette up/down three times to mix, and gently shake the plate.

41. Transfer plate to magnetic stand for 5 min.

42. Aspirate supernatant without disturbing bead pellet and discard.

43. Remove the plate from the magnetic stand.

44. Add 100 µl of nucleic acid wash solution to each sample, and shake at moderate speed for 1 min

45. Place plate on the magnetic stand for 5 min to capture the beads.

46. Aspirate supernatant without disturbing bead pellet and discard.

47. Remove the plate from the magnetic stand.

48. Repeat **steps 44–47** a second time using 100 µl of nucleic acid wash solution.

49. Elute cDNA by adding 30 µl of preheated elution solution (50–58 °C) to each sample.

50. Incubate at RT for 2 min.

51. Transfer plate to magnetic stand for 5 min.

52. Transfer supernatant to an appropriately labeled 1.5 ml tube and place on ice.

53. Quantitate the cRNA using the Nanodrop.

54. Fragmentation and labeling of single-stranded cDNA requires the Affymetrix GeneChip WT Terminal Labeling Kit (PN 900671).
    Prepare single-stranded DNA in a 0.2 ml tube as follows for each reaction.

Single-stranded DNA	5.5 µg (volume can vary)
RNase-free water	Up to 31.2 µl
Total volume	31.2 µl

55. Prepare the fragmentation master mix as follows for a single reaction. Adjust volume for multiple reactions as needed:

RNase-free water	10.0 µl
10× cDNA fragmentation buffer	4.8 µl
UDG, 10 U/µl	1.0 µl
APE 1, 1,000 U/µl	1.0 µl
Total volume	16.8 µl

56. Add 16.8 µl of the fragmentation mix to each sample prepared in **step 54**. Pipette to mix samples.

57. Incubate in a thermal cycler as below:
    37 °C 60 min; 93 °C 2 min; and 4 °C hold (at least 2 min).

58. Transfer 45 µl to a new 0.2 µl tube. As a QC procedure, remaining single-stranded DNA can be run on the Agilent Bioanalyzer using the RNA6000 LabChip kit II. Fragmented DNA size range should be 40–70 bp (Fig. 2).

59. Labeling of fragmented single-stranded DNA requires the Affymetrix GeneChip WT Terminal Labeling Kit (PN 900671). Prepare labeling reactions in 0.2 ml tube as follows for a single reaction:

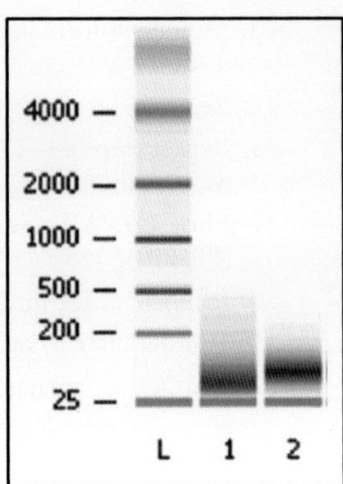

Fig. 2 Example of a well-fragmented cDNA before and after labeling. Bioanalyzer trace showing the fragmented samples before (*lane 1*) and after labeling (*lane 2*). *L* represents the ladder

Fragmented single-stranded DNA	45 μl
5× TdT buffer	12 μl
TdT	2 μl
DNA labeling reagent, 5 mM	1 μl
Total volume	60 μl

60. Flick mix, and spin down samples. Incubate in a thermal cycler as below:

37 °C	60 min
70 °C	10 min
4 °C	Hold (at least 2 min)

61. Hybridization onto gene chips: This procedure requires the use of the GeneChip Hybridization, Wash, and Stain kit (Affymetrix 900720). Prepare the hybridization cocktail in 1.5 ml tubes as follows (adjusted for sample number):
    The cocktail mix for the Human Gene 1.0 ST array is the 169 array format and Human Exon array 1.0 ST array is the 49 array format:

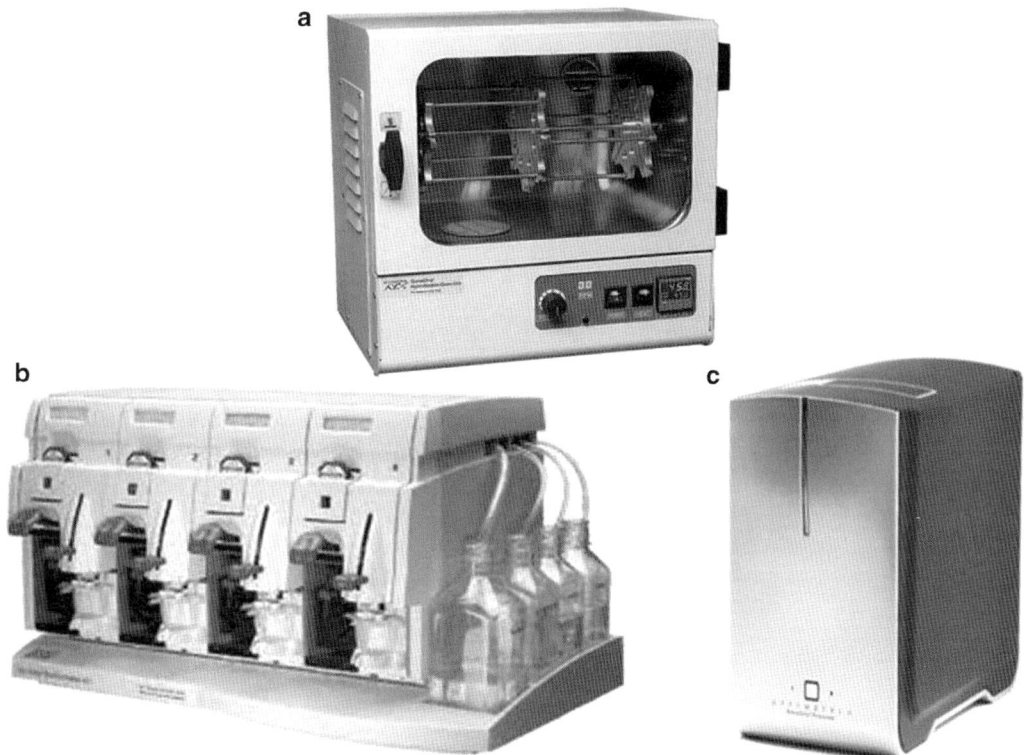

Fig. 3 Affymetrix instruments. (**a**) Hybridization oven. (**b**) Fluidics station. (**c**) Scanner

	Vol. for 1	Vol. for 1
Array format	49/64	169
Fragmented and labeled DNA target ~60.0 µl	27.0 µl	~25 ng/µl
Control Oligo B2 (3 nM) 50 pM	3.7 µl	1.7 µl
20× Eukaryotic hyb controls	11.0 µl	5.0 µl
2× Hybridization mix	110.0 µl	50.0 µl
DMSO	15.4 µl	7.0 µl
Nuclease-free water	Up to 220.0 µl	Up to 100.0 µl
Total volume	220 µl	100.0 µl

62. Denature cocktail for 5 min at 99 °C in heat block.

63. Cool for 5 min at 45 °C in the Hybridization Oven 640. Microcentrifuge at full speed for 1 min.

64. Inject the appropriate amount of denatured cocktail into the array.

65. Place arrays in the Hybridization Oven 640 (Fig. 3) and hybridize at 45 °C for 17 ± 1 h. Set the rotation speed at 60.

66. Washing and staining of hybridized gene chips: Aliquot the following reagents:
    600 μl Stain cocktail 1 into a labeled 1.5 ml amber tube.
    600 μl Stain cocktail 2 into a labeled 1.5 ml clear tube.
    800 μl Array holding buffer into a labeled 1.5 ml clear tube.

67. Prime the necessary Fluidics Station 450 (Fig. 3). Ensure that the appropriate wash A and wash B buffers have been installed in their appropriate location.

68. After 17 ± 1 h of hybridization remove arrays and remove cocktail to the original tube and store at $-80\,^{\circ}C$.

69. Refill probe array with wash A buffer as indicated by the array format.

70. Initiate Fluidics profile FS450_0001 or FS450_0007. Follow Fluidics Station prompts to install array and the stain cocktails and array holding buffer.

71. When fluidics is complete remove arrays and check for bubbles. If bubbles are present return array to module and engage. Array will be refilled with fresh array-holding buffer.

72. Engage wash blocks so that modules can prime. Follow prompts on fluidics monitors.

73. If all arrays are complete with fluidics, perform the fluidics shutdown protocol using Shutdown_450.

74. Scanning of the stained gene chips: Warm up Affymetrix Gene Chip Scanner 3000 for at least 10 min before scanning.

75. Clean array glass with water and a Kim wipe.

76. Install arrays into carousel of scanner (Fig. 3) starting with position 1 marked in red. Close scanner.

77. Select Run and Scanner from the menu bar or click the Start Scan in the tool bar. Select experiment name, and click Start button.

78. Each complete probe array image is stored in a separate data file identified by the experiment name and is saved with a data image file (.dat) extension.

79. Microarray data acquisition and analysis: This is a simplified version of analysis options using Affymetrix microarray data analysis Expression Console software. The technical manuals of the analysis software have extensive descriptions for choosing the various options, and the reader is advised to look into those websites. Describing them here is beyond the scope of this chapter (*see* **Note 5**).

80. Double-click on each *.dat file to open it.

81. Under the tab for "Image settings." choose autoscale and pseudocolor, and then click "OK."

82. Click on "Grid" tab, press "G," and select View → Grid from menu bar to superimpose grid lines on the scanned image.

83. Select Run → Analysis from menu bar.

84. At the end of each analysis of a *.dat file (~2 min), cell intensity data is computed and a *.cel file is created (single intensity value is computed for each probe cell).

85. QC metrics: A general visual inspection of the entire Gene Chip probe array should be performed after scanning (*see* **Note 6**). B2 oligos, which are spiked into the hybridization cocktail, should be used for checking the quality of hybridization grid alignment. The eukaryotic hybridization controls (EHCs) are a mixture of four biotin-labeled, antisense, fragmented noneukaryotic control cRNAs, which are spiked at staggered concentrations into the hybridization cocktail. Oligos complementary to these sequences are always spotted onto all gene chip arrays, and the EHCs thereby serve as hybridization controls. All four transcripts should maintain a maximum 1:2 ratio of signal intensities of the 5′ and 3′ probe sets. BioB should be present at least 50 % of the time, while bioC, bioD, and Cre should always be present, with increasing signal intensities.
The polyA spike in controls added during the labeling process should be a present call with increasing signal intensity in the order of lys, phe, thr, and dap (*see* **Notes 7** and **8**).

3.2.2 Preparation of Sequencing Libraries for RNA-Seq Using Illumina TruSeq Protocol

For Illumina libraries, total RNA will be processed according to the TruSeq Sample Preparation Kits (FC-122-1001, FC122-1002) following the manufacturer's directions and protocol (www.Illumina.com).

Optional: Ribosomal RNA-depleted material would also be a good starting material for the preparation of libraries for sequencing using the Illumina protocol. Ribo-depleted RNA can be generated using the Ribo-Zero Gold system (Epicentre Biotechnologies) according to the manufacturer's instructions.

Optional: In-line DNA controls supplied in the kit are intended for troubleshooting and are useful for identifying the specific mode of failure in this multistep library preparation process. Controls are added to the reactions just prior to their corresponding step in the protocol.

Completed libraries should be evaluated by DNA quantitation and Bioanalyzer analysis/agarose gel electrophoresis and then used for sequencing. Sequencing libraries constructed with barcodes allow multiplexing of 12 samples per lane, pooled to target 200 million clusters per channel and 100 million reads per library and distributed over multiple channels in flow cells to normalize for lane and run variability. Sequencing can be carried out on Illumina HiSeq 2000/2500 instruments following Illumina's recommended protocols.

1. Purification of mRNA: Dilute total RNA to 50 μl using nuclease-free H_2O in a new 96-well 0.3 ml non-skirted PCR plate.

2. Vortex the warmed RNA purification beads (RPB) vigorously to completely resuspend the Oligo-dT beads.

3. Add 50 μl of the RPB to each well of the plate. Gently pipette up and down to mix beads.

4. Seal plate with adhesive plate seal and incubate in thermal cycler with heated lid at 65 °C for 5 min followed by 4 °C hold.

5. Incubate plate at RT for 5 min to bind RNA to beads.

6. Transfer plate to magnetic stand for 5 min at RT.

7. Remove and discard the entire supernatant using multichannel pipette. Do not disturb pellets.

8. Remove plate from magnetic stand.

9. Wash beads by adding 200 μl of bead-washing buffer (BWB) to each well and pipette up and down to resuspend beads.

10. Return the plate to the magnetic stand for 5 min at RT.

11. Remove BWB from each well of plate using multichannel pipette and discard.

12. Add 50 μl of elution buffer (ELB) to each well of the plate. Pipette up and down to suspend beads.

13. Seal plate with adhesive seal.

14. Incubate the sealed plate in a thermal cycler using the following profile to elute mRNA:

    80 °C for 2 min followed by 25 °C hold.

15. Remove plate from the thermal cycler when 25 °C is reached and place at RT.

16. Add 50 μl of bead-binding buffer to each well of the plate to allow the RNA to rebind to the beads. Pipette up and down to thoroughly mix beads.

17. Incubate at RT for 5 min.

18. Keep the plate on magnetic stand at RT for 5 min.

19. Remove and discard supernatant with multichannel pipette without disturbing beads.

20. Remove the plate from magnetic stand, and wash beads with 200 μl BWB by pipetting up and down.

21. Keep the plate on magnetic stand at RT for 5 min.

22. Remove and discard the supernatant from each well of the plate.

23. Remove plate from magnetic stand.

24. Add 19.5 μl elute, prime, fragment (EPF) mix (which contains random hexamers for RT priming and serves as the first-strand cDNA synthesis buffer) to each well, and mix up and down with a multichannel pipette.

25. Seal the plate with an adhesive seal.

26. Incubate the sealed plate in a thermal cycler to elute, fragment, and prime the RNA at 94 °C for 8 min followed by 4 °C hold.

27. Remove the plate when 4 °C is reached and centrifuge briefly in plate adaptor at RT.

28. First-strand synthesis—Transfer the RNA bead plate (RBP) to the magnetic stand and incubate at RT for 5 min.

29. Remove the adhesive seal from the plate.

30. Transfer 17 μl of the supernatant (fragmented and primed mRNA) without disturbing the beads from each well of the RBP to the corresponding well of a new plate.

31. Prepare first-strand master mix by adding 50 μl SuperScript II to the thawed tube of first-strand buffer, and mix well. Centrifuge briefly.
    Note: The final first-strand master mix is a 1:7 ratio of SuperScript II to first-strand buffer.

32. Add 8 μl of supplemented first-strand master mix to each well of the plate. Gently mix by pipetting up and down.

33. Seal the plate with an adhesive seal and centrifuge briefly at 15 °C.

34. Incubate the plate in a thermal cycler using the following profile:
    25 °C 10 min; 42 °C 50 min; 70 °C 15 min; and 4 °C hold.

35. When the cycler reaches 4 °C remove plate and proceed immediately to the second-strand synthesis.

36. Synthesize second-strand cDNA: Prepare a fresh stock of 80 % EtOH. Warm AMPure beads to RT for 30 min. Vortex until well dispersed. Initiate thermal cycler profile, and preheat to 16 °C. Briefly centrifuge thawed second-strand master mix.

37. Add 25 μl of thawed second-strand master mix to each well of the plate using a multichannel pipette. Gently pipette up and down to mix thoroughly.

38. Seal plate with an adhesive seal.

39. Incubate the plate on a preheated thermal cycler with closed lid at 16 °C 1 h and hold at RT.

40. Vortex the AMPure XP beads until fully dispersed, and then add 90 μl of beads to each well of the plate containing 50 μl of ds cDNA. Pipette up and down gently to mix.

41. Incubate the plate at RT for 15 min.

42. Place the plate on the magnetic stand at RT for 5 min—make sure that beads are completely deposited on the side of the well.

43. Remove and discard 135 µl of the supernatant from each well without disturbing the beads using a multichannel pipette.

44. Leave the plate on the magnetic stand, and wash wells with 200 µl freshly prepared 80 % EtOH.

45. Incubate the plate at RT for 30 s. Remove EtOH using multichannel pipette. Repeat 80 % EtOH wash twice.

46. Let the plate stand at RT for 15 min to dry, and then remove plate from magnetic stand.

47. Add 52.5 µl resuspension buffer to each well of the plate. Mix up and down to completely resuspend the beads.

48. Incubate the plate at RT for 2 min.

49. Place the plate on the magnetic stand at RT for 5 min.

50. Transfer 50 µl of the supernatant containing the ds cDNA to a new 0.3 ml plate.

51. End repair process: Briefly mix and centrifuge the thawed end repair control tube.

52. Dilute the end repair control 1/100 in resuspension buffer (1 µl of end repair control + 99 µl of resuspension buffer).

53. Add 10 µl of diluted end repair control (or 10 µl resuspension buffer if control is not used) into each well of the IMP plate containing 50 µl of ds cDNA using a multichannel pipette.

54. Add 40 µl of end repair mix to each well of the plate using a multichannel pipette to obtain 100 µl total volume and mix by pipetting.

55. Seal the plate with adhesive seal.

56. Incubate the plate in a preheated thermal cycler with closed lid at 30 °C for 30 min using the following profile:
30 °C 30 min and 30 °C hold.

57. Remove the plate from thermal cycler.

58. Add 160 µl of AMPure XP beads to each well containing 100 µl of end repair mix using a multichannel pipette. Mix by pipetting.

59. Incubate the plate at RT for 15 min.

60. Transfer the plate to a magnetic stand for 5 min.

61. Remove and discard the supernatant from each well of the plate without disturbing the beads.

62. Add 200 µl of freshly prepared 80 % EtOH to each well.

63. Incubate the plate at RT for 30 s, and discard all supernatant from each well.

64. Repeat wash steps with 80 % EtOH.

65. Let the plate stand at RT for 15 min to dry, and remove the plate from the magnetic stand.

66. Resuspend the dried pellet in 17.5 µl resuspension buffer. Gently pipette up and down to mix thoroughly.

67. Incubate the plate at RT for 2 min.

68. Leave the plate on the magnetic stand for at least 5 min.

69. Transfer 15 µl of the supernatant from each well to a new PCR plate.

70. Adenylation process: Add 2.5 µl of diluted A-Tailing Control (or 2.5 µl resuspension buffer if control is not used) to each well of the plate using a multichannel pipette.

71. Add 12.5 µl of thawed A-Tailing Mix to each well of the ALP plate using a multichannel pipette. Pipette up and down to mix.

72. Incubate the plate in the preheated thermal cycler with closed lid at 37 °C for 30 min using the following profile:
37 °C—30 min and 37 °C—hold.

73. Ligation of adapters: Remove the appropriate RNA adapter index tubes (AR001–AR012, depending on the RNA adapter indexes being used) and one tube of stop ligase mix and ligase control from –20 °C and thaw at RT. Use of the ligase control is optional and can be replaced with the same volume of resuspension buffer.

74. Add 2.5 µl of DNA ligase mix to each well of the plate.

75. Dilute the ligase control 1/100 in resuspension buffer (1 µl ligase control + 99 µl resuspension buffer).

76. Add 2.5 µl of diluted ligase control (or 2.5 µl resuspension buffer if control is not used) to each well of the plate.

77. Add 2.5 µl of the appropriate thawed RNA adapter index (AR001–AR012) to each well of the plate.

78. Mix well, and seal the plate with an adhesive seal.
Note: When indexing libraries, Illumina recommends arranging samples that will be combined into a common pool in the same row. Each column should contain a common index (i.e., AR007). This will facilitate pipetting operations when dispensing indexed adapters and pooling indexed libraries later.

79. Incubate the plate in the preheated thermal cycler with closed lid at 30 °C for 10 min followed by holding at RT.

80. Add 5 µl stop ligase mix to each well of the plate, and mix by gently pipetting up and down with a multichannel pipette.

81. Add 42 µl of beads to each well of the plate using a multichannel pipette, and mix well.

82. Incubate the plate at RT for 15 min.

83. Place the plate on the magnetic stand at RT for at least 5 min.

84. Remove and discard supernatant from each well using a multichannel pipette.

85. Add 200 µl freshly prepared 80 % EtOH to each well without disturbing the beads.

86. Incubate the plate at RT for at least 30 s and discard all the supernatant from each well. Repeat ETOH wash twice.

87. Incubate the plate on the magnetic stand at RT for 15 min to dry residual EtOH.

88. Remove plate from magnetic stand, and resuspend the dry pellet in each well with 52.5 µl of resuspension buffer. Gently pipette up and down to mix.

89. Place the plate on the magnetic stand and incubate for at least 5 min.

90. Transfer 50 µl of clear supernatant from each well of the plate to the corresponding well of the new 0.3 ml plate.

91. Vortex the AMPure XP beads to disperse, and add 50 µl of beads into each well of the plate for a second cleanup.

92. Transfer 20 µl of clear supernatant from each well of the plate to the corresponding well of the new 0.3 ml PCR plate.

93. Enrichment of DNA fragments by PCR: Add 5 µl of thawed PCR primer cocktail to each well of the PCR plate using a multichannel pipette.

94. Add 25 µl of thawed PCR master mix to each well of the PCR plate using a multichannel pipette. Pipette up and down to mix thoroughly.

95. Incubate the PCR plate in the preheated thermal cycler with closed lid using the following profile:
1 cycle—98 °C 30 s.
15 cycles—98 °C 10 s, 60 °C 30 s, and 72 °C 30 s.
1 cycle—72 °C 5 min and 4 °C hold.

96. Add 50 µl of the AMPure XP beads to each well of the PCR plate containing 50 µl of PCR amplified library.

97. Incubate the PCR plate at RT for 15 min.

98. Transfer the PCR plate to the magnetic stand for at least 5 min making sure that the liquid clears.

99. Remove and discard the supernatant from each well using a multichannel pipette without disturbing the beads.

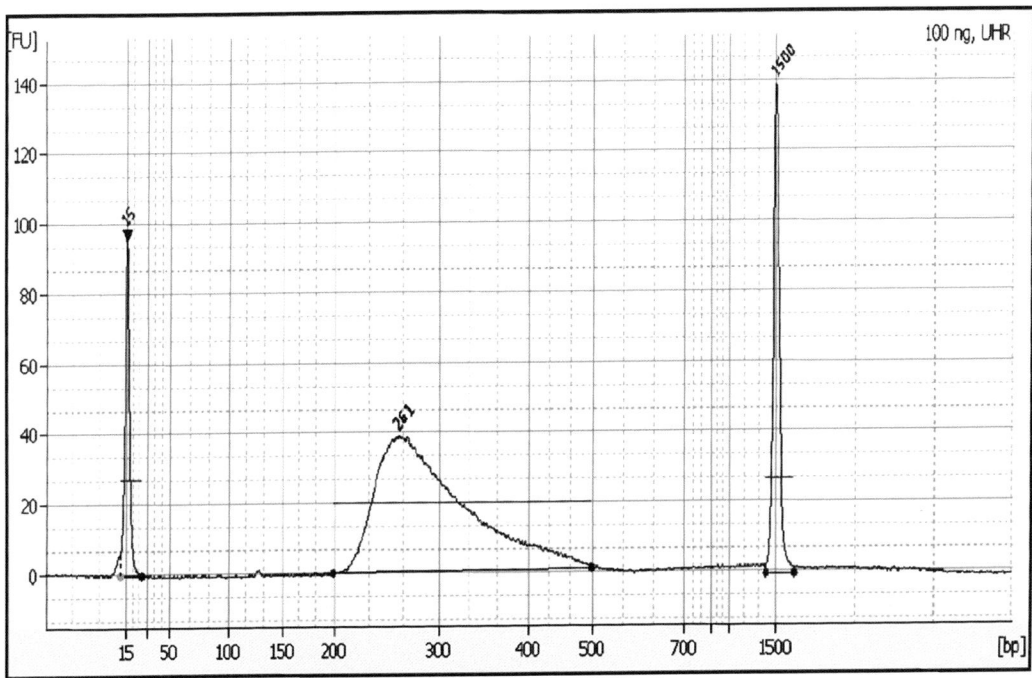

Fig. 4 Example of TruSeq RNA-seq library size distribution. *Lane 1* is the PCR-amplified DNA fragment ~260 bp

100. Add 200 µl of freshly prepared 80 % EtOH to each well without disturbing the beads.

101. Incubate the PCR plate at RT at least 30 s, and then remove and discard all the supernatant from each well. Repeat EtOH wash.

102. Incubate the PCR plate on the magnetic stand at RT for 15 min to dry residual EtOH.

103. Remove plate from magnetic stand, and resuspend the dry pellet in each well with 32.5 µl of resuspension buffer. Gently pipette up and down to mix.

104. Incubate the PCR plate at RT for 2 min.

105. Place the PCR plate on the magnetic stand at RT for at least 5 min.

106. Transfer 30 µl of clear supernatant from each well of the PCR plate to the corresponding well of the new 0.3 ml plate.

107. Quantitation of library: Determine the concentration of each amplified library using the Nanodrop or the Qubit fluorometer.

108. Perform QC of the amplified library by running 1 µl of each sample on the Agilent 2100 Bioanalyzer using the Agilent DNA 1000 Chip (Fig. 4).
     Note: The final product should be a band at approximately 260 bp for a single-read library (Fig. 5).

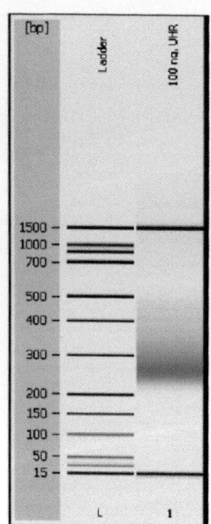

L - Ladder
Lane 1 is the PCR amplified DNA fragment ~ 260bp

Fig. 5 Example of a PCR-amplified library

109. Calculate the nM concentration of each library utilizing the Ambion RNA and DNA concentration calculator available at http://www.ambion.com/techlib/append/concentration_calculator.html.

    Note: Libraries can also be quantitated by KAPA library quant Kits (www.kapabiosystems.com). Though time consuming and a little expensive, this method is more accurate and can rule out primer dimers giving false readings in Nanodrop or Qubit measurements.

110. Normalization and pooling of libraries: This process would prepare DNA templates that are needed for cluster generation in Illumina c-Bot. Dilute each library (if run as a non-multiplexed sample) to 10 nM concentration using EB buffer (10 mM Tris–HCL pH 8.5) supplemented with 0.1 % Tween-20 (*see* **Note 9**).

111. Multiplexed DNA libraries need to be normalized to 10 nM in the diluted cluster plate (DCT) and then pooled in equal volumes in the pooled DCT plate.

112. Determine the number of samples to be combined together for each pool. Transfer 10 μl of each normalized library to one well of a new PCR plate. The total volume in each well should be 10× the number of combined sample libraries and will be 20–240 (2–24 libraries).

113. Gently pipette up and down to mix the libraries.

    Note: Libraries diluted to 10 nM in EB buffer supplemented with 0.1 % Tween-20 are safe for long-term storage at –20 °C.

Hiseq – 2000

cBOT

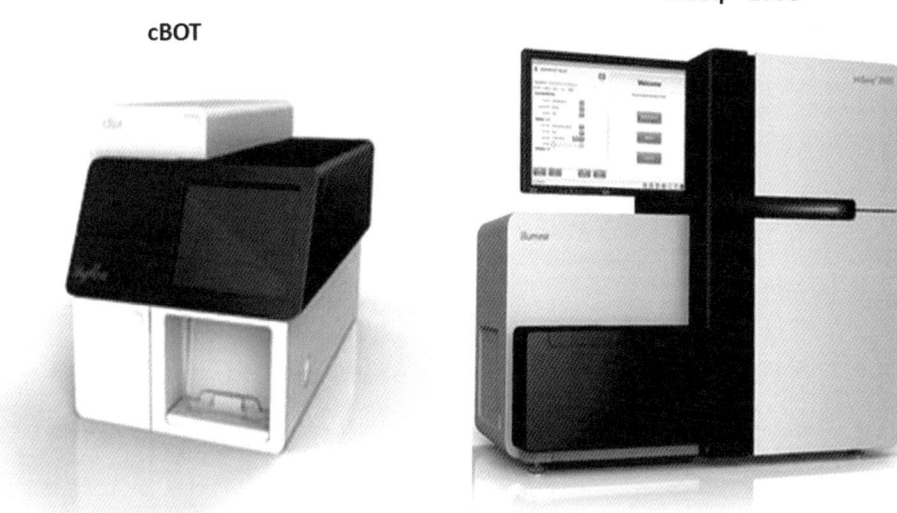

Fig. 6 Illumina sequencing equipment

114. Process the libraries in cBot (Fig. 6) that loads the flow cell and begins the process of clustering required for initiating the sequencing protocol following Illumina's protocol (www. illumina.com).

115. The clustered flow cell is then loaded onto the Illumina HiSeq 2000/2500 for sequencing (Fig. 6) following the manufacturer's directions (www.illumina.com. (*See* **Note 10.**)
Quality metrics: The Illumina HiSeq system provides real-time information on read quality as it is being generated. Important metrics to review the sequencing data include pass filter rate (ideally above 85 %), total number of reads per lane (ideally 150–200 millions per lane), and alignment to the genome which can vary with library preparation, RNA quality, and sample type but should be in the range of 70–80 % (*see* **Note 11**).

4   Notes

1. The quality of RNA is very critical in both microarray and RNA-seq experiments, as impurities have an adverse effect on the amplification and the labeling efficiency. The A260:A280 ratio for RNA samples of acceptable purity should be between 1.8 and 2.1. Sample integrity is also very important and is determined by the entire electrophoretic trace of the RNA sample using the Agilent Bioanalyzer, including the presence or the

absence of degradation products. The RNA integrity number (RIN) is a software tool for estimating the integrity of total RNA samples on a quantitative scale of 1 (worst)–10 (best). The RIN for samples of acceptable integrity should be >7. Use of degraded RNA can result in low yield, over-representation of the 3′ ends of the RNA molecules, or failure of the protocol during preparation of sequencing libraries.

2. The DNase treatment of RNA is crucial. In addition to removing the DNA, it purifies the RNA and reduces background.

3. The protocols described here are optimized in the author's laboratory and adapted from protocols in Affymetrix's and Illumina's user manuals. Methodologies and instrumentation frequently evolve over time especially with RNA-seq. If newer versions of the recommended kits and instrumentation become available, refer to the updated user manuals from the manufacturer.

4. The Affymetrix Gene Chip WT amplification protocol should be applied to samples with RNA concentration greater than 50 ng. For small-size samples less than 20 ng, Nugen's SPIA amplification protocol can be used (www.nugeninc.com).

5. The analysis of microarray data is complex and needs an experienced statistician/bioinformatics expert to look at the quality of the data generated and perform downstream analysis. Standardized analysis methods are frequently used for identification of differentially expressed genes, determination of p-values and false discovery rates, as well as clustering, generation of heat maps, running class prediction algorithms, and pathway analysis for the identified genes and gene sets.

6. A general visual inspection of the entire Gene Chip probe array should be performed after scanning. B2 oligos, which are spiked into the hybridization cocktail, should be used for checking the quality of hybridization grid alignment.

7. The EHCs are a mixture of four biotin-labeled, antisense, fragmented noneukaryotic control cRNAs, which are spiked at staggered concentrations into the hybridization cocktail. Oligos complementary to these sequences are always spotted onto all gene chips arrays, and the EHCs thereby serve as hybridization controls. All four transcripts should maintain a maximum 1:2 ratio of signal intensities of the 5′ and 3′ probe sets. BioB should be present at least 50 % of the time, while bioC, bioD, and Cre should always be present, with increasing signal intensities.

8. The polyA spike in controls added during the labeling process, i.e., added to target RNA, should be a present call with increasing signal intensity in the order of lys, phe, thr, and dap.

9. A decision that must be made at the start of every sequencing project is the read length and multiplexing by pooling bar-

coded libraries. For simple RNA expression profiling, the objective is to detect an mRNA and determine relative abundances. This can be done with single reads of 50–100 bp in length. Paired-end sequencing for RNA expression profiling is highly recommended to improve the alignment quality.

10. The last major issue to note is that of "read depth." This term is calculated by the total number of reads multiplied by the fraction (%) of reads that align to the genome. The key point is that deep sequencing generates a huge amount of data, but the important data is only that which contributes to your experimental objective. For example, we can generate a billion reads with one RNA-seq sample in a single run. The opportunity to do multiple samples in a single lane using barcodes allows considerable cost savings and increased efficiency. So the real question is how many reads are actually needed to determine the global gene expression profile. Generally, ten million aligned reads give results that are comparable to profiling with the latest generation of Affymetrix microarrays, and so 12 different samples can be run in a single lane for RNA-seq in the sequencer. For the profiling of alternative splicing the read depth needs to be considerably greater, at least in the 30 million aligned read range.

11. Similarly, sequencing generates millions of short reads for each sampled individual. Combining these short reads into mRNA transcripts and then using the transcripts to detect differentially expression require programming skills, considerable computing power, statistical capability, and bioinformatics expertise. Rapid evolution of software in this field has made it possible to align, map, and detect differentially expressed, alternatively spliced genes with their sequence information. The readers need to evaluate the advantages and disadvantages of each of these tools before applying them to their studies.

Acknowledgements

The author gratefully acknowledges the Genomics Core Facility in intramural research in NHLBI and the extramural research program in NIA for their support and resources during the preparation of this chapter.

References

1. Esrefoglu M (2013) Role of stem cells in repair of liver injury: experimental and clinical benefit of transferred stem cells on liver failure. World J Gastroenterol 19:6757–6773

2. Weiss DJ (2013) Stem cells, cell therapies and bioengineering in lung biology and diseases: comprehensive review of the recent literature 2010–2012. Ann Am Thorac Soc 10:S45–S97

3. Volk SW, Theoret C (2013) Translating stem cell therapies: the role of companion animals in regenerative medicine. Wound Repair Regen 21:382–394

4. Preda MB, Valen G (2013) Evaluation of gene and cell-based therapies for cardiac regeneration. Curr Stem Cell Res Ther 8:304–312

5. Hansson EM, Lendahl U (2013) Regenerative medicine for the treatment of heart disease. J Intern Med 273:235–245

6. Szkolnicka D, Zhou W, Lucendo-Villarin B, Hay DC (2013) Pluripotent stem cell-derived hepatocytes: potential and challenges in pharmacology. Annu Rev Pharmacol Toxicol 53: 147–159

7. Cohen KS, Cheng S, Larson MG et al (2013) Circulating CD34(+) progenitor cell frequency is associated with clinical and genetic factors. Blood 121:e50–e56

8. Bentzinger CF, Wang YX, von Maltzahn J, Rudnicki MA (2012) The emerging biology of muscle stem cells: implications for cell-based therapies. Bioessays 35:231–241

9. Ding L, Morrison SJ (2013) Haematopoietic stem cells and early lymphoid progenitors occupy distinct bone marrow niches. Nature 495:231–235

10. Miao W, Xufeng R, Park MR et al (2013) Hematopoietic stem cell regeneration enhanced by ectopic expression of ROS-detoxifying enzymes in transplant mice. Mol Ther 21: 423–432

11. Geiger H, de Haan G, Florian MC (2002) The ageing haematopoietic stem cell compartment. Nat Rev Immunol 13:376–389

12. Liu P, Barb J, Woodhouse K, Taylor JG, Munson PJ, Raghavachar N (2011) Transcriptome profiling and sequencing of differentiated human hematopoietic stem cells reveal lineage-specific expression and alternative splicing of genes. Physiol Genomics 43:1117–1134

13. Rieger MA, Schroeder T (2012) Hematopoiesis. Cold Spring Harb Perspect Biol 4:pii:a008250

14. Cain CJ, Manilay JO (2012) Hematopoietic stem cell fate decisions are regulated by Wnt antagonists: comparisons and current controversies. Exp Hematol 41:3–16

15. Seke Etet PF, Vecchio L, Bogne Kamga P, Nchiwan Nukenine E, Krampera M, Nwabo Kamdje AH (2012) Normal hematopoiesis and hematologic malignancies: role of canonical Wnt signaling pathway and stromal microenvironment. Biochim Biophys Acta 1835:1–10

16. Sood R, Liu P (2012) Novel insights into the genetic controls of primitive and definitive hematopoiesis from zebrafish models. Adv Hematol 2012:830703

17. Zhan Y, Xu Y, Lew AM (2012) The regulation of the development and function of dendritic cell subsets by GM-CSF: more than a hematopoietic growth factor. Mol Immunol 52:30–37

18. Tsujioka T, Matsuoka A, Tohyama Y, Tohyama K (2012) Approach to new therapeutics: investigation by the use of MDS-derived cell lines. Curr Pharm Des 18:3204–3214

19. Nagasawa T (2012) Regulation of immune cell production by bone marrow niches]. Seikagaku 84:163–167

20. Kokkaliaris KD, Loeffler D, Schroeder T (2012) Advances in tracking hematopoiesis at the single-cell level. Curr Opin Hematol 19:243–249

21. Rappold I, Ziegler BL, Kohler I et al (1997) Functional and phenotypic characterization of cord blood and bone marrow subsets expressing FLT3 (CD135) receptor tyrosine kinase. Blood 90:111–125

22. McNiece I, Briddell R, Stoney G et al (1997) Large-scale isolation of CD34+ cells using the Amgen cell selection device results in high levels of purity and recovery. J Hematother 6:5–11

23. Hicks C, Wong R, Manoharan A, Kwan YL (2007) Viable CD34+/CD133+ blood progenitor cell dose as a predictor of haematopoietic engraftment in multiple myeloma patients undergoing autologous peripheral blood stem cell transplantation. Ann Hematol 86: 591–598

24. El-Badri NS, Hakki A, Saporta S et al (2006) Cord blood mesenchymal stem cells: Potential use in neurological disorders. Stem Cells Dev 15:497–506

25. Gao Z, Fackler MJ, Leung W et al (2001) Human CD34+ cell preparations contain over 100-fold greater NOD/SCID mouse engrafting capacity than do CD34– cell preparations. Exp Hematol 29:910–921

26. Riviere C, Subra F, Cohen-Solal K et al (1999) Phenotypic and functional evidence for the expression of CXCR4 receptor during megakaryocytopoiesis. Blood 93:1511–1523

27. Huss R (1996) Applications of hematopoietic stem cells and gene transfer. Infusionsther Transfusionsmed 23:147–160

28. Sutherland DR, Yeo EL, Stewart AK et al (1996) Identification of CD34+ subsets after glycoprotease selection: engraftment of CD34+Thy-1+Lin– stem cells in fetal sheep. Exp Hematol 24:795–806

29. Jensen K, Brusletto BS, Aass HC, Olstad OK, Kierulf P, Gautvik KM (2013) Transcriptional profiling of mRNAs and microRNAs in human bone marrow precursor B cells identifies subset- and age-specific variations. PLoS One 8:e70721

30. Liu F, Lu J, Fan HH et al (2006) Insights into human CD34+ hematopoietic stem/progenitor

cells through a systematically proteomic survey coupled with transcriptome. Proteomics 6: 2673–2692

31. Eckfeldt CE, Mendenhall EM, Flynn CM et al (2005) Functional analysis of human hematopoietic stem cell gene expression using zebrafish. PLoS Biol 3:e254

32. McKinney-Freeman S, Cahan P, Li H et al (2012) The transcriptional landscape of hematopoietic stem cell ontogeny. Cell Stem Cell 11:701–714

33. Gabut M (2012) Alternative splicing: a new mechanism controlling stem cell pluripotency. Med Sci (Paris) 28:372–374

34. Kramer S (2011) Developmental regulation of gene expression in the absence of transcriptional control: the case of kinetoplastids. Mol Biochem Parasitol 181:61–72

35. Raghavachari N, Barb J, Yang Y et al (2012) A systematic comparison and evaluation of high density exon arrays and RNA-seq technology used to unravel the peripheral blood transcriptome of sickle cell disease. BMC Med Genomics 5:28

36. Kandpal RP, Rajasimha HK, Brooks MJ et al (2012) Transcriptome analysis using next generation sequencing reveals molecular signatures of diabetic retinopathy and efficacy of candidate drugs. Mol Vis 18:1123–1146

37. Driver AM, Penagaricano F, Huang W et al (2013) RNA-Seq analysis uncovers transcriptomic variations between morphologically similar in vivo- and in vitro-derived bovine blastocysts. BMC Genomics 13:118

38. Hackett NR, Butler MW, Shaykhiev R et al (2012) RNA-Seq quantification of the human small airway epithelium transcriptome. BMC Genomics 13:82

39. Wang Z, Gerstein M, Snyder M (2009) RNA-Seq: a revolutionary tool for transcriptomics. Nat Rev Genet 10:57–63

40. McGettigan PA (2013) Transcriptomics in the RNA-seq era. Curr Opin Chem Biol 17:4–11

41. Vijay N, Poelstra JW, Kunstner A, Wolf JB (2012) Challenges and strategies in transcriptome assembly and differential gene expression quantification. A comprehensive in silico assessment of RNA-seq experiments. Mol Ecol 22:620–634

42. Tariq MA, Kim HJ, Jejelowo O, Pourmand N (2011) Whole-transcriptome RNAseq analysis from minute amount of total RNA. Nucleic Acids Res 39:e120

Chapter 8

Measuring MicroRNA Expression in Mouse Hematopoietic Stem Cells

Wenhuo Hu and Christopher Y. Park

Abstract

MicroRNAs (miRNAs) are important regulators of diverse biologic processes. In the hematopoietic system, miRNAs have been shown to regulate lineage fate decisions, mature immune effector cell function, apoptosis, and cell cycling, and a more limited number of miRNAs has been shown to regulate hematopoietic stem cell (HSC) self-renewal. Many of these miRNAs were initially identified as candidate regulators of HSC function by comparing miRNA expression in hematopoietic stem and progenitors cells (HSPCs) to their mature progeny. While the measurement of miRNA expression in rare cell populations such as HSCs poses practical challenges due to the low amount of RNA present, a number of techniques have been developed to measure miRNAs in small numbers of cells. Here, we describe our protocol for measuring miRNAs in purified mouse HSCs using a highly sensitive real-time quantitative PCR strategy that utilizes microfluidic array cards containing pre-spotted TaqMan probes that allows the detection of mature miRNAs in small reaction volumes. We also describe a simple data analysis method to evaluate miRNA expression profiling data using an open-source software package (HTqPCR) using mouse HSC miRNA profiling data generated in our lab.

Key words Hematopoietic stem cell, MicroRNA, FACS, qPCR, HTqPCR, TaqMan, R/bioconductor

1 Introduction

MicroRNAs (miRNAs) are small, evolutionarily conserved noncoding RNAs approximately 22 nucleotides (nts) in length that exert their biological effects by negatively regulating the stability and translational efficiency of multiple target mRNAs by binding the 3′-untranslated region (3′-UTR) of target mRNAs [1]. Following posttranscriptional processing by endonucleases, mature miRNAs are incorporated into a protein–RNA complex called the RNA-induced silencing complex (RISC), which potentiates interactions between 6–8 nts "seed sequences" in miRNAs and their near-complementary "seed match" sequences in the 3′-UTRs of their target mRNAs [2]. Typically, such base pairing is imperfect, resulting in translational suppression, but in the presence of perfect base

Kevin D. Bunting and Cheng-Kui Qu (eds.), *Hematopoietic Stem Cell Protocols*, Methods in Molecular Biology, vol. 1185, DOI 10.1007/978-1-4939-1133-2_8, © Springer Science+Business Media New York 2014

pairing endonucleolytic mRNA cleavage is triggered [3]. In silico analyses predict that each miRNA targets hundreds of mRNAs on average; thus, the ~1,000 miRNAs in the human genome likely regulate approximately 30 % of the estimated ~30,000 mRNAs in humans, reflecting widespread regulation of gene expression networks by miRNAs [4].

miRNA features such as their short length and ability to inhibit targets via interactions that rely on base pairing make miRNAs attractive targets for therapeutic manipulation since stable nucleic acid miRNA analog mimics and inhibitors (e.g., locked nucleic acids, LNAs) can be efficiently delivered to tissues in vivo such as the liver [5, 6] as well as cancer cells [7, 8] when administered in vivo for relatively short periods of time. However, since long-term in vivo experiments (4–6 months) are required to evaluate the function of hematopoietic stem cells (HSCs), such studies present significant practical problems for studying HSC miRNA function under normal physiological states. Therefore, ectopic expression of miRNAs using viral transduction techniques is commonly used to evaluate miRNA function in HSCs [9–12]. Prior studies have shown that cloning miRNAs with their adjacent sequences into lenti- and retroviral constructs can aid in generating high-expressing constructs [13].

While miRNAs have been shown to regulate a variety of cellular processes, a number of studies have definitively demonstrated that miRNAs regulate HSC function in humans and mice (reviewed in ref. 14). These studies have largely relied on the development of methods to identify miRNAs that are differentially expressed between HSCs and their downstream progeny in either the mouse [10, 12, 15–18] or the human hematopoietic system [11]. While initial efforts to measure miRNAs in rare HSC populations used partially purified cell populations to increase cell number and RNA yield [19], for the past several years it has become possible to use increasingly more sensitive techniques to measure absolute or relative miRNA abundance using sequence-specific PCR-based assays or solid-state (hybridization-based) arrays. In addition, the advent of next-generation sequencing techniques has allowed investigators the ability to readily identify novel miRNAs while simultaneously measuring absolute and relative abundance of different known miRNA species; however, the ability to apply RNA-sequencing techniques to small numbers of cells is still an emerging area (reviewed in ref. 20). Both solid-state arrays and qPCR strategies have been improved by utilizing chemically modified (e.g., LNA oligonucleotides) or stem–loop-based probes (e.g., TaqMan) that can improve complementarity matches or probe binding by increasing the Tm, leading to the development of both sensitive and specific assays; however, even these assays have potential pitfalls. For example, while stem–loop primer-based TaqMan qPCR assays can efficiently detect known miRNA species in even single cells [21], they can only measure one form of an individual miRNA. Thus, these assays may not accurately detect the various mature miRNAs that can be generated following posttranscriptional modification at the 3′ or the 5′ end of miRNAs (including nucleotide additions and deletions), which, in turn, can affect miRNA stability or function [22–26].

While numerous approaches are now available to perform wide-scale measurements of miRNA expression and abundance including small RNA-sequencing, miRNA hybridization arrays (e.g., Affymetrix, Agilent, and Illumina), and qPCR-based approaches including TaqMan (Life Technologies) and QuantiMir (System Biosciences) [20], we have chosen to use the TaqMan miRNA Array card technology using stem–loop primers to measure miRNA expression in both mouse and human HSCs since we have found that it can routinely measure miRNA expression in small cell numbers (<1,000). Despite the potential pitfalls of stem–loop primers, we and many other investigators have chosen to perform miRNA profiling studies using this method since it is relatively rapid, obviates the need to perform RNA ligations, and bypasses the need to separately validate miRNA expression results generated from miRNA hybridization arrays. We use this approach when our goal is not to identify novel miRNAs, but rather to rapidly identify differentially expressed miRNAs among known, annotated miRNAs. Herein, we describe our protocol for miRNA expression profiling studies in mouse HSCs as well as our basic workflow for analyzing the generated miRNA expression data using the open-source software package, HTqPCR [27].

2   Materials

Prepare all solutions using ultrapure water (prepared by purifying deionized water to attain a specific resistance of 18 MΩ cm at 25 °C) and analytical grade reagents; water for resuspending RNA and all reactions must be treated with DEPC to inactivate RNases. Prepare and store all reagents at room temperature (unless indicated otherwise). Diligently follow all waste disposal regulations when disposing waste materials.

2.1 Reagents and Antibodies for HSC Purification

1. FACS buffer: Sterile phosphate-buffered saline (PBS, without Mg^{2+} or Ca^{2+}) supplemented with 2 % fetal calf serum.

2. Ammonium chloride K [ACK (potassium bicarbonate)] red cell lysis solution (150 mM NH_4CL, 10 mM $KHCO_3$) containing 10 mM EDTA.

3. Anti-mouse c-Kit magnetic beads (Miltenyi Biotec, cat# 130-091-224) to enrich for c-Kit+ cells.

4. LS columns (Miltenyi Biotec, cat# 130-042-401) and QuadroMACS separator (Miltenyi Biotec, cat# 130-090-976).

5. Antibodies to identify mouse HSCs include the following: a lineage antibody-containing cocktail (Lin) in PECy5 [containing antibodies against CD3 (clone: 145-2C11), CD4 (GK1.5), CD8 (53-6.7), B220 (RA3-6B2), Ter-119 (Ter-119), Gr-1 (RB6-8C5), Mac-1 (M1/70)], c-Kit in APC-Cy7 (2B8), Sca-1 in Pacific Blue (E13-161.7), CD34 in PE (RAM34), Slamf1 in allophycocyanin (TC15-12F12.2), CD16/32 in Alexa Fluor 700 (93). All these antibodies are commercially available from either BD Biosciences or eBioscience.

6. 1,000× Propidium iodide (PI) (1 mg/ml).

7. Trypan blue for counting viable cells.

2.2 miRNA Purification, Measurement, and Data Analysis

1. RNA purification reagents include Trizol reagent (Life Technologies, cat# 15596-026), chloroform, isopropyl alcohol, 75 % ethanol (in DEPC-treated water), RNase-free water. Use a benchtop centrifuge with temperature control for RNA preparation.

2. miRNA array cards and primer pools. The following reagents are available from Life Technologies: TaqMan rodent miRNA A array V2.0 (cat# 4398967) and B array v3.0 (cat# 4444899) cards each with 384 wells (each well contains dried primers and probes); Megaplex primer pools of rodent pools A (cat# 4401090) and pools B v3.0 (cat# 4444752) for RT reactions; Megaplex pre-amp primers for rodent pool A (cat# 4399203) and pool B v3.0 (4444308) for pre-amplification.

3. TaqMan miRNA reverse transcription kit (Life Technologies, cat# 4366596) for reverse transcription (RT), TaqMan PreAmp master mix (Life Technologies, cat# 4391128) for pre-amplification, and TaqMan Universal PCR master mix (Life Technologies, cat# 4324018) for quantitative polymerase chain reaction (qPCR).

4. Qubit 2.0 Fluorometer (Life Technologies, cat# Q32866) to quantitate the small quantity of RNA typically isolated from small numbers of purified HSCs (1–1.5 ng per 1,000 cells).

5. TaqMan miRNA array card sealer (Life Technologies, cat# 4331770).

6. Applied biosystems ViiA™ 7 Real-Time PCR System: This system has the ability to hold heating blocks for 96-well plates, 384-well plates, and TaqMan Array Micro-Fluidic Cards. We describe the use of TaqMan Array Micro-Fluidic Cards and the ViiA™ 7 Real-Time PCR System equipped with the TaqMan Array Block here.

7. Install the R language program for Mac OS, Linux, or Windows operating systems by following the online manual (r-project.org).

Installation of the HTqPCR software package should be performed inside the R environment as shown below. Online manuals (r-project.org) such as "An Introduction of R" and "R Data Import/Export" are suggested for readers to help perform analyses. The codes are after the symbol ">" which are provided by the R environment:

```
> source("http://bioconductor.org/biocLite.R")
> biocLite("HTqPCR")
```

---

3 Methods

3.1 Bone Marrow Cell Preparation

1. Dissect each mouse donor to obtain long bones, including tibias and femurs. We also routinely harvest the entire spine and bilateral pelvises in order to maximize cell yield (*see* **Note 1**). After cleaning the muscles from the bones, crush the bones using a mortar and pestle in 5–10 ml of staining media. Filter the cells with a 70 μm nylon cell strainer into 50 ml conical tubes.

2. Spin the cells down at $300 \times g$ for 10 min to pellet them.

3. (Optional step) Add 1 ml ACK lysis buffer to the cell pellet from each mouse, and resuspend the cells using a pipetteman. Leave the cells on ice for 10 min. Add >10 ml staining buffer to dilute the ACK lysis buffer. Spin at $300 \times g$ for 10 min to pellet the cells. Remove the supernatant, and resuspend the cell pellet in 1 ml of staining buffer.

4. Add mouse anti-c-Kit magnetic beads as per manufacturer's instructions (or, in order to preserve beads, add about 10 μl beads for the total bone marrow cells from each mouse). Mix the cells and beads well using a pipetteman, ensuring that the cells are resuspended, and incubate on ice for 15 min.

5. Add 5 ml FACS buffer to the cells.

6. Magnetic bead purification should be performed as per manufacturer's instructions. Briefly, prime the LS column with 5 ml FACS buffer after mounting onto the QuadroMACS separator. Apply the resuspended cells to the primed LS column. Following loading of the column with cells, wash the column with 5 ml FACS buffer, and then repeat the wash step 1 additional time. Finally, remove the column from the magnetic separator and then add another 5 ml buffer to the column. Press the provided plunger into the column, and elute the cells into a FACS tube or a 15 ml conical.

7. Spin at $300 \times g$ for 10 min to pellet the cells. Resuspend in an appropriate volume of staining media to achieve a cell density of approximately 20 million cells/ml.

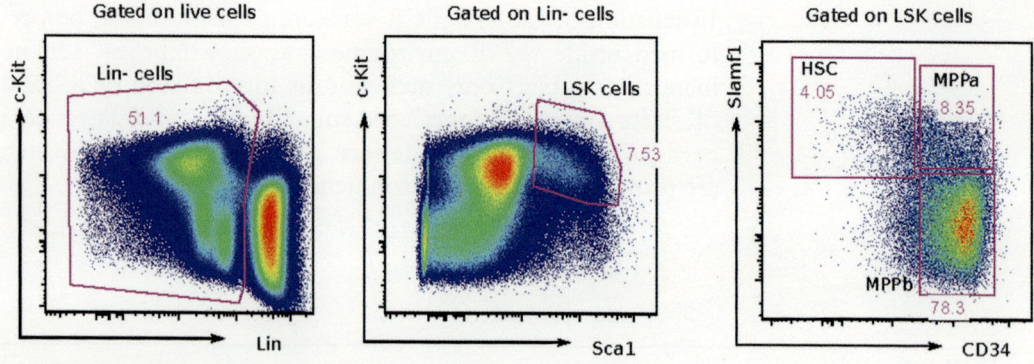

Fig. 1 Gating strategies for identifying mouse HSCs and multipotent progenitors following enrichment using mouse c-Kit magnetic beads. Immunophenotypic definitions of HSPC populations described include the following: HSC, Lin– c-Kit+ Sca1+ Slamf1+ CD34–; MPPa, Lin– c-Kit+ Sca1+ Slamf1+ CD34+; MPPb, Lin– c-Kit+ Sca1+ Slamf1– CD34+

3.2 FACS Purification of HSCs

1. Stain the c-Kit-enriched cells with the following antibodies: Lineage cocktail (all in PECy5), c-Kit (APC-Cy7), Sca1 (Pacific Blue), CD16/32 (Alexa Fluor 700), Slamf1 (allophycocyanin), CD34 (phycoerythrin). Incubate on ice for at least 45 min in order to optimize staining for CD34, allowing the best separation between CD34-positive and -negative cells (*see* **Notes 2** and **3**).

2. Cell sorting: Gating strategies and stem and progenitor cell fractions are shown in Fig. 1 using a FACS Aria 2 or 3 (BD) cell sorter and FACS Diva software. We routinely double-sort cells without moving the gates in order to achieve purities of >90 % (*see* **Note 4**). Each adult male mouse (>10 weeks old) yields approximately 3,000–3,500 double-sorted HSCs (Lin– c-Kit+Sca1+ Slamf1+ CD34–).

3.3 RNA Preparation

1. Spin down cells after the second FACS-sort, and remove supernatant carefully.

2. Add 500 μl Trizol, and then mix well by vortexing for 15 s. Add 200 μl chloroform, and then mix well again (*see* **Note 5**).

3. Centrifuge the samples at $13,000 \times g$ for 10 min at 4 °C. Remove the aqueous phase (top layer) containing total RNA into a new tube without disturbing the middle layer (protein).

4. Add 250 μl isopropanol to the aqueous phase (this volume should be half the Trizol volume originally added), and then mix by inversion. Centrifuge at $13,000 \times g$ for 10 min at 4 °C.

5. Wash the pellet with 500 μl 75 % ethanol, and spin the pellet at $13,000 \times g$ for 5 min. Discard the supernatant. Air-dry or vacuum dry the pellet, and then add 20 μl DEPC-treated RNase-free water to dissolve the RNA pellet (*see* **Note 6**).

6. RNA can be quantitated by Qubit 2.0 as per manufacturer's instructions.

Table 1
RT buffer mixture for TaqMan PCR

RT reaction mix components	Volume for one sample (µl)	Volume for ten samples (µl)
MegaPlex RT Primers (10×)	0.80	9.00
dNTPs with dTTP (100 mM)	0.20	2.25
MultiScribe Reverse transcriptase (50 U/µl)	1.5	16.88
10× RT Buffer	0.80	900
MgCl$_2$ (25 mM)	0.90	10.12
RNase inhibitor (20 U/µl)	0.10	1.12
Nuclease-free water	0.20	2.25
Total	4.50	50.62

3.4 TaqMan Reverse Transcription Reactions

ABI Mouse miRNA Array Cards (Life Technologies, cat# 4398979) can be purchased as a paired set which includes two 384-well plates (A and B) that have been pre-spotted with TaqMan probes for 641 unique mature mouse miRNA species and controls. The MegaPlex reverse transcription primers and Mega PreAmp pre-amplification primers have, respectively, A and B pools for each of these plates. Total RNA prepared from a single mouse (about 3,000 HSCs) ranges between 3 and 5 ng based on our experience. If the total amount of starting RNA is greater than 350 ng (typically from non-HSC populations used as reference controls), reverse transcription products can be used for qPCR reactions without the need for a pre-amplification step.

1. Prepare the RT buffer mixture by combining the components listed in Table 1.
   Following addition of all components, briefly mix and spin. We typically prepare master mixes for ten samples that include 12.5 % excess volume to account for losses during pipetting.

2. Pipette 4.5 µl RT mixture into the PCR tubes or plates, and add 3 µl of RNA samples into each tube or well. Prepare A and B pools separately.

3. Seal the tubes or the plates, and then spin briefly. Incubate the samples on ice while programming the PCR machine for the following thermal cycling conditions.

4. Set up the thermal cycling conditions for the PCR machine block as in Table 2.

5. Load the samples, and run the reaction. This run takes approximately 2 h and 30 min.

Table 2
Thermal cycling conditions for PCR

Stage	Temperature (°C)	Time
Cycle (40 cycles)	16	2 min
	42	1 min
	50	1 sec
Hold	85	5 min
Hold	4	∞

Table 3
Master mix for TaqMan PCR

PreAmp reaction mix components	Volume for one sample (μl)	Volume for ten samples (μl)
TaqMan PreAmp Master Mix (2×)	12.5	140.62
MegaPlex PreAmp Primers (10×)	2.5	28.13
Nuclease-free water	7.5	84.37
Total	22.5	253.12

3.5 Pre-amplification of miRNAs

A pre-amplification step is used to increase the sensitivity of the PCR reactions when starting with limited numbers of cells, i.e., less than 350 ng RNA for each reaction. We routinely use this step when profiling miRNAs from double-FACS-purified HSCs from a single mouse.

1. Thaw the buffers, and combine them; mix and spin briefly after that. The master mix volumes for ten samples include 12.5 % excess volume for each component in order to account for loss during pipetting (Table 3).

2. Pipette 22.5 μl into each PCR tubes or plates. Add 2.5 μl RT product. Seal the tubes or the plates, and spin briefly. Incubate on ice before running the reaction.

3. Program the PCR machine as shown in Table 4.

4. Load the samples, and run the reaction, which takes approximately 2 h. At the end of the reaction, add 75 μl of 0.1 mM TE pH 8.0 to each well or tube to stop the reaction.

3.6 qPCR Reaction

1. Thaw and mix reagents, and combine them in 1.5 ml tubes as shown in Table 5.
   Perform a quick spin after mixing the components. The recommended volume for each reagent reflects 12.5 % excess volume to account for volume losses during pipetting.

Table 4
PCR parameter for TaqMan PCR

Stage	Temperature (°C)	Time
Hold	95	10 min
Hold	55	2 min
Hold	72	2 min
Cycle (12 cycles)	95 60	15 s 4 min
Hold	99.9	10 min
Hold	4	∞

Table 5
Quantitative PCR amplification reagents required

Component	Volume for one array (µl)
TaqMan Universal PCR Master Mix No AmpErase UNG, 2×	450
Diluted PreAmp product	9
Nuclease-free water	441
Total	900

2. Pipette 100 µl of the reaction mix into each port of the TaqMan miRNA array card. Centrifuge for $300 \times g$ for 2 min without the brake on, and then repeat. Seal the plates using an array card sealer.

3. Set up the PCR on the ViiA 7 Real-Time PCR System. Create a new document, and check the following options: relative quantification ($\Delta\Delta$Ct), 384-well TaqMan array cards, and standard reaction procedure. Then import the rodent SDS files from the CD that is included with the array cards for plates A and B. Load the TaqMan array card, and then initiate the program. The total reaction time is approximately 2 h.

4. Export only the result table as a text, which can be directly imported into the R environment.

3.7 Data Analysis

Once the raw miRNA expression data is acquired, analysis is performed using the R/bioconductor platform using the open-source HTqPCR software package designed for large-scale qPCR data analyses [27]. Below, we use data generated in our lab to demonstrate the individual steps of data analysis. This experiment was designed to identify differences in miRNA expression in two

Table 6
Example of metadata table for analysis of miRNA expression data using the R program

	File	Group
1	ctrl1.csv	Ctrl
2	ctrl2.csv	Ctrl
3	ctrl3.csv	Ctrl
4	ctrl4.csv	Ctrl
5	ctrl5.csv	Ctrl
6	treat1.csv	Drug
7	treat2.csv	Drug
8	treat3.csv	Drug
9	treat4.csv	Drug
10	treat5.csv	Drug

populations of mouse HSCs, which we have designated as control (ctrl) and treatment (treat) groups. The miRNA expression profiling data for each sample was generated from total RNA isolated from approximately 2,000 FACS-sorted HSCs from a single mouse. There were five samples included from the ctrl and treatment groups.

1. Open the R program, and assign the working directory to the folder where the exported text files are using the function set work directory as shown below. All the following steps in the data analysis will read data from, or write data to, this folder:

   ```
   > setwd("path/name/to/data/")
   ```

2. Make a text file, referred to as "metadata" (data about data), that describes each of the samples. For example, Table 6 shows how for our data we have created a metadata file "target.txt" with the following content. Each row describes one sample generated from plates A and B.

3. Load the software packages needed for miRNA expression analysis by typing in the functions below. "HTqPCR" is designed for high-throughput qPCR assays across multiple conditions [27]. Package "gplots" contains various R programming tools for plotting data [28]. The function shown below is provided by this package to draw a heatmap. The information after "#" in the following code represents the comments for the code before this symbol in the same line, so this information is not necessary to execute the function:

   ```
   > library(HTqPCR)
   > library(gplots) # for cluster analysis,
     heatmap.2( )
   ```

4. Read sample information:

```
read.delim("target.txt", header=T) -> files
```

5. There are several ways to import the data exported from the ViiA7 qPCR software into the R environment. Each exported data file contains information including gene names and their Ct values as well as the metadata about the experiments. Furthermore, since two files (one each for plates A and B) make up each sample, it is necessary to delete the metadata and combine the paired files. In general, the expression data can be directly imported into R in order to extract the information needed, and then the data for plates A and B can be combined for each sample. One simple method to organize the data is to use Microsoft Excel before importing the data into R software. In Excel, first delete the unnecessary information in the description lines for each of the files, leaving only the 384 rows representing the wells associated with miRNA probes. All columns can be retained since the columns can be assigned in the program (see next step). The data file for each sample can then be generated by copying the 384 lines from plates A and B together. The final file names for each sample should be kept the same as shown in the file of "target.txt" such as ctrl1.csv, treat1.csv et al.

6. Import the Ct values for each miRNA into the R environment using the code below. Here, the files represent the file names generated from Subheading 3.7, **step 5**; the header parameter is assigned as "False", which indicates for the program that there is no head line in these files; n.features represent the gene numbers (768 genes for each sample from plates A and B); the information assigned to column.info links the column numbers (which corresponds to the column numbers in the data file) to their data type such as feature (same as gene name, column 4), position (column 1), and Ct values (column 11):

```
> raw = readCtData(files=as.character(files$
file),
    header=False,
    n.features=768,
    column.info=list(feature=4, position=1, Ct=11))
```

7. The gene names in the data file provided by ABI contain trailing numbers. For ease of reading, we frequently delete the trailing numbers (optional). For example, change "mmu-let-7b-000378" to "mmu-let-7b":

```
> tmp = sub("-\\d+$", "", featureNames(raw))
> tmp2 = sub("-\\d+_mat$", "", tmp)
> tmp3 = sub("#$", "", tmp2)
> featureNames(raw) = tmp3
```

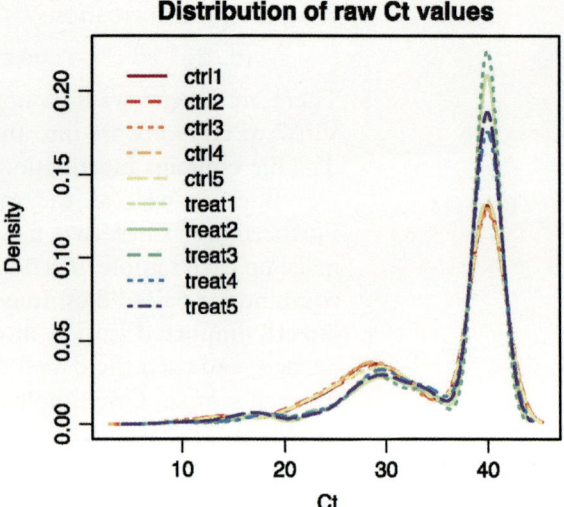

Fig. 2 Density plot of Ct values for miRNA expression using raw sample data. Each sample was evaluated using 768 unique TaqMan-based probes that detect 641 unique miRNAs as well as a subset of snoRNAs and U6

8. Check the quality of the raw data by examining the distribution of Ct values for all miRNAs using a histogram as follows. The resulting histogram is shown in Fig. 2:

```
> plotCtDensity(raw, main='Distribution of
raw Ct values', col=cols)
```

9. Perform a principal component analysis (PCA). A PCA analysis is performed in order to help identify the most important variables that account for the differences among sample groups. We routinely run this type of analysis on all our samples. This analysis can be run by typing in the code below. The result is presented in Fig. 3:

```
> plotCtPCA(raw, features=F, pch=2)
```

10. Identify miRNAs that are expressed in HSCs. The criteria for inclusion of miRNA expression data are somewhat subjective, but we generally consider a miRNA as detectable (and therefore evaluable) if it is expressed in more than 25 % of samples in which it was measured; applying this criterion results in the inclusion of 218 miRNAs (28 % of the total measured). The distribution of Ct values after eliminating non-detectable miRNAs is shown in Fig. 4:

```
> f = filterCtData(raw, remove.category =
"Undetermined", n.category=0.25*nrow(files))
> nrow(exprs(f))/nrow(exprs(raw)) # 0.2838542
> plotCtDensity(f, main='Distribution of Ct
values\nfor, expressed microRNAs', col=cols)
```

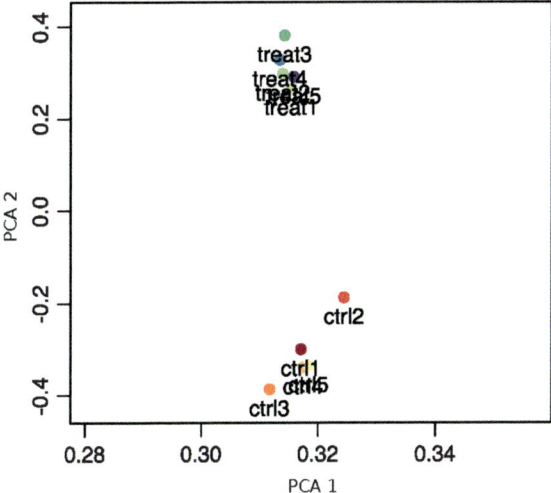

Fig. 3 PCA analysis of the raw Ct values reveals differences between samples representing control and treatment conditions

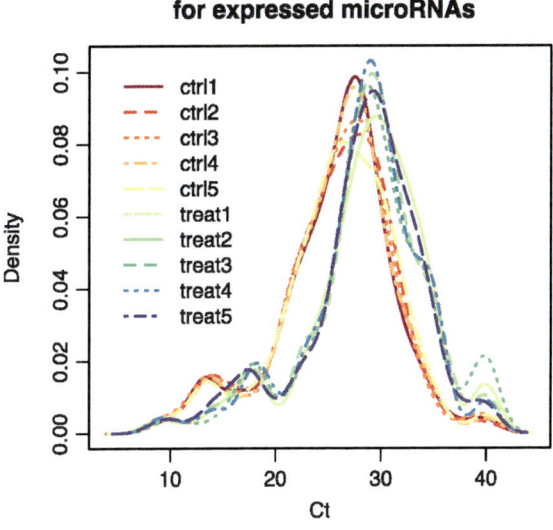

Fig. 4 Ct density plot of miRNAs detected across samples. Expressed mRNAs are defined by their mean Ct value when detected in more than 25 % samples tested, resulting in a detection rate of about 28 % of total miRNAs measured

11. Normalize the data across the various samples using the quantile method (*see* **Note 7**). The distribution of Ct values after quantile normalization is shown in Fig. 5:

```
> f.qt=normalizeCtData(f, norm="quantile")
> plotCtDensity(f.qt, main='After normali-
zation')
```

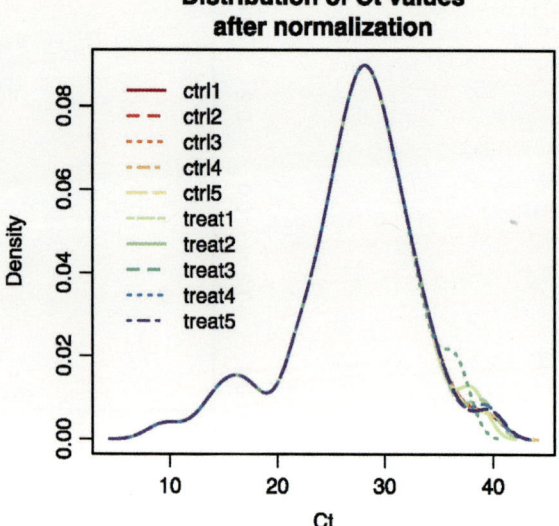

Distribution of Ct values after normalization

Fig. 5 MiRNA Ct density plot following quantile normalization. This example demonstrates an equivalent distribution of miRNA gene expression levels

12. Identify miRNAs that are differentially expressed at statistically significant levels by applying a *t* test (*see* **Note 8**). Here in the ttestCtData function, the parameter "groups" indicate the control and treated samples; "calibrator" is assigned as "ctrl" which indicates the control group as baseline values; "p.adjust" indicates the method of *p* value adjustment; other parameters are provided for the *t* test. The results were in ttest variable, which contains information about fold change, *p* values, and adjusted *p* values. Then the results were exported by the write. csv function into a file named "gene-list-ttest.csv":

```
> ttest = ttestCtData(f.qt, groups = files$
group, calibrator='ctrl', alternative = "two.
sided",
    paired = FALSE, replicates = TRUE, sort =
TRUE, stringent = TRUE, p.adjust = "BH")
> write.csv(file='gene-list-ttest.csv', ttest)
```

Use the function colnames to show the information included in the results ttest variable. The text below this function represents the results following execution of this function:

```
> colnames(ttest)
[1] "genes" "feature.pos" "t.test"
[4] "p.value" "adj.p.value" "ddCt"
[7] "FC" "meanCalibrator" "meanTarget"
[10] "categoryCalibrator" "categoryTarget"
```

The top ten differential miRNAs identified in our data set are shown here:

```
> ttest[1:10,c('genes', 'FC', 'p.value',
'adj.p.value')]
   genes FC p.value adj.p.value
   203   snoRNA135   0.25621911   7.579055e-13
1.504506e-10
   205   U6   snRNA   0.12617037   1.453629e-12
1.504506e-10
   171 mmu-miR-706 0.34257220 1.268243e-08
8.750876e-07
   204   snoRNA202   0.07579676   6.024003e-08
3.117421e-06
   206   U87   0.14726842   4.306654e-07
1.782955e-05
   35 mmu-miR-126-3p 0.05519303 1.151795e-06
3.973691e-05
   195   rno-miR-489   6.78985140   1.779365e-06
5.261836e-05
   158   mmu-miR-652   0.06423129   2.240539e-06
5.448856e-05
   99 mmu-miR-2183 5.78661636 2.616545e-06
5.448856e-05
   22   Mamm   U6   0.43218623   2.632297e-06
5.448856e-05
```

13. Prepare the data for cluster analysis using normalized Ct values:

```
> exprs(f.qt) -> all.exprs # export Ct
values
> fdr <- 0.01
> ttest.sig <- as.character(ttest[ttest$
adj.p.value < fdr, 'genes']) # gene names
> exprs(f.qt)[row.names(all.exprs)   %in%
ttest.sig,] -> sig.exprs
> all.exprs[apply(all.exprs, 1, sd)>0,]
-> all.exprs
```

14. View the cluster analysis showing the entire set of detectable miRNAs using the codes provided below. The code "all.exprs" ensures that the Ct values are appropriately assigned such that miRNAs with higher Ct values are given negative values since high Ct values correspond to lower expression levels; the code "row" signifies that the expression levels for each miRNA are normalized across all samples measured; the color scheme (col parameter) is set with the code "greenred", which means that miRNAs with higher expression levels will be depicted in red in the heatmap, while those with lower expression will be in green. Since the expression levels are scaled for each "row" or miRNA, the color in the heatmap only allows visualization of

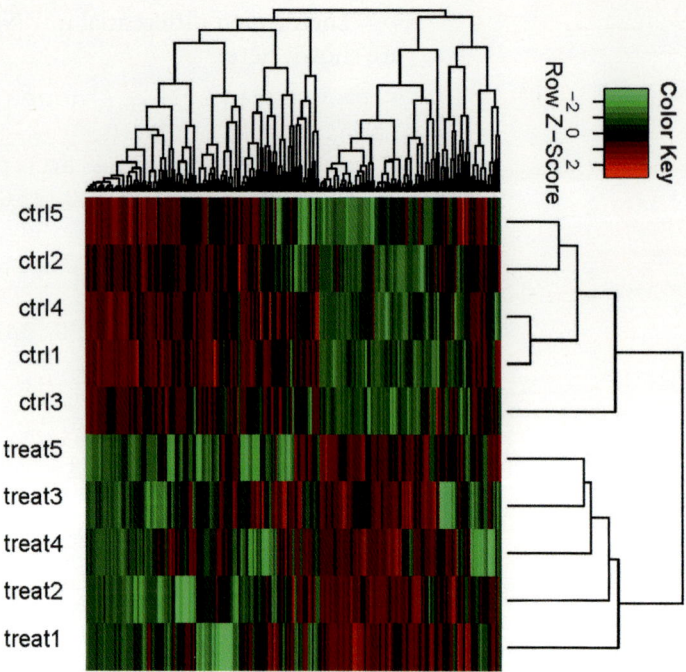

Fig. 6 Cluster analysis of HSC miRNA profiling data set including detectable genes following normalization

differences in miRNA expression level from the mean expression for any individual sample. Thus, the same shade of red or green for different miRNAs does not necessarily correspond to the same absolute differences in expression. The "heatmap.2" function is provided by the software package gplots with the detailed manual included in a PDF file. Also, a question mark with the function name will print the manual for this function too, such as "> ?heatmap.2". The result is presented in Fig. 6:

```
> heatmap.2(-all.exprs,
         scale='row',
         Colv=T,
         key=T,
         symbreaks=T,
         trace="none",
         col=greenred,
         density.info = "none",
         margin=c(4,2), labRow = '',
         hclust=function(x)hclust(x,method=
         "complete"),
         distfun=function(x) as.dist(1-cor
         (t(x), method='pearson')))
```

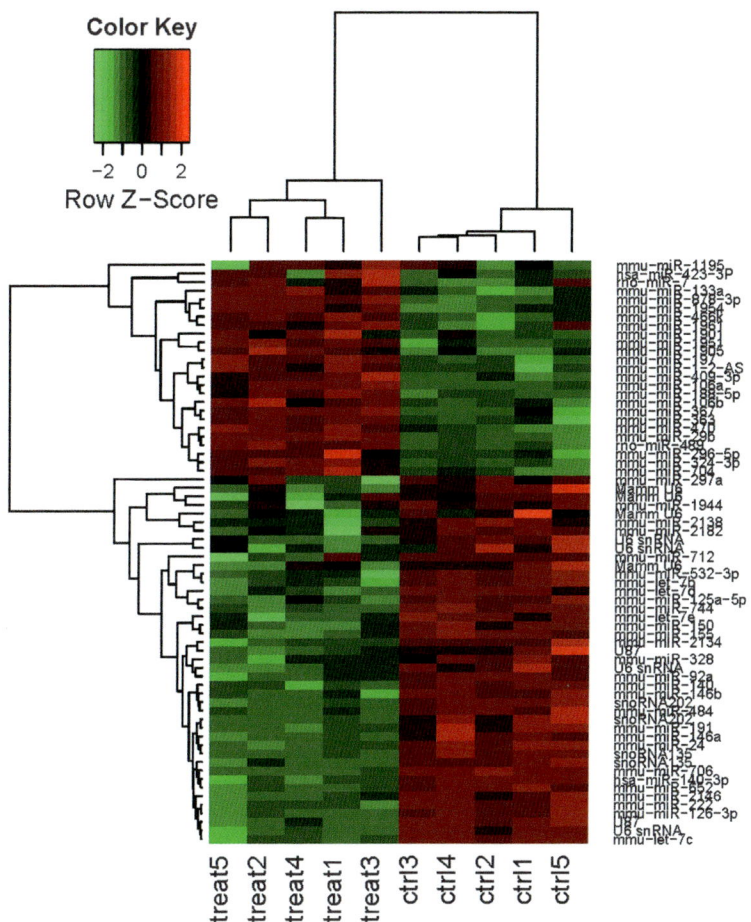

Fig. 7 Cluster analysis of differentially expressed miRNAs in HSCs identified after treatment. FDR < 0.01

15. Perform a clustering analysis for the miRNAs that exhibit significant changes in expression after drug treatment. The result is shown in Fig. 7:

```
> heatmap.2(-sel.rows[1:50,],
            scale='row',
            Colv=T,
            key=T,
            symbreaks=T,
            trace="none",
            col=greenred,
            density.info = "none",
            margin=c(4,7),
            hclust=function(x) hclust(x,method=
            "complete"),
            distfun=function(x) as.dist(1-cor
            (t(x), method='pearson')))
```

4   Notes

1. The number of harvested bone marrow cells (and hence HSCs) from donor mice can be increased by crushing the spine and pelvises in addition to the long bones. Since cells may aggregate during processing due to release of genomic DNA from lysed or damaged cells, DNase (10 U/ml or 1 µg/ml) should be added to the FACS buffer in all subsequent steps to reduce the formation of insoluble aggregates.

2. Antibodies against mouse CD34 typically do not stain cells brightly; however, staining can be improved by incubating cells with antibody for more than 45 min.

3. Whenever a new lot of an antibody is used, it should be re-titered to achieve comparable staining results.

4. Double-FACS-sorting HSCs significantly increases the purity of HSC preparations. While the first sort typically achieves an HSC purity of 60–70 % (calculated as a percentage of total events collected), a second sort can routinely result in >95 % purity; therefore, this approach is recommended.

5. The RNeasy Mini Kit (Qiagen) can be used instead of Trizol for RNA preparation if the investigator wishes to avoid using phenol-containing reagents.

6. Do not allow the RNA pellet to dry completely following the final EtOH wash because this makes it more difficult to redissolve the RNA.

7. There are several methods to normalize HTqPCR data. "deltaCt" calculates the standard deltaCt values by subtracting the mean of the chosen controls from all other treatment groups. "scale.rankinvariant" sorts features from each sample based on Ct values and identifies a set of features that remain rank invariant, i.e., those whose ordering is constant, and the average of these rank-invariant features is then used to scale the Ct values for each array individually. "norm.rankinvariant" also identifies rank-invariant features between each sample and a reference and then uses these features to generate a normalization curve individually for each sample by smoothing. "geometric.mean" calculates the geometric mean of all Ct values below Ct.max in each sample and scales the Ct values accordingly. The effect of normalization can be demonstrated by the function plotCtDensity.

8. The Limma software package uses a linear model to identify differentially expressed genes and is widely used to analyze mRNA expression microarray data [29]. HTqPCR also provides a function, "limmaCtData", to identify differentially expressed miRNAs using this linear model. Similar results can usually be obtained by this method compared to those obtained using statistical t tests.

Acknowledgments

This work was supported by a New Investigator Award in Bone Marrow Failure from the Department of Defense as well as an R01 award (1R01CA164120-01A1) from the NIH/NCI (to CYP).

References

1. Bartel DP (2004) MicroRNAs: genomics, biogenesis, mechanism, and function. Cell 116:281–297

2. Winter J, Jung S, Keller S, Gregory RI, Diederichs S (2009) Many roads to maturity: microRNA biogenesis pathways and their regulation. Nat Cell Biol 11:228–234

3. Jones-Rhoades MW, Bartel DP, Bartel B (2006) MicroRNAS and their regulatory roles in plants. Annu Rev Plant Biol 57:19–53

4. Filipowicz W, Bhattacharyya SN, Sonenberg N (2008) Mechanisms of post-transcriptional regulation by microRNAs: are the answers in sight? Nat Rev Genet 9:102–114

5. Elmen J, Lindow M, Schutz S et al (2008) LNA-mediated microRNA silencing in non-human primates. Nature 452:896–899

6. Elmen J, Lindow M, Silahtaroglu A et al (2008) Antagonism of microRNA-122 in mice by systemically administered LNA-antimiR leads to up-regulation of a large set of predicted target mRNAs in the liver. Nucleic Acids Res 36:1153–1162

7. Cogoi S, Zorzet S, Rapozzi V, Geci I, Pedersen EB, Xodo LE (2013) MAZ-binding G4-decoy with locked nucleic acid and twisted intercalating nucleic acid modifications suppresses KRAS in pancreatic cancer cells and delays tumor growth in mice. Nucleic Acids Res 41:4049–4064

8. Nagai Y, Miyazawa H, Huqun TT et al (2005) Genetic heterogeneity of the epidermal growth factor receptor in non-small cell lung cancer cell lines revealed by a rapid and sensitive detection system, the peptide nucleic acid-locked nucleic acid PCR clamp. Cancer Res 65:7276–7282

9. Chaudhuri AA, So AY, Mehta A et al (2012) Oncomir miR-125b regulates hematopoiesis by targeting the gene Lin28A. Proc Natl Acad Sci U S A 109:4233–4238

10. Guo S, Lu J, Schlanger R et al (2010) MicroRNA miR-125a controls hematopoietic stem cell number. Proc Natl Acad Sci U S A 107:14229–14234

11. Lechman ER, Gentner B, van Galen P et al (2012) Attenuation of miR-126 activity expands HSC in vivo without exhaustion. Cell Stem Cell 11:799–811

12. Ooi AGL, Sahoo D, Adorno M, Wang Y, Weissman IL, Park CY (2010) MicroRNA-125b expands hematopoietic stem cells and enriches for the lymphoid-balanced and lymphoid-biased subsets. Proc Natl Acad Sci U S A 107:21505–21510

13. Hwang HW, Wentzel EA, Mendell JT (2007) A hexanucleotide element directs microRNA nuclear import. Science 315:97–100

14. Chung SS, Hu W, Park CY (2011) The role of MicroRNAs in hematopoietic stem cell and leukemic stem cell function. Ther Adv Hematol 2:317–334

15. Arnold CP, Tan R, Zhou B et al (2011) MicroRNA programs in normal and aberrant stem and progenitor cells. Genome Res 21:798–810

16. Petriv OI, Kuchenbauer F, Delaney AD et al (2010) Comprehensive microRNA expression profiling of the hematopoietic hierarchy. Proc Natl Acad Sci U S A 107:15443–15448

17. Han Y-C, Park CY, Bhagat G et al (2010) microRNA-29a induces aberrant self-renewal capacity in hematopoietic progenitors, biased myeloid development, and acute myeloid leukemia. J Exp Med 207:475–489

18. O'Connell RM, Chaudhuri AA, Rao DS, Gibson WS, Balazs AB, Baltimore D (2010) MicroRNAs enriched in hematopoietic stem cells differentially regulate long-term hematopoietic output. Proc Natl Acad Sci U S A 107:14235–14240

19. Georgantas RW 3rd, Hildreth R, Morisot S et al (2007) CD34+ hematopoietic stem-progenitor cell microRNA expression and function: a circuit diagram of differentiation control. Proc Natl Acad Sci U S A 104:2750–2755

20. Pritchard CC, Cheng HH, Tewari M (2012) MicroRNA profiling: approaches and considerations. Nat Rev Genet 13:358–369

21. Chen C, Ridzon DA, Broomer AJ et al (2005) Real-time quantification of microRNAs by stem-loop RT-PCR. Nucleic Acids Res 33:e179

22. Katoh T, Sakaguchi Y, Miyauchi K et al (2009) Selective stabilization of mammalian micro RNAs by 3′ adenylation mediated by the cytoplasmic poly(A) polymerase GLD-2. Genes Dev 23:433–438

23. Kozomara A, Griffiths-Jones S (2011) miRBase: integrating microRNA annotation and deep-sequencing data. Nucleic Acids Res 39:D152–157

24. Cloonan N, Wani S, Xu Q et al (2011) MicroRNAs and their isomiRs function cooperatively to target common biological pathways. Genome Biol 12:R126

25. Jones MR, Quinton LJ, Blahna MT et al (2009) Zcchc11-dependent uridylation of microRNA directs cytokine expression. Nat Cell Biol 11:1157–1163

26. Wyman SK, Knouf EC, Parkin RK et al (2011) Post-transcriptional generation of miRNA variants by multiple nucleotidyl transferases contributes to miRNA transcriptome complexity. Genome Res 21:1450–1461

27. Dvinge H, Bertone P (2009) HTqPCR: high-throughput analysis and visualization of quantitative real-time PCR data in R. Bioinformatics 25:3325–3326

28. Warnes GR, Bolker B, Bonebakker L, et al (2013) gplots:various R programming tools for plotting data, R package version 2.13.0 Cran.r.project.org/web/packages/gplots/gplots.pdf

29. Smyth GK (2005) Bioinformatics and computational biology solutions using R and bioconductor. Springer, New York

Chapter 9

DNA Methylation Profiling of Hematopoietic Stem Cells

Amber Hogart Begtrup

Abstract

DNA methylation is a key epigenetic mark that is essential for properly functioning hematopoietic stem cells. Determining where functionally relevant DNA methylation marks exist in the genome is crucial to understanding the role that methylation plays in hematopoiesis. This chapter describes a method to profile DNA methylation by selectively enriching methylated DNA sequences that are bound in vitro by methyl-binding domain (MBD) proteins. The MBD-pulldown approach selects for DNA sequences that have the potential to be "read" by the endogenous machinery involved in epigenetic regulation. Furthermore, this approach is feasible with very small quantities of DNA, and is compatible with the use of any downstream high-throughput sequencing approach. This technique offers a reliable, simple, and powerful tool for exploration of the role of DNA methylation in hematopoietic stem cells.

Key words DNA methylation, Methyl-binding domain protein, MBD-pulldown, Methylation enrichment, High-throughput sequencing, Epigenetics

1 Introduction

Epigenetics refers to covalent modifications of DNA and histone proteins that are heritable and reversible. Epigenetic modifications contribute to the cell-type specific gene expression differences that drive differentiation and cell fate specification. DNA methylation in mammals occurs primarily in the context of CpG dinucleotides and is essential for early development [1]; however, little is known about how DNA methylation contributes to cell specification. Hematopoietic stem cells require DNA methylation for proper differentiation into the various lineages of the blood [2, 3]. Profiling DNA methylation in hematopoietic stem cells allows for identification of specific sequences and genes that are marked by methylation and ultimately assists in understanding the role of epigenetics in cellular identity [4].

A variety of approaches exist to identify the location of DNA methylation marks on a genomic scale; however, all methods have limitations. For a comprehensive review of current approaches that

Kevin D. Bunting and Cheng-Kui Qu (eds.), *Hematopoietic Stem Cell Protocols*, Methods in Molecular Biology, vol. 1185, DOI 10.1007/978-1-4939-1133-2_9, © Springer Science+Business Media New York 2014

are applicable to the genome-wide scale, please refer to Harris et al. [5]. The approach described in this chapter is based upon an in vitro protein pull-down technique. The proteins that bind to DNA methylation marks within the cell are called methyl-binding domain (MBD) proteins. The MBD proteins represent a small family of proteins that specifically bind to methylated CpG sites and interact with a host of other key proteins that regulate chromatin organization and gene expression [6]. One advantage to this approach is the ability to profile DNA methylation marks that are relevant for the proteins that read the DNA methylation and translate the patterns into altered chromatin and gene expression. Additionally, this approach is amenable to very small quantities of genomic DNA. Finally, because the protein-pulldown involves minimal manipulation, the enriched DNA is of high quality and is well suited for downstream procedures such as high-throughput sequencing.

Recent advances in DNA sequencing technology have allowed for investigation of the entire genome of an organism to be feasible. Using massively parallel or next-generation sequencing to identify the sequences marked by DNA methylation therefore permits unprecedented characterization of the role that methylation plays in hematopoietic stem cells. This chapter describes the procedures to isolate genomic DNA from primary cells and selectively enrich the methylated DNA sequences. Specific instruction regarding performing next-generation sequencing is not included in this chapter, but can be found elsewhere [7]. Additionally, this chapter does not address the informatics component of analyzing the methylation profiles; however, two recent papers demonstrate the types of analyses that can be performed with the genomic DNA obtained from the MBD-enrichment protocol [4, 8].

2   Materials

2.1   Primary Enriched HSCs

The exact quantity of cells required will depend on the downstream methodology used for profiling the methylation-enriched DNA sequences. If high-throughput sequencing will be utilized, 3–5 µg of genomic DNA is generally sufficient (obtained from 1 to 2×10^6 cells). Due to the limited abundance of HSCs in the bone marrow, a large number of primary cells must be isolated and enriched in order to obtain sufficient quantities for whole-genome sequencing. Enrichment strategy for HSCs will vary depending on current knowledge of cellular surface markers and the desired characteristics of the cell population. Reagents listed below are one option for enrichment of HSCs from mice and represent the method described previously [9, 10].

1. Bone marrow from tibias and femurs from 20 mice typically yields 10^9 white blood cells, leading to approximately $3–5 \times 10^5$ enriched HSCs.

2. Needles, small gauge (25G) and 3 mL syringe.

3. ACK lysing buffer.

4. Lineage-specific antibodies that recognize the following cell surface antigens: CD8a, CD4, CD11b, Ly-6G/Ly-6C, CD45R, and Ter119.

5. BioMag magnetic beads coated with antibody-specific IgG (corresponding to antibodies in **item 3** above).

6. Fluorescently labeled antibodies for flow cytometry (i.e., anti-mouse Sca1 and anti-mouse CD117).

7. 1× Phosphate Buffered Saline (PBS) with 0.5 % Tween®20.

8. Hemocytometer and light microscope for counting cells.

9. Magnetic tube holder to secure beads.

10. Cell strainer, Falcon® 40 µM.

11. BD FACS Aria flow cytometer (BD Biosciences).

2.2 Genomic DNA Isolation and Methylation Enrichment

1. Gentra Puregene DNA purification kit (Qiagen, Cat. No 158745).

2. Nanodrop 2000 UV spectrophotometer (Thermo Scientific).

3. Probe sonicator.

4. Agarose gel, 100 bp DNA ladder, ethidium bromide.

5. Methyl Collector™ Ultra kit (Active Motif, Cat. No 55005).

6. Small rotating shaker.

7. Qiagen MinElute Reaction Cleanup kit (Qiagen, Cat No 28204).

2.3 Quantitative PCR Instrumentation for Quality Assurance

1. Positive control oligonucleotides to amplify known methylated genomic sequence.

2. Negative control oligonucleotides to amplify unmethylated genomic sequence.

3. SYBR® Select Master Mix (Life Technologies, Cat. No 4472937).

4. Optical 96-well reaction plate (Applied Biosystems, Cat No 4483354).

5. Applied Biosystems 7500 Real-Time PCR System (Applied Biosystems).

3 Methods

3.1 Isolation and Enrichment of Primary HSCs

1. Starting with freshly sacrificed mice, remove skin and muscle from the femur and tibias.

2. Using sharp scissors cut the ends of the bones to create an open tube for flushing marrow.

3. Prepare 10 mL of 1× phosphate buffered saline (PBS) in a conical tube.

4. Using a small (22G or 25G) gauge needle with a 3 mL syringe, flush the open bones with 1× PBS to dislodge the marrow (*see* **Note 1**).

5. Homogenize the bone marrow cells in the 1× PBS by passing through the needle 4–5 times.

6. Add 3×–5× total volume, i.e., 40 mL, of ACK lysis buffer to 10 mL of suspended cells to lyse red blood cells. Gently pipet up and down to mix cells.

7. Incubate cells in ACK lysis solution on ice for 10 min, then pellet cells by centrifugation for 5 min at 4 °C.

8. Remove supernatant and resuspend cell pellet in 5–10 mL of cold 1× PBS. (*See* **Notes 2** and **3**).

9. Count cells on a hemocytometer to determine the quantity of reagents required for lineage depletion. (*See* **Note 4**).

10. Incubate cells with lineage-specific antibodies on ice for 30 min.

11. Prepare magnetic beads by thoroughly resuspending, aliquoting in a clean tube, and washing with an equal volume of cold 1× PBS/0.5 % Tween. (*See* **Note 5**).

12. Combine the antibody-incubated cell suspension with the washed magnetic beads, vortex briefly, and incubate on ice for 10 min.

13. Place tube(s) on the magnet and incubate for 5 min at 4 °C.

14. Remove supernatant to clean tube(s) and place back on magnet for 5 min at 4 °C.

15. Remove supernatant to a clean tube(s) and centrifuge cells for 5 min at 4 °C to pellet lineage-depleted cells.

16. Remove supernatant and resuspend pelleted cells in 1–2 mL of 1× PBS.

17. Count lineage depleted cells on a hemocytometer to ensure adequate lineage subtraction. (*See* **Note 6**).

18. Remove approximately $1–2 \times 10^4$ cells for each control cell population desired for flow cytometry (i.e., unstained control, and each single color antibody).

19. Incubate cells with fluorescently labeled antibodies (i.e., PE-anti-mouse Sca1 and APC-anti-mouse c-kit (BD Biosciences)) and incubate on ice for 30 min. (*See* **Note 7**).

20. Wash cells with approximately 10 mL of cold 1× PBS, spin and resuspend in 1–2 mL of 1× PBS.

21. Prior to placing cells on a flow cytometer, filter out any clumps by passing cells through a cell strainer.

22. Stained primary cells are enriched for HSCs and other multipotent progenitors and are ready to be further enriched by flow cytometry. The protocol for cell sorting depends on the instrument available and is beyond the scope of this procedure.

3.2 Isolation of Genomic DNA from HSCs

1. After enriched cells have been recovered from cell sorting, vortex to resuspend and combine into one tube to pellet cells.

2. Spin cells for 5–10 min at 4 °C and remove supernatant. Cell pellets are typically visible even when the total yield of cells is on the order of 2–5×10^5.

3. Resuspend cell pellet in residual liquid by vortexing and then add 150 µl of cell lysis solution, and pipet up and down to thoroughly mix the pelleted cells. Transfer to a 1.5 mL eppendorf tube for subsequent processing. (*See* **Note 8**).

4. Add 1.5 µl of RNase A solution to the cells and mix by pipetting. Incubate at 37 °C for 5 min and then immediately place on ice.

5. Add 100 µl of protein precipitation solution and vortex vigorously for 20–30 s.

6. Spin tube in a 4 °C centrifuge for 5 min at $13,000$–$16,000 \times g$. The protein pellet should be visible as white precipitate along the edge of the tube. If the protein pellet is loose, it may be necessary to repeat the centrifugation.

7. Carefully remove the supernatant to a clean 1.5 mL eppendorf tube containing 300 µl of isopropanol. Mix by inverting until a visible strand of DNA becomes visible.

8. Spin the DNA at $13,000$–$16,000 \times g$ at 4 °C for 5 min. The DNA should be visible as a small white pellet. Carefully remove the supernatant by pipetting off the isopropanol.

9. Add 300 µl of 70 % ethanol and invert tube until pellet becomes dislodged. Spin at $13,000$–$16,000 \times g$ for 5 min.

10. Remove ethanol supernatant, being careful not to dislodge the DNA pellet. If the pellet does appear loose, use a small tip to pipet away the remaining ethanol. Allow the clean pellet to dry inverted for 2–3 min to eliminate residual ethanol.

11. Add 50 µl of DNA hydration solution to the pellet and pipet up and down and vortex to resuspend DNA pellet. Overnight incubation at room temperature or 1 h incubation at 55 °C will ensure complete hydration of DNA. (*See* **Note 9**).

12. Using a Nanodrop UV spectrophotometer, quantify the yield of genomic DNA. Prior to proceeding with the methylated DNA enrichment ensure that a minimum of 1 µg of total HSC genomic DNA are available.

Table 1
Components for the methylated DNA binding reaction

Reagent	Volume per reaction (µl)
Complete Binding Buffer	70
Fragmented genomic DNA (100–500 ng)	10
Magnetic Beads	10
His-MBD2/MBD3L1 protein complex	10

3.3 MBD-Enrichment of Methylated DNA

1. Fragment genomic DNA through mechanical shearing with sonication. A minimum concentration of 20 ng/µl is needed and a minimum volume of 100 µl in a 1.5 mL eppendorf tube. If necessary, adjust total volume by adding additional DNA hydration solution. (*See* **Notes 10** and **11**).

2. Check DNA fragment size by running a small aliquot of DNA (2–4 µl) on a 1 % agarose gel with a 100 bp DNA ladder. The predominant smear of genomic DNA should be between 200–400 bp.

3. Prepare the Binding Reaction components for the desired number of reactions according to the order listed in Table 1. Binding reactions occur in 0.5 mL strip PCR tubes provided with the kit. The binding reaction is tolerant to a range of DNA quantities (100–500 ng). If the final concentration of fragmented DNA is 10–50 ng/µl, then the DNA can be added without dilution to an individual binding reaction. If the DNA concentration exceeds 50 ng/µl then it is recommended that the DNA be diluted with water when adding to the binding reaction.

4. Repeat the binding reaction setup for up to 8 tubes. Ensure that the tubes are kept on ice while pipetting and that magnetic beads are mixed thoroughly by pipetting before adding to each binding reaction tube. Mix each reaction thoroughly by pipetting up and down and completely cap the PCR tubes. (*See* **Note 12**).

5. Incubate the binding reactions at 4 °C for 1 h while on a rotating shaker. During this incubation gently flick the tubes to resuspend the bead slurry at 15 min intervals.

6. After the 1 h incubation spin briefly or flick tubes to ensure the liquid is removed from the cap. Place the tubes on the magnetic stand assembled from the kit, to pellet the beads.

7. Remove the supernatant to clean tubes and save for further analysis. (*See* **Note 13**).

8. Wash beads four times with 200 µl of binding buffer (provided by kit). Holding the tubes away from the magnet, gently pipet

the liquid into the tube and thoroughly resuspend the beads with pipetting. Be careful to avoid generating bubbles. Place the tubes on the magnet to pellet the beads and remove and discard of the supernatant.

9. After the final wash add 100 μl of complete elution buffer (created by adding 2 μl of provided proteinase K to 98 μl of Elution Buffer EM (1 for each reaction). Incubate the beads in the elution buffer at 50 °C for 30 min. During this incubation, ensure that the beads are resuspended in the elution buffer every 10 min.

10. Warm the Proteinase K stop solution at 37 °C during the last 10 min of the elution incubation.

11. Flick or quick spin the PCR tubes to remove liquid from the cap and place reaction tubes on the magnetic stand to pellet the beads. Carefully remove the supernatant to a clean tube. Depending on the total number of tubes used for the binding reaction, the supernatants from multiple binding reactions can be combined into a single 1.5 ml eppendorf tube or split into different tubes.

12. Add 2 μl of proteinase K stop solution per 100 μl of eluted DNA. Thoroughly mix by pipetting.

13. Proceed to purification of the methylation-enriched and supernatant DNA samples with column purification. 5 volumes of Buffer PB are added to every 1 volume of the binding reaction, or supernatant fraction. For the bound fraction 500 μl of PB are added to 100 μl of eluted DNA.

14. Apply the sample to the column and centrifuge for 1 min. Remove flow-through.

15. Wash bound sample with 750 μl of PE and centrifuge for 1 min. Remove flow-through and spin for 1 additional minute.

16. Place column in a clean 1.5 mL tube and elute with 30 μl of water. Let sit for 2–3 min at room temperature and spin for 1 min to elute DNA. (*See* **Note 14**).

17. Remove 1–2 μl of DNA and check concentration and yield from the methylation-enriched and supernatant samples on a Nanodrop spectrophotometer.

18. Dilute with water an aliquot of the supernatant sample to approximately the same concentration as the undiluted methylation-enriched sample.

3.4   Quality Control for Methylation Enrichment

1. Prepare a quantitative PCR master mix for at least one positive control primer set and one negative control primer set in the methylation enriched and supernatant samples. Refer to Tables 2 and 3 for PCR setup instructions and Table 4 for example primers. (*See* **Notes 15** and **16**).

Table 2
Components for quantitative PCR setup

Reagent	Volume per reaction (µl)
Water	9.5
2× Sybr Master Mix	12.5
Primer For + Rev	2
DNA	1

Table 3
Quantitative PCR standard cycle conditions

		40 Cycles	
Step	Initial denaturation	Denature	Anneal/extend
Temperature	95 °C	95 °C	60 °C
Time	10 min	15 s	1 min

2. Using an optical-grade 96-well plate, perform quantitative PCR with standard reaction conditions. Approximately 10 ng of DNA is required for each PCR amplification. Include a non-template control to ensure PCR product specificity.

3. Perform qPCR with standard cycling conditions. Ensure that fluorescence readings are measured at each cycle and a melting curve analysis was performed. (*See* **Note 17**).

4. Determine if methylation enrichment was achieved by calculating the ΔCT. (*See* **Note 18**) Fig. 1 shows representative qPCR data for three positive control methylated sequences. (*See* **Notes 19** and **20**).

5. Methylation enriched DNA is ready to be used in any number of downstream procedures, such as library preparation for next-generation sequencing.

4   Notes

1. In our experience, several factors contribute to the overall yield of cells from mice. First, younger mice will often produce a larger number of HSCs than comparable older mice. Second, if the marrow can be expelled cleanly with one push, the yield of cells will be higher. Success of flushing the marrow will be improved by cutting bones with very sharp scissors, and away from joints and by using a needle that fits tightly into the bone.

Table 4
qPCR primers for quality control

Organism	Gene	Methylation status	Primer forward	Primer reverse	Size (bp)
Mouse	*Snrpn*	Methylated	TAGCTGCCTTTTGGCAGGA	CGCAATGGCTCAGGTTTGT	135
Mouse	*H19*	Methylated	GACCCACAGCATTGCCATT	TTGGACGTCTGCTGAATCAGT	135
Mouse	*Rasgrf1*	Methylated	TGGAATTCTGGGGACTCTTCA	AACAGCAATAGCGGTAGCCA	136
Mouse	*Actb*	Unmethylated	ACACTGGCACAGCCAACTTTA	TCATCAAATGCCCACACCG	140
Human	*GAPDH*	Unmethylated	TACTAGCGGTTTTACGGGCG	AAAAGAAGATGCGGCTGACT	187
Human	*SNRPN*	Methylated	TATCCTGTCCGCTCGGCATT	GAACTGCAATCACCCTGATGT	176

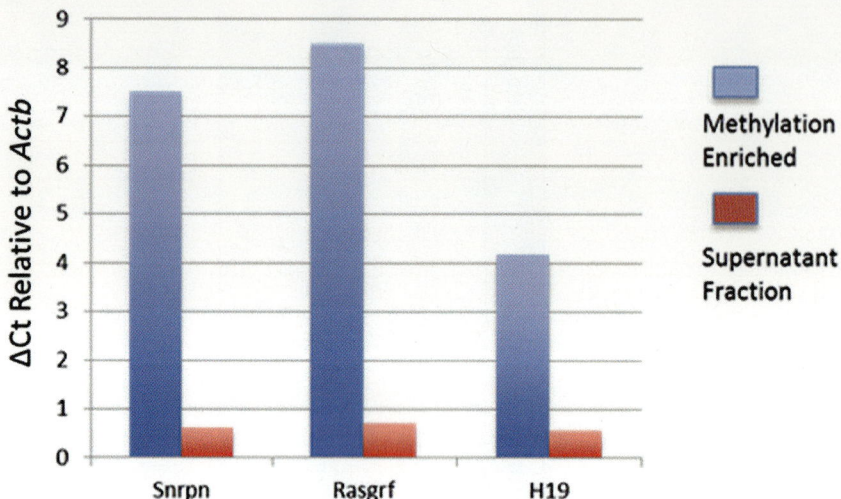

Fig. 1 Quantitative PCR Quality Control for Methylation Enrichment. Amplification of three known methylated regions of the mouse genome representing the imprinting control regions for *Snrpn*, *Rasgrf1*, and *H19* was compared to the amplification of the unmethylated CpG island promoter of *Actb*. Crossing point values for the methylated control genes were normalized to the unmethylated control gene for the Methylation-enriched (*blue*) and Supernatant fractions (*red*), and the differences in crossing points (ΔCT) are plotted. All three genes demonstrate significant enrichment of methylated sequences in the methylation-enriched sample relative to the supernatant

2. Incomplete RBC lysis is obvious if the cell pellet is pink instead of white. If this is seen, a higher volume of ACK lysis solution should be used.

3. The volume of PBS used to resuspend the cell pellet is arbitrary and can be reduced to any volume that will sufficiently allow for complete resuspension of cells. As antibody incubation will occur immediately after resuspending and counting cells, use of a smaller volume (such as 4–5 mL) may increase the efficiency of antibody binding.

4. The quantity of lineage antibodies used for depletion can be empirically determined based upon the concentration of each antibody and the total number of cells. When depleting 10^9 bone marrow cells, 50–100 μg of each depletion antibody was used; however, this may be in excess of what is needed for depletion. If resources are limited, it may be possible to reduce the quantity of antibody used for this step.

5. Magnetic bead incubations are performed at 4 °C to preserve the primary cells. Store all PBS solutions and magnetic beads at 4 °C to ensure the cells remain cold during processing.

6. Typically using this procedure, greater than 90 % of total cells are depleted. If lineage depletion does not lead to significant reduction in total cell count, it is likely that an insufficient quantity of cell-surface antibodies were utilized.

7. As with the lineage-specific cell surface antibodies, the total quantity of antibody used for staining cells for cell sorting will be empirically determined based upon the concentration of the antibody, background fluorescence, and affinity of the antibodies.

8. Lysed samples are stable at room temperature in cell lysis solution. It is important that primary cells are processed rapidly to reduce the amount of cellular degradation; however, once the cells are lysed, the suspension is stable for at least 2 years on the bench.

9. The final volume used to resuspend the DNA depends on the total yield of cells from the sorting procedure. Aim to keep the concentration at least 20 ng/μl, as more concentrated genomic DNA will lead to more uniform fragment size during the sonication. Typically 50 μl will be an appropriate volume if the yield is 0.5×10^6 however, a volume of 100 μl may be appropriate for higher cell yields.

10. Uniformly fragmented genomic DNA is crucial for the success of the downstream enrichment; however, the sonication program that will produce the desired fragments must be empirically determined. Factors including the concentration and volume of DNA, temperature, duration, and intensity of the sonication pulses will influence the resulting fragmentation. Using an abundant DNA source, modify the duration and intensity of sonication pulses until the ideal settings are identified to produce consistently fragmented DNA.

11. Digestion of genomic DNA with restriction enzymes is an alternative approach to sonication fragmentation. This method is not recommended for whole-genome profiling, however, due to the introduction of fragment bias due to the sequence-specific cutting by restriction enzymes. If used, restriction enzyme digestion also requires additional purification steps to ensure the elimination of the restriction enzymes after digestion. Please see the user manual for the Methyl Collector™ Ultra kit for details.

12. Repeated freeze–thaw cycles may lead to degradation in the quality of the His-MBD2/MBD3L1 protein complex. To preserve the integrity of the protein pulldown, it is recommended to aliquot the protein into several tubes to avoid repeated thawing.

13. Although there is some loss of total DNA during the purification step, purifying the total enriched fraction and total supernatant will provide some measure of the efficiency of the methylated enrichment process. One spin column can be used to purify several binding reactions, as the total binding capacity of the column is approximately 5 μg.

14. The elution volume for the methylation-enriched DNA may be modified depending on the requirements for next-generation sequencing. To ensure that enough DNA is available for quality control as well as the downstream sequencing procedure, a minimum of 30 µl elution is recommended.

15. Quantitative PCR amplification of known methylated sequences relative to known unmethylated sequences ensures that the enrichment procedure was successful before proceeding to genome-wide sequencing. If performing genome-wide methylation profiling in mouse or human cells, the imprinting center for several well-studied imprinted genes serves as a nice positive control, as both a highly methylated and unmethylated chromosomes are present in every cell.

16. When designing qPCR primers, the optimal product size is between 125 and 200 bp.

17. Unincorporated primers can sometimes generate crossing point values in non-template controls, or in samples with little template. To ensure the crossing point value represents a unique and specific product the melting curve analysis should be performed. If primer amplification is specific, one peak of fluorescence will be observed as the products are heated, generally in the range of 70–90 °C.

18. Calculating the enrichment using the ΔCT method involves subtracting the crossing point values for a control gene from the gene of interest. The unmethylated control amplicon is subtracted from the methylated control amplicon (ΔCT). This method assumes that the primer efficiency between the two amplicons is similar, which may not be correct. Therefore, to ensure an accurate representation of the enrichment, it is recommended that at least two different target regions are tested.

19. As PCR amplification occurs on a logarithmic scale, the ΔCT values represent $\log_2$ differences in abundance between the methylation-enriched and supernatant fractions. ΔCT of 8 represents approximately a 250-fold increase in enrichment.

20. If methylated enrichment is not seen, the most likely reason is a problem with the binding buffer. Low and high stringency buffers are provided. Repeating the protocol with a high stringency buffer may improve the methylated enrichment.

References

1. Li E, Bestor TH, Jaenisch R (1992) Targeted mutation of the DNA methyltransferase gene results in embryonic lethality. Cell 69:915–926

2. Broske AM, Vockentanz L, Kharazi S, Huska MR, Mancini E et al (2009) DNA methylation protects hematopoietic stem cell multipotency from myeloerythroid restriction. Nat Genet 41:1207–1215

3. Trowbridge JJ, Snow JW, Kim J, Orkin SH (2009) DNA methyltransferase 1 is essential for and uniquely regulates hematopoietic stem and progenitor cells. Cell Stem Cell 5:442–449

4. Hogart A, Lichtenberg J, Ajay SS et al (2012) Genome-wide DNA methylation profiles in hematopoietic stem and progenitor cells reveal overrepresentation of ETS transcription factor binding sites. Genome Res 22:1407–1418

5. Harris RA, Wang T, Coarfa C et al (2010) Comparison of sequencing-based methods to profile DNA methylation and identification of monoallelic epigenetic modifications. Nat Biotechnol 28:1097–1105

6. Bogdanovic O, Veenstra GJ (2009) DNA methylation and methyl-CpG binding proteins: developmental requirements and function. Chromosoma 118:549–565

7. Pomraning KR, Smith KM, Bredeweg EL et al (2012) Library preparation and data analysis packages for rapid genome sequencing. Methods Mol Biol 944:1–22

8. Deaton AM, Webb S, Kerr AR et al (2011) Cell type-specific DNA methylation at intragenic CpG islands in the immune system. Genome Res 21:1074–1086

9. Nemeth MJ, Curtis DJ, Kirby MR et al (2003) Hmgb3: an HMG-box family member expressed in primitive hematopoietic cells that inhibits myeloid and B-cell differentiation. Blood 102:1298–1306

10. Orlic D, Fischer R, Nishikawa S et al (1993) Purification and characterization of heterogeneous pluripotent hematopoietic stem cell populations expressing high levels of c-kit receptor. Blood 82:762–770

Chapter 10

Metabolic Characterization of Hematopoietic Stem Cells

Fatih Kocabas, Junke Zheng, Chengcheng Zhang, and Hesham A. Sadek

Abstract

An important feature of stem cells is their maintenance in their respective hypoxic niche. Survival in this low-oxygen microenvironment requires significant metabolic adaptation. We demonstrated that mouse HSCs utilize glycolysis instead of mitochondrial oxidative phosphorylation to meet their energy demands. We have adapted various tools for characterization of the metabolic properties of hematopoietic stem cells (HSCs). These techniques include flow cytometric profiling of HSCs based on mitochondrial potential and NADH fluorescence as well as measurement of ATP content, oxygen consumption rate, and glycolytic flux in purified HSCs.

Key words Hematopoietic stem cells, HSC, Stem cell metabolism, Reactive oxygen species, ROS, Glycolysis, ATP content, Oxygen consumption, NADH, Glycolytic flux, Lactate

---

1 Introduction

Hematopoietic stem cells (HSCs) are able to self-renew and provide lifelong supply of blood cells. They are primarily located in the endosteal regions of the bone marrow with unique vasculature and limited perfusion of oxygen, which results in very low oxygen tension, also known as hypoxic niche (reviewed in ref. 1). The understanding of HSC function and metabolic adaptation to their respective hypoxic niche in the bone marrow has important implications in HSC biology. We have recently demonstrated that HSCs in the bone marrow preferentially utilize anaerobic glycolysis instead of mitochondrial oxidative phosphorylation [1–4]. HSCs demonstrate lower rates of oxygen consumption and lower ATP content with a significant increase in rates of lactate production. This metabolic phenotype is associated with stabilization of the master metabolic regulator hypoxia-inducible factor-1α (Hif-1α). With hundreds of downstream targets, Hif-1α results in stimulation of glycolysis and inhibition of oxidative phosphorylation. Genetic manipulation of Hif-1α and its transcriptional regulator Meis1 in HSCs outlined the link between metabolic and redox regulation of HSCs and their stemness [2] (Fig. 1).

Kevin D. Bunting and Cheng-Kui Qu (eds.), *Hematopoietic Stem Cell Protocols*, Methods in Molecular Biology, vol. 1185, DOI 10.1007/978-1-4939-1133-2_10, © Springer Science+Business Media New York 2014

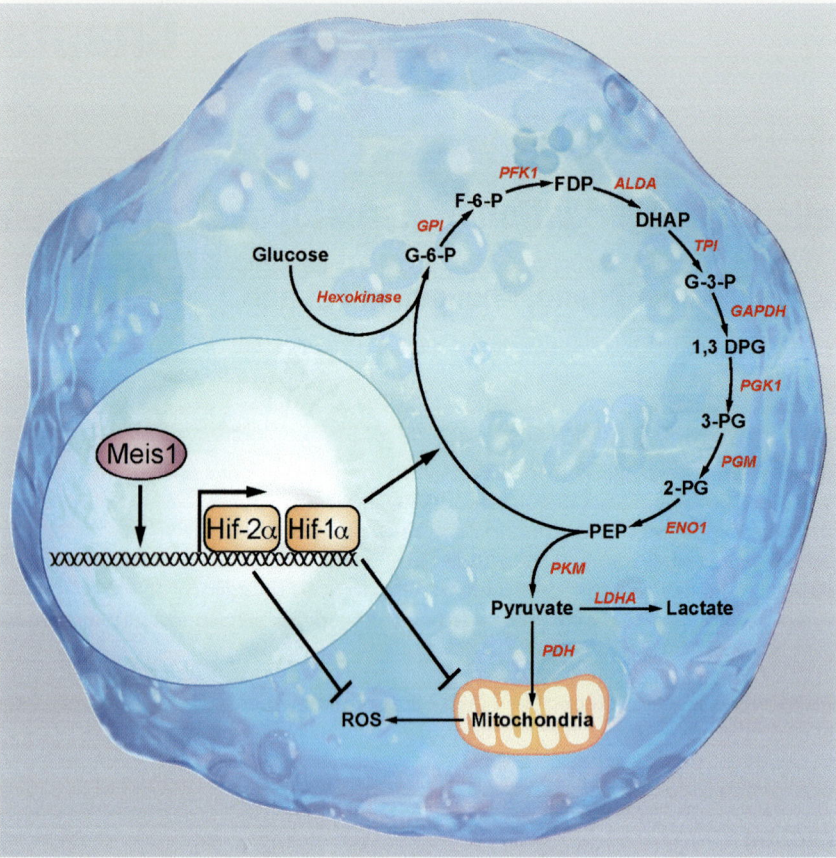

Fig. 1 Schematic of hypoxic HSC depicting metabolic and redox regulation of HSCs, which are integral for normal HSC function

HSCs are responsible for generating over a billion cells daily [5]. However, their frequency is very low with estimates of 1 long-term (LT) HSC out of 30,000 mouse bone marrow mononuclear cells [6]. Isolation of LT-HSCs by fluorescence-activated cell sorter (FACS) provides tools to study HSC gene expression, repopulation ability following bone marrow transplantations, and stemness with serial transplantation. One of the dilemmas of studying HSC metabolism is the isolation of limited number of HSCs from bone marrow and lack of sensitive metabolic assays to measure HSC function. Thus, we developed metabolic assays that facilitate robust measurement of HSC metabolism. To this end, we initially pooled bone marrow cells from a large number of mice (up to 62 animals) to isolate a sufficient number of LT-HSCs for metabolic assays [4]. This allowed us to determine ATP content, glycolysis, and oxygen consumption rate in HSCs. In addition, we developed a flow cytometric profile of the unfractionated bone marrow using fluorescent dyes that stains cells based on their mitochondrial potential and content, and we found that a vast majority of HSCs are located in low mitochondrial potential gate. It is even possible to enrich for HSCs simply by metabolic profiling of cells with low mitochondrial content. Further refinements of our metabolic assays allowed us to markedly increase the sensitivity of our assays whereby a full metabolic profile could be scaled down to pooled purified HSCs from 5 to 7 mice [2, 3]. This chapter provides details of metabolic assays to study mouse HSC function.

2 Materials

2.1 Bone Marrow Cell Isolation and Flow Cytometry Components

1. Isoflurane.
2. Phosphate-buffered saline, 1×.
3. RPMI 1640 medium.
4. RPMI medium supplemented with 20 % FBS (Hyclone, Defined Fetal Bovine Serum, SH30070.03).
5. RBC lysis solution (150 mM NH_4Cl, 10 mM $KHCO_3$, and 0.1 mM EDTA).
6. Lineage cocktail: Anti-CD3, anti-CD5, anti-B220, anti-Mac-1, anti-Gr-1, anti-Ter119.
7. FC blocker (anti-mouse CD16/32) and streptavidin-PE/Cy5.5.
8. HSC markers: Anti-Sca-1-FITC, anti-Kit-APC, anti-Flk-2-PE, and anti-CD34-PE or Sca-1-PE/Cy5.5, C-Kit-APC, anti-CD34-PE, and anti-Flk-2-PE.

2.2 Metabolic Assay Components

1. Mitotracker dyes (MitoTracker Green, Cat#M-7514; MitoTracker deep red 633, Cat#M22426; or MitoTracker Red CMXRos, Cat#M-7512 from Life Technologies).
2. Antimycin A (AMA) solution (Sigma, A0149).
3. 5-(and-6)-Carboxy-2′,7′-dichlorofluorescein diacetate (carboxy-DCFDA) (Cat# C-369, Invitrogen).
4. ATP Bioluminescence Assay Kit HS II (Roche) (Cat# 11699709001).
5. Fluorescence plate reader.
6. BD Oxygen Biosensor System 96- and 384-well formats.
7. Sodium sulfate (100 mM in water).
8. Borosilicate tubes.
9. D-[1-6-^{13}C]-glucose (Cambridge Isotope Labs, Cat# CLM-2717-0, Cambridge Isotope Laboratories, Inc).
10. 1:500 Dilution of 20 % w:v solution of Na[$^{13}C_3$]-lactate.
11. 1:500 Dilutions of 20 % w:v stock solutions of unlabeled sodium lactate and sodium L-[3-^{13}C]-lactate.

12. Tri-Sil (Thermo Scientific, Prod#TS-49001 Tri-Sil HTP reagent, 50 mL).

13. Glycolytic flux medium: Base DMEM supplemented with 10 % dialyzed fetal calf serum and ^{13}C-enriched nutrient of choice. DMEM powder (Sigma D5030) 8.3 g/L, L-glutamine 0.58 g/L (4 mM), NaHCO$_3$ 3.7 g/L 42.5 mM, HEPES 5.9 g 25 mM, Pen/Strep (10 mL from 100× stock), NaOH (9 mL from 1 N stock), add fetal calf serum to 10 % (Hyclone, Defined Fetal Bovine Serum, SH30070.03). Adjust pH to 7.4. Bring final volume to 1 L with mQ water, and filter sterilize. Add desired D-[1-6-^{13}C]-glucose (1-^{13}C, 99 %; 6-^{13}C, 97 %+, Cat# CLM-2717-0, Cambridge Isotope Laboratories, Inc) or D-[1-^{13}C]-glucose, 10 mM final concentration. (Remember to use dialyzed serum in experiments with isotopically labeled nutrients.)

3   Methods

3.1   Flow Cytometric Isolation of HSCs

All antibodies were purchased from BD PharMingen.

1. Euthanize 8–12-week-old mouse (C57BL/6 mice) using isoflurane chamber.

2. Cut the edges of the bones, and flush with PBS using 25G needles.

3. Complete the solution up to 50 mL with PBS (washing step).

4. Centrifuge at $1,200 \times g$ for 10 min.

5. Resuspend in 1 mL of media without FBS, and add 5 mL of RBC lysis buffer (see **Note 1**).

6. Incubate maximum for 5 min at room temperature (RT).

7. Add 25 mL of media with FBS (RPMI medium supplemented with 20 % FBS).

8. Centrifuge at $1,200 \times g$ for 10 min.

9. Aspirate the supernatant and resuspend in 0.5 mL of PBS supplemented with 2 % FBS.

10. Count the number of cells (see **Note 2**).

11. Add FC blocker (1:200 dilution) (anti-mouse CD16/32) on ice for 15 min.

12. Stain bone marrow cells with a biotinylated lineage cocktail. Add 5 μL biotinlyted Lin$^+$ antibody cocktail (1:100) (anti-CD3, anti-CD5, anti-B220, anti-Mac-1, anti-Gr-1, anti-Ter119; Stem Cell Technologies) to each 0.5 mL of cells. Incubate on ice for 15 min.

13. Add 10 mL PBS + 2 % FBS. Spin at $1,200 \times g$ for 5 min. Discard supernatant.

14. Resuspend in 225 μL of PBS + 2 % FBS.

15. Add 25 μL (1:100) streptavidin-conjugated microbeads. Incubate on ice for 15 min.

16. Add 10 mL volume PBS + 2 % FBS. Spin at $1,200 \times g$ for 5 min. Remove the supernatant.

17. Resuspend cells in 5 mL PBS + 2 % FBS.

18. Use autoMACS separator (Miltenyi Biotec) to perform magnetic lineage depletion.

19. Right before loading, strain cells through 70 μM strainer.

20. Collect unbound bone marrow cells (lineage-negative cells) (*see* **Notes 3** and **4**).

21. Spin down Lin⁻ cells at $1,200 \times g$ for 5 min.

22. Resuspend in 200 μL PBS + 2 % FBS.

23. Stain lineage-negative cells with 1 μL streptavidin-PE/Cy5.5 (1:200 dilution), 1 μL anti-Sca-1-FITC (1:200 dilution), 1 μL anti-Kit-APC (1:200 dilution), 1 μL anti-Flk-2-PE (1:200 dilution), and 1 μL anti-CD34-PE (1:200 dilution) antibodies. Mix by pipetting.

24. Incubate on ice for 15 min.

25. Add 10 mL volume PBS + 2 % FBS. Spin at $1,200 \times g$ for 5 min. Remove the supernatant.

26. Resuspend the pellet in 3 mL of PBS + 2 % FBS.

27. Isolate Lin⁻Sca1⁺Kit⁺Flk2⁻CD34⁻ cells (LT-HSCs) using FACS.

28. Perform metabolic studies within 24 h of LT-HSC isolation (*see* **Note 5**).

3.2 Flow Cytometric Profiling and Separation of Cells Based on Mitochondrial Activity

1. Collect cells from mouse bone marrow (*see* Subheading 3.1).

2. Wash collected bone marrow cells in 50 mL of PBS.

3. Spin down at $1,200 \times g$ for 10 min.

4. Resuspend the pellet with 1 mL of RPMI medium without FBS.

5. Add 5 mL of RBC lysis buffer, and incubate for 5 min at RT.

6. Add 25 mL of RPMI medium supplemented with 20 % FBS.

7. Spin down at $1,200 \times g$ for 10 min.

8. Discard supernatant, and resuspend the cells with 10 mL of RPMI medium supplemented with 20 % FBS.

9. Count the cells.

10. Prepare 10 mL of cells with 2,000,000 cells/mL for staining.

11. Add mitotracker dye (MitoTracker Green, MitoTracker deep red 633, MitoTracker Red CMXRos, or MitoTracker Red CM-H2XRos) with final concentration of 200 nM (*see* **Note 6**).

12. Incubate the cells at 37 °C water bath for 15 min.

13. Spin down cells at $1,200 \times g$ for 10 min.

14. Discard supernatant, and resuspend the cells with RPMI medium supplemented with 20 % FBS.

15. Adjust cell concentration to 10,000,000 cells/mL (*see* **Note 7**).

16. Carry out flow cytometric separation of high and low MP cells by separating cells in the low MP gate (6–9 %) and an equivalent number of cells with high mitochondrial potential.

3.3 Determination of Mitochondrial Source of NADH

1. Collect mouse bone marrow cells and stain with Mitotracker Deep Red (*see* Subheading 3.1).

2. Treat high and low mitochondrial potential cells with 2 mM AMA for 5 min at 37 °C (*see* **Note 8**).

3. Profile for NADH fluorescence versus Mitotracker Deep Red using flow cytometer.

4. Measure endogenous NADH fluorescence flow cytometrically at 37 °C with a UV laser (Ex: 350 nm, Em: 460 nm, Moflo analyzer, Cytomation) as described previously [7].

3.4 Measurement of Reactive Oxygen Species

1. Isolate mouse bone marrow cells, and collect Lin⁻ cells (*see* Subheading 3.1).

2. Incubate Lin⁻ cells with and without 1 μM of carboxy-DCFDA for 30 min in 37 °C water bath in the dark.

3. Spin down cells at $440 \times g$ for 5 min.

4. Resuspend in 200 μL PBS + 2 % FBS

5. Stain for LT-HSC surface markers: Sca-1-PE/Cy5.5, C-Kit-APC, CD34-PE, and Flk2-PE. Mix by pipetting.

6. Incubate on ice for 15 min.

7. Add 10 mL volume PBS + 2 % FBS. Spin at $440 \times g$ for 5 min. Remove the supernatant.

8. Resuspend the pellet in 3 mL of PBS + 2 % FBS.

9. Assay DCFDA fluorescence (ROS content) by flow cytometer.

3.5 Determination of ATP Content in HSCs

1. Isolate at least 150×10^3 LT-HSCs from mouse bone marrow using FACS (*see* Subheading 3.1) (*see* **Note 9**).

2. Centrifuge cells at $1,200 \times g$ for 10 min, and remove any medium.

3. Use ATP Bioluminescence Assay Kit HS II (Roche) in accordance with the manufacturer's recommendations.

4. Prepare 50 μL of ATP standards with concentration of ATP ranging from 10^{-6} to 10^{-12} M.

5. Lyse at least 50,000 LT-HSCs in 50 μL of lysis solution provided in the kit for 5 min at room temperature, and perform

measurement right away by adding 50 µL of luciferase reagent using Fluostar Optima plate reader (BMG Labtech) (*see* **Note 10**).

6. Finally, normalize relative light units with protein concentration (*see* **Note 11**) and determine ATP content/million cells by analysis of ATP standard curve.

3.6 Oxygen Consumption Assays in HSCs

1. Isolate at least 30×10^4 LT-HSCs from mouse bone marrow using FACS (*see* Subheading 3.1).

2. Use BD Oxygen Biosensor System to determine the rate of oxygen consumption in freshly isolated LT-HSCs.

3. Plate equal numbers of cells (10×10^4 cells/well) in 50 µL of RPMI medium supplemented with 20 % FBS in the provided 384-well plate (BD Oxygen Biosensor System, CA, USA) (*see* **Note 12**).

4. Use culture media lacking cells as a negative control and sodium sulfate (100 mM in water) as a positive control.

5. Seal with regular RT-PCR transparent cover to prevent air exchange prior to measurement.

6. Set the microplate reader temperature to 37 °C (Fluostar Optima plate reader, BMG Labtech).

7. Put the plate into incubated microplate reader, and measure oxygen consumption rate up to 6 h (10–15-min intervals) (excitation = 485 nM and emission = 630 nM) (*see* **Note 13**).

3.7 Culture of HSCs in Glycolytic Flux Medium (GC/MS Method)

1. Isolate at least 150×10^3 LT-HSCs ($n = 3$) from mouse bone marrow using FACS (*see* Subheading 3.1) and collect in PBS.

2. Centrifuge cells at $1,200 \times g$ for 10 min, and remove any medium (*see* **Note 14**).

3. Culture at least 50,000 LT-HSCs in 40 µL of glycolytic flux medium supplemented with 10 mM D-[1-6-^{13}C]-glucose (Cambridge Isotope Labs) to allow up to all of the glucose-derived lactate pool to be labeled on C-3 in 96-well plate (round bottom) overnight (12–16 h) at 37 °C tissue culture incubator (*see* **Note 15**).

4. Next day, pellet the cells and collect the supernatant for gas chromatography-mass spectrometry. Store samples at –80 °C until GC/MS analysis.

3.8 Preparation of Samples for GC/MS Analysis

1. Vortex medium samples.

2. Pipette 25 µL into a borosilicate tube.

3. Add 5 µL of internal standard (*see* **Note 16**).

4. Add 1 mL methanol; vortex.

5. Add 1 mL chloroform; vortex.

6. Add 1 mL mQ water; vortex (*see* **Note 17**).

7. Spin at $440 \times g$ for 5 min to separate phases. The aqueous phase will be on top and will be separated from the organic phase by a film of protein. Using a glass transfer pipette, transfer the upper (aqueous) phase into a screw-topped glass tube (*see* **Note 18**).

8. Dry down the aqueous phase at 42 °C (*see* **Note 19**).

9. Add a few drops of methylene chloride, and dry for a few more minutes to completely remove the residual moisture (*see* **Note 20**).

10. Add 100 μL Tri-Sil, and derivatize at 42 °C for 30 min (*see* **Note 21**).

11. Transfer to auto-injector vials, and analyze by GC/MS.

12. Determine the lactate abundance by monitoring m/z at 117 (unenriched), 118 (lactate containing ^{13}C from glucose), and 119 (internal standard).

4   Notes

1. We recommend to keep RBC lysis buffer at 37 °C before use.

2. Expected yield of bone marrow cell isolation from one mouse is around 5×10^7 cells.

3. Evaluate the cells by flow cytometry to ensure adequate lineage depletion. Expected yield is >95 % lineage depletion.

4. Count the Lin⁻ using hemocytometer or cell counter. Lin⁻ cells should be 10 % of original cells.

5. Keep LT-HSC on ice until use for metabolic assays.

6. Mitotracker dyes are dissolved in DMSO and stored as 1 mM, 5,000× stocks at −20 °C.

7. The optimal cell density for FACS is 1×10^7 cells/mL.

8. AMA is an electron transport chain complex III inhibitor. Prepare AMA stock solution at 1 M in DMSO.

9. It is necessary to collect LT-HSCs, which is at least enough for triplicate measurements. Increase the number of animals accordingly, and pool bone marrow cells.

10. Measurement settings are as follows: Inject at least 50 μL of luciferase substrate, take measurements after 1-s delay, and perform 10-s measurement of luciferase activity.

11. Do not discard samples after ATP luciferase assay, and perform Bradford protein assay. Use protein content to ensure loading of equal number of cells.

12. 96-Well plate BD Oxygen Biosensor System also works fine, but it requires the use of more cells ($>20 \times 10^4$ cells/well in 200 μL of RPMI medium supplemented with 20 % FBS).

13. We found that emission = 600 nM could also be used. There may be a drop in the fluorescence signal at the beginning due to an equilibration of oxygen content in the well. Keep in mind that when they breathe, it glows.

14. It is essential to remove any remaining source of FBS and glucose, and perform additional washes with PBS when it is required.

15. 10 mM of D-[1-^{13}C]-glucose (from Cambridge Isotope Labs) could also be used, but this will allow up to only half of the glucose-derived lactate pool to be labeled on C-3, which may decrease the signal intensity at least by half. If you happened to use D-[1-^{13}C]-glucose, increase cell density at least twice.

16. Internal standard is as follows: 1:500 dilution of 20 % w:v solution of Na[^{13}C$_3$]-lactate. This is equivalent to 17.9 nM of standard.

17. Use boiled mQ water.

18. Because of the internal standard, it is not necessary to collect the entire aqueous phase. Try to collect about 75 % of it, but do not take any of the interphase or the organic phase. Collected aqueous phase is usually around 1.5 mL.

19. It takes about 1 h.

20. This is an optional step, and most of the time it is not required when you dry samples for at least 1 h.

21. Use only glass pipettes, and avoid inclusion of any water or humidity.

Acknowledgements

This work was supported by grants from the American Heart Association (Grant-in-Aid), the Gilead Research Scholars Program in Cardiovascular Disease, Foundation for Heart Failure Research, NY, and the NIH (1R01HL115275-01).

References

1. Zhang CC, Sadek HA Sadek (2013) Hypoxia and metabolic properties of hematopoietic stem cells. Antioxid Redox Signal

2. Kocabas F (2012) Meis1: at the crossroads between metabolic and cell cycle regulation. UT Southwestern Electronic Theses and Dissertations 1–226

3. Kocabas F, Zheng J, Thet S, Copeland NG, Jenkins NA, DeBerardinis RJ et al (2012) Meis1 regulates the metabolic phenotype and oxidant defense of hematopoietic stem cells. Blood 120(25): 4963–4972. doi:10.1182/blood-2012-05-432260

4. Simsek T, Kocabas F, Zheng J, DeBerardinis RJ, Mahmoud AI, Olson EN et al (2010) The dis-

tinct metabolic profile of hematopoietic stem cells reflects their location in a hypoxic niche. Stem Cell 7(3):380–390. doi:10.1016/j.stem.2010.07.011

5. Shizuru JA, Negrin RS, Weissman IL (2005) Hematopoietic stem and progenitor cells: clinical and preclinical regeneration of the hematolymphoid system. Annu Rev Med 56:509–538

6. Zhang CC, Lodish HF (2005) Murine hematopoietic stem cells change their surface phenotype during ex vivo expansion. Blood 105(11):4314–4320

7. Chance B, Thorell B (1959) Localization and kinetics of reduced pyridine nucleotide in living cells by microfluorometry. J Biol Chem 234: 3044–3050

Chapter 11

Nanoproteomic Assays on Hematopoietic Stem Cells

Heath L. Bradley, Himalee Sabnis, Deborah Pritchett, and Kevin D. Bunting

Abstract

Dysregulation of cytokine signaling pathways is associated with benign and malignant hematologic disorders. Improvements in therapy rely on understanding the biology of the pathways and the proteins involved. Studying these pathways in patient samples is challenging as samples are difficult to obtain, contain fewer cells, and are heterogeneous in nature. To address some of these difficulties, we have utilized the technique of microcapillary electrophoresis. Using the NanoPro 1000 system (ProteinSimple) which is built on an automated, capillary-based immunoassay platform, we have developed rapid and quantitative assays for specific proteins from relatively small sample sizes. The NanoPro provides precise and quantitative data of the phosphorylation states of a specific protein of interest. We describe our experience with NanoPro assay development and optimization with specific application toward understanding aberrant cytokine signaling in human leukemia cells.

Key words Hematopoietic stem cells, Signal transduction, Proteomic, Nanoimmunoassay, Capillary electrophoresis

1 Introduction

Analysis of hematopoietic stem cells (HSCs) has been primarily accomplished by the use of flow cytometry, a powerful technique developed more than 30 years ago for analysis of single hematopoietic cells and which has become an essential technique for understanding the heterogeneity of HSC populations based on cell surface marker expression. Despite its steadily growing popularity for clinical research on blood lineages, the flow cytometry technique is limited by cell surface marker compatibility when combined with methods for intracellular staining of cytoplasmic or nuclear proteins. Some cell surface markers are difficult to keep on the cell surface during the detergent solubilization needed to get antibodies into the cell to detect intracellular proteins [1, 2]. A newer technique called mass cytometry [3] promises to be a major tool for analysis of HSC at the single-cell level and alleviates compensation issues, but instrumentation is not yet widely

Kevin D. Bunting and Cheng-Kui Qu (eds.), *Hematopoietic Stem Cell Protocols*, Methods in Molecular Biology, vol. 1185, DOI 10.1007/978-1-4939-1133-2_11, © Springer Science+Business Media New York 2014

available and the analysis of the large amounts of data obtained requires significant and often customized bioinformatic resources. Additional techniques such as reverse-phase protein array (RPPA) [4], a high-throughput antibody-based technique that provides a comprehensive proteomic analysis in a sorted population of cells, are feasible for biomarker discovery, but the turnaround time of RPPA is long and not compatible with the short-term analysis needed for monitoring biomarkers in patients and informing treatment decisions. Furthermore, assays of mRNA by multiplexed real-time quantitative PCR have general utility but cannot always predict changes at the protein level, and classic proteomic assays such as traditional mass spectroscopy and immunohistochemistry lack the sensitivity and are not practical for analysis of rare hematopoietic cell populations. Understanding changes in signaling molecules at the protein level is critical for monitoring the effects of drugs on their targets; e.g., phosphorylation or other posttranslational modifications that change the charge of the protein; for identifying and monitoring new biomarkers in patient samples; and for determining relative protein expression levels or stoichiometry among interacting proteins. In order to be useful, these analyses need to be rapid and provide multiple levels of information.

We have developed assays on the NanoPro 1000 system (ProteinSimple), which is based on a well-established technique of isoelectric focusing coupled with capillary electrophoresis. The NanoPro enables a rapid and quantitative analysis of specific proteins in relatively small numbers of cells due to the inherent sensitivity of microcapillary electrophoresis [5, 6]. The NanoPro provides precise and quantitative data for the phosphorylation states of specific proteins of interest. This system is built on an automated platform whereby proteins are focused to their isoelectric point (pI) to resolve the various modification states of proteins, immobilized, and then probed with specific antibodies. Signal is generated by a horseradish peroxidase (HRP)-conjugated chemiluminescence system and captured by a scientific-grade camera within the NanoPro platform. The signal intensity and data are then automatically expressed in an electropherogram. There are several features of the NanoPro 1000 that make it a unique tool to look at signaling within rare cell populations like stem cells. It is capable of detecting multi-phosphorylation states of the same protein while using the same antibody, which is impossible to accomplish by Western blot or intracellular flow cytometry. The conceptual advantage of this approach is that the instrument is capable of analyzing dozens of different proteins within a single sample or it can analyze a single protein in up to 96 samples per run.

There are important steps to take in design and validation of new assays. This automated system permits analysis of numerous proteins within very few cells using the same antibodies used in traditional Western blots. The absolute amount of protein needed for these assays is very low (nanogram range), but there are still some bottlenecks in regard to getting this amount of protein per capillary due to the fluidics of the system. We present several features of the sample preparation that we have optimized in our work and describe how to run and interpret data generated on the NanoPro. In this chapter we describe our experience with this new technology and discuss the strengths and weaknesses.

The NanoPro is ideally suited for the serial analysis of clinical specimens from patients receiving therapy. Several reports have already documented the power of the approach in a range of solid and liquid malignancies [7–10]. We have initially used the NanoPro 1000 to standardize the assays for important proteins involved in the survival pathways in acute myeloid leukemia (AML), but the assays can be applied to any hematopoietic cell population that can be enriched by either CD34+ isolation or multiparameter flow cytometric sorting. We focus this chapter on the process of NanoPro analysis using limiting numbers of hematopoietic cells. We also point out where optimization of assays can be done to increase the number of cells per capillary, to boost the signal strength, and to refine the peak profiles.

2 Materials

1. The cell source of human bone marrow, umbilical cord blood, or mobilized peripheral blood will depend on the investigator and the specific questions being asked. We describe our experience with analysis of human bone marrow from AML patients.

2. Phosphate-buffered saline (PBS without calcium and magnesium).

3. 50 ml Conical tubes.

4. 15 ml Conical tubes.

5. Bradford Dye protein reagent.

6. 1.7 ml Microcentrifuge tubes.

7. BD Stain buffer (BSA) (BD Biosciences 554657).

8. 0.5 % BSA in PBS.

9. NanoPro assay kit (ProteinSimple p/n CBS2001)—all reagents from ProteinSimple.

   (a) Capillaries-Charge Separation for Peggy or NanoPro 1000 (5-Pack) (p/n CBS701).

   (b) NanoPro 1000 Plate Kit, 5-pack of 384-well plates with lids (p/n 040-901).

   (c) Anolyte (200 ml) (p/n 040-337).

   (d) Catholyte (200 ml) (p/n 040-338).

    (e) Wash concentrate (200 ml) (p/n 041-108).

    (f) Luminol (2 ml) (p/n 040-652).

    (g) Peroxide (2 ml) (p/n 040-653).
       (All the above items can also be ordered a la carte.)

10. Bicine/Chaps lysis buffer (ProteinSimple p/n 040-764).

11. 50× DMSO inhibitors (phosphatase inhibitor cocktail) (ProteinSimple p/n 040-510).

12. 25× Aqueous inhibitors (protease inhibitor cocktail) (ProteinSimple p/n 040-482).

13. Goat-anti-rabbit secondary antibody, HRP conjugate.

14. Goat-anti-mouse secondary antibody, HRP conjugate.

15. Amplified Rabbit Secondary Antibody Detection kit (ProteinSimple p/n 041-126).

16. Amplified Mouse Secondary Antibody Detection kit (ProteinSimple p/n 041-127).

17. Premix G2, pH 4–8 (nested) separation gradient, contains pH 2–4 plug (ProteinSimple p/n 040-972).

18. pI Standard ladder 3: 4.9, 6.0, 6.4, 7.0, 7.3 (ProteinSimple p/n 040-646).

19. pI Standard, 5.5 (ProteinSimple p/n 040-028).

20. Antibody Diluent (ProteinSimple p/n 040-309) (is supplied with all secondary antibodies in sufficient quantity that we have not had to order individually).

---

3 Methods

The key to obtaining high-quality data is always in the sample preparation. We first describe the methods for obtaining protein from cells and for transferring the protein to the capillaries. In between there are important steps and considerations regarding controls.

3.1 Sample Preparation

3.1.1 Sample Prep for Normal Hematopoietic Stem Cells and Leukemic Blasts

1. Dilute bone marrow aspirate sample 1:3–1:4 with PBS (*see* **Note 1**). Overlay 15 ml of the diluted sample on 10 ml of Ficoll-Paque™ in a sterile 50 ml conical tube (be careful not to disrupt the layers).

2. Centrifuge tubes at $800 \times g$ at room temperature for 15 min in a swinging bucket rotor centrifuge without brake (or at least deceleration speed set to a minimum).

3. Carefully remove the tubes from the centrifuge, and aspirate the top layer.

4. Carefully collect the cells at the interphase between the two layers which will contain either HSCs or leukemic blasts with a

pipettor, and transfer this layer to a new 15 ml centrifuge tube. Be careful not to take more than 5 ml of material per tube to be washed.

5. Wash the cells isolated from the interphase with 10 ml of ice-cold PBS (*see* **Note 1**) at least twice. To remove Ficoll from the cells centrifuge at $500 \times g$ at 4 °C in a swinging bucket rotor centrifuge (*see* **Note 2**).

6. Proceed with antibody staining for isolation of HSC or leukemic blasts.

3.1.2 Flow Cytometry Staining and Sorting of HSC and Leukemic Blast Populations

1. Resuspend the cells from the isolation above in BD Stain buffer (100 µl/10^8 cells), and mix well.

2. Add antibodies for HSC staining or as appropriate to isolate leukemic blasts depending on the specific experiment and the patient to be analyzed. Incubate the tubes on ice for 30 min.

3. Wash cells with 5 ml 0.5 % BSA in PBS and centrifuge for 10 min at $500 \times g$ at 4 °C in a swinging bucket rotor centrifuge.

4. Resuspend cells in appropriate volume of 0.5 % BSA in PBS for flow sorting.

5. Sort appropriate population to be analyzed, and proceed with cell processing as below.

3.1.3 Sample Prep for Cell Lines and Sorted HSC and Leukemic Populations

1. Pellet cells by centrifugation ($300 \times g$ for 5 min). The amount of cells required depends on the source.

2. Aspirate supernatant, wash with 1 ml ice-cold PBS, and transfer cells to pre-chilled 1.7 ml microcentrifuge tubes.

3. Spin cells at $500 \times g$ for 5 min to pellet.

4. Lyse cell pellets in Bicine/Chaps Buffer (ProteinSimple p/n 040-764) with added 1× DMSO inhibitors and 1× aqueous inhibitors. According to protein content, ProteinSimple provides a recommended lysis volume based on cell number but does not recommend lysis in less than 10 µl, as per its NanoPro Cell Lysis Kit product insert (*see* **Note 3**).

5. Lyse on ice for 30–60 min with occasional vortexing (every 20–30 min).

6. Clarify lysate by centrifuging at $14,000 \times g$ for 10 min at 4 °C.

7. Quantitate protein using Bradford Dye protein reagent.

8. Aliquot lysate, snap-freeze, and store at –80 °C.

3.2 Choosing Isoelectric Point Standards

1. Determine the theoretical isoelectric point (pI) of the protein of interest by using a pI calculator such as http://scansite.mit. edu/calc_mw_pi.html, http://web.expasy.org/compute_pi/, or http://www.phosphosite.org/homeAction.do;jsessionid=0 7CF04E16BB8B3457E12521E1FEA4D3F.

2. According to the theoretical pI select an ampholyte premix from ProteinSimple and its corresponding pI standard ladder. For most of our applications we use a 5–8 premix with a 2–4 pH plug with ladder 3 and individual pI standard 5.5 (*see* Subheading 2).

3. It may be necessary to supplement pI ladders with individual pI standards to increase reproducibility of the determined pI, as calculation by the algorithm of the software, for a given peak. For example, pI ladder 3 includes standards of 4.9, 6.0, 6.4, 7.0, and 7.3. If the protein of interest falls near the midway between the 4.9 and 6.0 standards, it is advisable to add the 5.5 individual pI standard in order for the software to more robustly calculate the pI.

3.3 Ampholyte Premix and Antibody Selection

1. To make the ampholyte premix + pI ladder, add the pI ladder of your choice (supplied by ProteinSimple at a 60× concentration in relation to the concentration loaded in the assay capillary) at a 1:45-fold dilution. For example to obtain 150 μl of ampholyte premix 5–8 with a 2–4 plug + pI standard ladder 3, add 3.3 μl of the ladder to 146.7 μl of premix. In order to spike individual pI standards into your existing ladder we typically dilute 1:5 in the ladder before taking the ladder/individual standard mix and spiking it into the ampholyte premix at 1:60.

2. Using Bicine/Chaps buffer with inhibitors, dilute the sample to four times the final concentration to be loaded in the capillary. For example, if using 0.2 mg/ml total protein per capillary, the users dilute their sample to 0.8 mg/ml in the Bicine/Chaps buffer with inhibitors. Then take the diluted sample, and add 1 part sample with 3 parts ampholyte premix + pI ladder. The individual standards within the pI ladder are labeled with a fluorescent dye (*see* **Note 4**).

3. Antibodies detecting native epitopes are optimal choices for the NanoPro 1000 platform (*see* **Note 5**). A 1:100 antibody dilution in antibody diluent is often a good starting concentration, but optimal working dilutions can range from 1:25 to 1:1,000 depending on the antibody affinity and abundance of protein expression.

4. Secondary antibodies are typically HRP-labeled anti-species antibodies and can be purchased directly from ProteinSimple for rabbit or mouse primary antibodies. Detection occurs with the addition of a luminol–peroxide substrate (*see* **Note 6**).

5. The user then loads samples, diluted antibodies, and detection reagent in a 384-well plate. Samples and reagents are only loaded to 12 consecutive wells across rows of the plate. The NanoPro 1000 uses sets of 12 capillaries per assay cycle, and up to 8 cycles per run can be programmed, for a total of 96 possible separate conditions.

6. Load 10–12 µl/well of sample (diluted in premix + pI ladder). Load 20 µl/well for antibodies and other reagents. Loading the luminol detection reagent far from the secondary antibody to limit potential background carryover can also be a good practice. A user will typically load this reagent in row P on the plate to allow distance from the secondary reagent.

3.4 Running Samples

1. After creating a template in Compass software (*see* below in Subheading 3.5) and saving the template, click the start button and the program prompts the user to load the reagents and capillaries (make certain that an adequate number of capillary rows is available for the assay). The NanoPro 1000 can accommodate two boxes of capillaries, allowing the user to efficiently utilize partially filled boxes.

2. Typical assay settings load sample (0.4 µl) into the capillaries, and then transport the capillaries to a separation chamber where one capillary end is submerged in anolyte and the other capillary end submerged in catholyte before applying constant voltage for a set time (typically 2,100 MicroWatts for 40 min) to complete isoelectric focusing. Following focusing, an immobilization step uses an 80–100-s UV light exposure to cross-link the sample proteins to the walls of the capillaries (*see* **Note 7**).

3. Then, the NanoPro 1000 proceeds with an immunoassay, automating steps typical of a Western blot. Following a wash, capillaries are loaded with primary antibody and moved to an incubation tray. Sets of 12 capillaries, composing 1 cycle of an assay, follow this pattern in a staggered fashion until a maximum of 8 cycles (96 capillaries) have undergone focusing and incubated with antibody.

4. Incubation with primary antibody can be modified but typically is 120 min. After primary antibody incubation, the NanoPro 1000 washes the capillaries and then loads the secondary antibody (typical incubation for this step is 60 min). The NanoPro 1000 then proceeds with either a tertiary antibody incubation (*see* **Note 6**) or detection using the luminol–peroxidase substrate. An internal CCD camera captures image exposures at various time intervals specified by the user when setting up the initial template. We typically gather data with 30-, 60-, 120-, 240-, 480-, and 960-s exposures.

3.5 Data Analysis

1. The data is saved to the internal computer in the NanoPro 1000 as well as to the external computer used to interface with the NanoPro 1000 for backup purposes.

2. Data is analyzed in Compass software (ProteinSimple).

3. Prior to the assay the user specifies a plate map in the Compass assay tab to identify the location for samples and various reagents (Fig. 1).

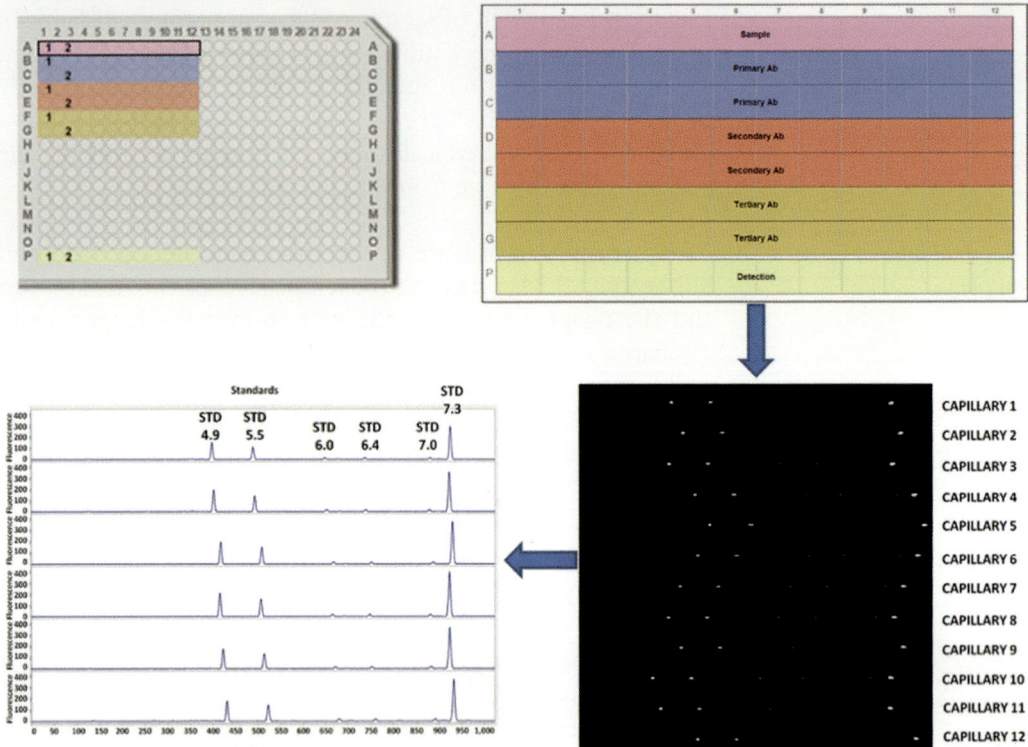

Fig. 1 NanoPro1000 assay flow chart (clockwise representation). *Top* shows a 384-well plate set up in the Compass® software for a 2-cycle run (24 total data points) and the placement of sample (*pink*), primary antibody (*blue*), secondary anti-species antibody (*red*), (optional) tertiary antibody (*dark yellow*), and luminol detection reagent (*light yellow*). *Bottom* shows the fluorescent image captured by the NanoPro 1000 camera for the separated pI standards of all 12 capillaries (samples are detected through chemiluminescence on a different wavelength). Following acquisition, the computer represents the data as electropherograms (*bottom left*) showing the pI location of the standards. The standards are used to determine the pI of the signal obtained by the specific antibody complex via relative position

4. Data analysis proceeds by first clicking on the edit tab, selecting Analysis, and specifying the standards settings to reflect the pI ladder used in the assay. The user can also name specific detected peaks based on pI determination and can select the exposure to use for analysis and displaying the Images section.

5. Next, the user would return to the Analysis section and cycle through the samples to visually check pI standard assignment. Typically the first and last standards in the ladder have a higher intensity and are used by Compass for pI determination as well as for image registration (set points that align images of the capillary prior to and after movement during the assay). If the standards are not correctly assigned by the software, the user can manually adjust them by right clicking on any individual peak.

6. The user then returns to the sample view to analyze the electropherograms and to determine if the peaks are at the expected pI. The user can also overlay or stack individual capillaries for comparison (for example if there is a drug-treated sample vs. a non-treated sample), analyze changes in the peak profiles, and also obtain the areas under the peaks for quantitative analysis.

3.6 Validation of Peaks

One of the most important things to consider when analyzing the electropherograms is whether the detected peaks represent the phospho-forms of the proteins to be studied. For example a good way to validate a peak is to use an antibody that recognizes all phosphorylated forms of a protein, such as the Total 4EBP-1 antibody from Cell Signaling Technology (cs9644) that detects a multiple peak profile on an electropherogram. Cell Signaling Technology also has a phospho-specific antibody to the serine 65 isoform of 4EBP-1 (cs9451) that provides a single peak profile on the electropherogram. If the user then loads these antibodies in separate capillaries with the same sample lysate and aligns the output electropherograms, the peak representing the serine 65 phosphorylated form can then be identified.

Other ways to validate peaks include using a drug to inhibit protein expression, using a growth factor or cytokine to stimulate expression, or using a phosphatase to reduce or eliminate phosphorylation. In our hands we have used various treatments of cells in culture with small molecules such as Dasatinib to knock down p-STAT5, the mTORc1/mTORc2 inhibitor AZD8055 to inhibit 4EBP-1 phosphorylation (Fig. 2), and the MAP kinase inhibitor U0126 to inhibit ERK phosphorylation. We have also used lambda phosphatase on cell lysates to demonstrate knockdown of p-STAT5 signals.

4 Notes

1. Keeping the phosphoproteins of a clinical sample stable can be problematic due to lag time from collection to processing in the lab. Both phosphatases and kinases have been shown to be active in patient tissue samples ex vivo. For Ficoll separation purposes it is best not to keep the sample on ice due to temperature-sensitive gradient effects, but it is recommended to process the fresh sample as soon as possible. After performing the Ficoll step at room temperature, it is recommended that the user perform washes in ice-cold PBS supplemented with protease, phosphatase, and kinase inhibitors [11]. Patient samples have much lower protein amount per cell than do cell lines. For example we have detected as much as 100-fold differences in protein content per cell.

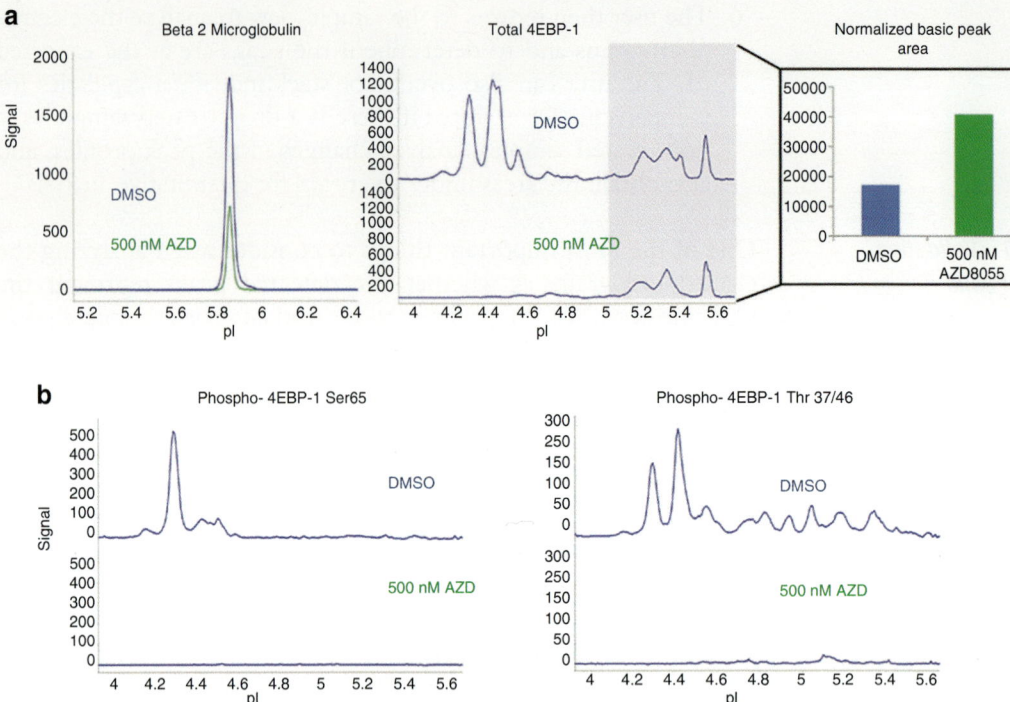

Fig. 2 Validation of 4EBP-1 using a primary sample. Primary bone marrow samples treated for 24 h with either DMSO or the mTORC1/2 inhibitor AZD 8055 and analyzed using the NanoPro 1000 instrument show a specific response in ablation of the "acidic" or the phosphorylated peaks. (**a**) Using beta 2 microglobulin antibody as a loading control it was possible to quantitate the difference in amount of sample loaded in order to normalize the "basic" non-phosphorylated peaks and represent the difference in a bar graph format using area under the curve as our readout. (**b**) Electropherograms of two of the phospho forms of 4EBP-1 shown with DMSO (*top*) or AZD 8055 (*bottom*) treatment

2. Optional—Between the washes the user may want to count the cells via trypan blue exclusion or some other method at this point.

3. Our experience has found that the lysis volume is a potential bottleneck for primary bone marrow samples due to the low yield of protein per cell compared with tissue culture cells. For example, we typically lyse 5×10^6 cells from the MV4-11 human AML cell line in 150 μl of lysis buffer and obtain approximately between 2 and 6 mg/ml of protein. However, we have previously lysed 1.7×10^7 primary AML cells from PBMC in 50 μl of lysis buffer and obtained 5 mg/ml of protein. When lysing samples for low cell number analysis remember that you need to start with a good number of more cells than the final cells/capillary. Things that should be considered are lysis volume (it is recommended that you use no less than 10 μl of Bicine/Chaps buffer with inhibitors). After lysis you

will dilute your sample further, by fourfold with your viscous ampholyte premix (A.P.), and finally the NanoPro 1000 loads 400 nl of volume per capillary. So for example, if you lyse 5×10^4 cells in 10 µl of lysis buffer, then dilute to 40 µl final volume (by adding 30 µl A.P.), and the NanoPro 1000 loads 0.4 µl, you would be loading 500 cells/capillary.

4. Be sure to make up enough for 2–4 extra samples as the G2 premix supplied by ProteinSimple is very viscous and difficult to pipet accurately.

5. Selection of antibodies for the NanoPro can be challenging. Since the preferred lysis method selects for native protein, it is best to choose antibodies developed for immunoprecipation or generated against a native epitope. ProteinSimple has many application briefs and an antibody database on its website (http://www.proteinsimple.com) for many applications. Check these resources first. Many antibodies, especially those developed for traditional Western blotting and denatured linearized epitopes, are simply not compatible with the NanoPro 1000. This can be a rate-limiting step to developing new assays.

6. ProteinSimple has developed a biotin–streptavidin method of amplifying signal from potentially weak detection. This utilizes a biotinylated anti-species antibody and a tertiary streptavidin-horse radish peroxidase (SA-HRP) conjugate (p/n 041-126 and 041-127). This amplification step can increase background peaks, and in particular, we have often seen peaks at 5.9, 6.4, 6.5, and 6.7, possibly due to endogenous biotin expression. One should consider these background peaks if using amplification, as it may interfere with the data. Typical incubation time with the SA–HRP conjugate is recommended for 10 min; however in our hands an incubation time of 120 min significantly increases signal strength for certain antibodies (Fig. 3).

7. Note that the longer the time, the potential for degradation of epitopes. However, too short of UV immobilization and the protein will not be efficiently captured on the capillary walls. Therefore, UV immobilization time may need optimization in specific assays.

Acknowledgements

The authors thank the Aflac Cancer and Blood Disorders Center Leukemia/Lymphoma Research Program, Children's Healthcare of Atlanta, the Emory + Children's Pediatric Flow Cytometry Core Facility, and the Cure Childhood Cancer Foundation for supporting the NanoPro assay development.

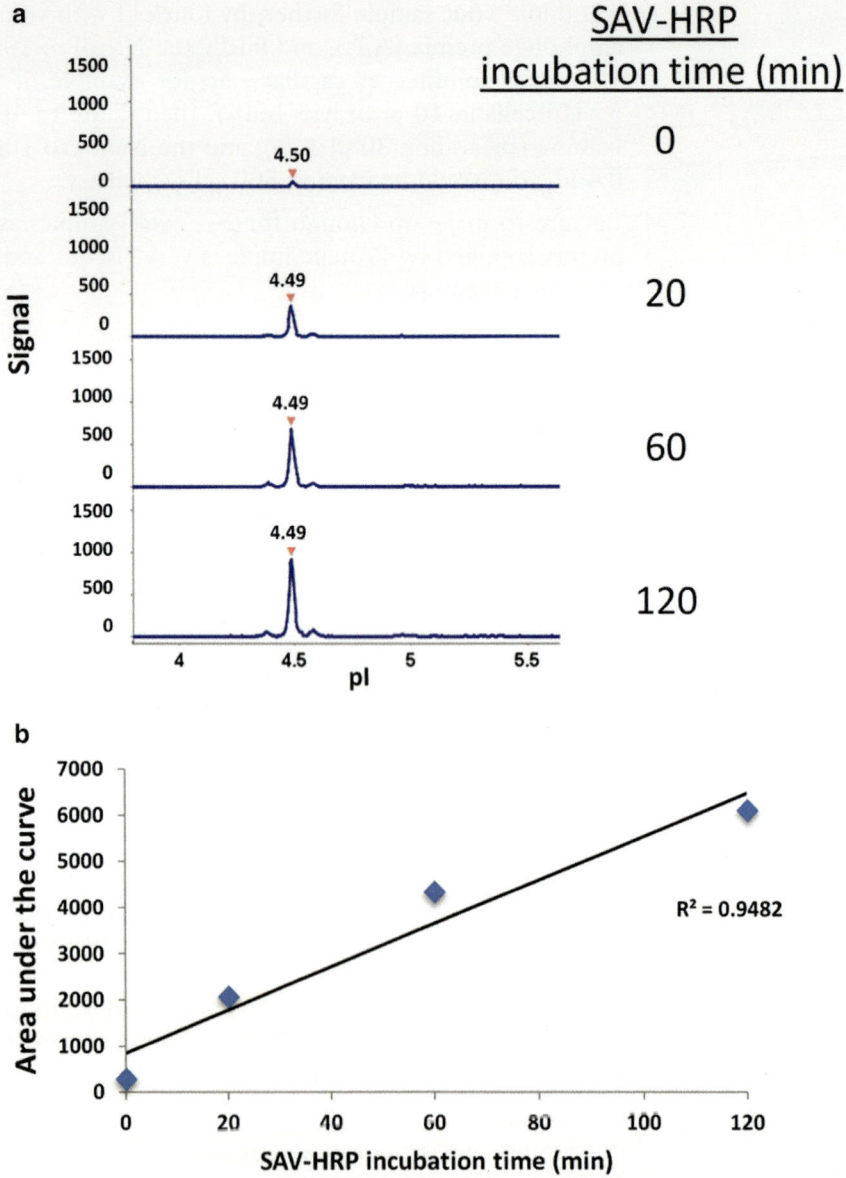

Fig. 3 Amplification of p4EBP-1 Ser 65. (**a**) Electropherograms of MV4-11 AML cell lysates detected with p4EBP-1 antibody (Cell Signaling) showing a stepwise increase in exposure time to the tertiary SAV–HRP antibody. (**b**) Area under the curve for the individual peaks plotted versus time shows a linear increase of signal over time

References

1. Kalaitzidis D, Neel BG (2008) Flow-cytometric phosphoprotein analysis reveals agonist and temporal differences in responses of murine hematopoietic stem/progenitor cells. PLoS One 3:e3776

2. Han L, Wierenga AT, Rozenveld-Geugien M, van de LK, Vellenga E, Schuringa JJ (2009) Single-cell STAT5 signal transduction profiling in normal and leukemic stem and progenitor

cell populations reveals highly distinct cytokine responses. PLoS One 4:e7989

3. Bendall SC, Simonds EF, Qiu P, Amir E, Krutzik PO, Finck R, Bruggner RV, Melamed R, Trejo A, Ornatsky OI, Balderas RS, Plevritis SK, Sachs K, Pe'er D, Tanner SD, Nolan GP (2011) Single-cell mass cytometry of differential immune and drug responses across a human hematopoietic continuum. Science 332:687–696

4. Wong SC, Chan CM, Ma BB, Lam MY, Choi GC, Au TC, Chan AS, Chan AT (2009) Advanced proteomic technologies for cancer biomarker discovery. Expert Rev Proteomics 6:123–134

5. Knittle JE, Roach D, Horn PB, Voss KO (2007) Laser-induced fluorescence detector for capillary-based isoelectric immunoblot assay. Anal Chem 79:9478–9483

6. O'Neill RA, Bhamidipati A, Bi X, Deb-Basu D, Cahill L, Ferrante J, Gentalen E, Glazer M, Gossett J, Hacker K, Kirby C, Knittle J, Loder R, Mastroieni C, Maclaren M, Mills T, Nguyen U, Parker N, Rice A, Roach D, Suich D, Voehringer D, Voss K, Yang J, Yang T, Vander Horn PB (2006) Isoelectric focusing technology quantifies protein signaling in 25 cells. Proc Natl Acad Sci U S A 103:16153–16158

7. Fan AC, Deb-Basu D, Orban MW, Gotlib JR, Natkunam Y, O'Neill R, Padua RA, Xu L, Taketa D, Shirer AE, Beer S, Yee AX, Voehringer DW, Felsher DW (2009) Nanofluidic proteomic assay for serial analysis of oncoprotein activation in clinical specimens. Nat Med 15:566–571

8. Chen JQ, Lee JH, Herrmann MA, Park KS, Heldman MR, Goldsmith PK, Wang Y, Giaccone G (2013) Capillary isoelectric-focusing immunoassays to study dynamic oncoprotein phosphorylation and drug response to targeted therapies in non-small cell lung cancer. Mol Cancer Ther 12:2601–2613

9. Iacovides DC, Johnson AB, Wang N, Boddapati S, Korkola J, Gray JW (2013) Identification and quantification of AKT isoforms and phosphoforms in breast cancer using a novel nanofluidic immunoassay. Mol Cell Proteomics 12:3210–3220

10. Kentsis A, Reed C, Rice KL, Sanda T, Rodig SJ, Tholouli E, Christie A, Valk PJ, Delwel R, Ngo V, Kutok JL, Dahlberg SE, Moreau LA, Byers RJ, Christensen JG, Vande WG, Licht JD, Kung AL, Staudt LM, Look AT (2012) Autocrine activation of the MET receptor tyrosine kinase in acute myeloid leukemia. Nat Med 18:1118–1122

11. Espina V, Edmiston KH, Heiby M, Pierobon M, Sciro M, Merritt B, Banks S, Deng J, VanMeter AJ, Geho DH, Pastore L, Sennesh J, Petricoin EF III, Liotta LA (2008) A portrait of tissue phosphoprotein stability in the clinical tissue procurement process. Mol Cell Proteomics 7:1998–2018

Part IV

In Vitro Assays and Differentiation

Chapter 12

Hematopoietic Differentiation of Pluripotent Stem Cells in Culture

Jason A. Mills*, Prasuna Paluru*, Mitchell J. Weiss, Paul Gadue, and Deborah L. French

Abstract

This chapter describes a two-dimensional "monolayer" system for differentiating human pluripotent stem cells (PSCs) into "primitive" hematopoietic progenitor cells (HPCs) resembling those produced in vivo by the early embryonic yolk sac. This experimental system utilizes defined conditions without serum or feeder cells. Cytokines are added sequentially to stimulate the formation of mesoderm and its subsequent patterning to hematopoietic progenitors. The HPCs produced by this protocol have multi-lineage potential (erythroid, megakaryocyte, and myeloid) and can be isolated as a homogeneous population for use in standard hematopoietic studies including liquid expansion to mature lineages and colony assays. In addition, the HPCs can be cryopreserved for distribution or analysis at later times. The HPCs generated by this protocol have been used successfully to better define intrinsic variation in hematopoietic potential between different PSC lines and to model human hematopoietic diseases using patient-derived induced pluripotent stem cells.

Key words Hematopoietic progenitor cells, Embryonic stem cells, Induced pluripotent stem cells, Primitive hematopoiesis

1 Introduction

The advent of pluripotent stem cell (PSC) technologies has created excitement for biomedical research and regenerative medicine. The derivation of human embryonic stem cell (ESC) and patient-specific induced pluripotent stem cell (iPSC) lines creates powerful systems for studying tissue development and disease [1]. The capacity of PSCs to form three germ layers (endoderm, ectoderm, mesoderm) and their derivative differentiated tissues provides remarkable new opportunities for modeling normal and pathological development in vitro and, potentially, to generate transplantable tissues for regenerative therapies. In vitro, PSC differentiation

---

*Author contributed equally with all other contributors.

Kevin D. Bunting and Cheng-Kui Qu (eds.), *Hematopoietic Stem Cell Protocols*, Methods in Molecular Biology, vol. 1185, DOI 10.1007/978-1-4939-1133-2_12, © Springer Science+Business Media New York 2014

resembles the ontogeny of embryogenesis and the most optimal in vitro differentiation methodologies utilize defined growth conditions that seek to reproduce the microenvironments in which tissues of interest are formed in vivo [2]. With these concepts in mind, normal and patient-derived PSCs provide outstanding models for understanding hematopoietic development and associated diseases [3].

Hematopoietic development in PSC differentiation cultures resembles events that occur during embryogenesis, at least to some extent. Several spatiotemporally distinct waves of hematopoiesis occur in mammalian embryos, as defined mainly in mice but also documented to occur in humans [4–7]. The first wave, termed "primitive," occurs in the yolk sac and produces erythrocytes, myeloid cells, and megakaryocytes. A second wave of yolk sac hematopoiesis produces "definitive" progenitors that colonize the fetal liver transiently. Independently, definitive hematopoietic progenitors arise from hemogenic-endothelium within several vascular beds of the developing embryo, including the dorsal aorta. These progenitors colonize the fetal liver and give rise to hematopoietic stem cells capable of repopulating the hematopoietic system after transplantation. Primitive and definitive hematopoietic progenitors are distinguished by the types of globin genes expressed in their erythroid progeny [8], characteristic patterns of gene expression, distinct transcription factor requirements, and T lymphocyte potential uniquely in the definitive lineage [9]. The first wave of hematopoiesis to arise in PSC differentiation cultures most closely resembles the primitive yolk sac-derived lineage. Differentiation of PSCs for more prolonged times and/or under certain defined conditions gives rise to hematopoietic progenitors capable of producing T cells and erythrocytes expressing definitive-type globins [10, 11]. However, it has not yet been possible to generate bone fide hematopoietic stem cells from PSCs consistently or reproducibly. The extent to which PSCs faithfully recapitulate the developmental stages of human hematopoiesis is a topic of active investigation. In this regard, in vitro protocols to generate hematopoietic stem cells from PSCs consistently and reliably are a "holy grail" yet to be achieved [12].

This chapter outlines a detailed protocol to differentiate human PSCs into hematopoietic progenitor cells (HPCs) most closely resembling the first wave of hematopoiesis in the embryo, i.e., primitive. Previous methods for in vitro hematopoietic differentiation of PSCs include co-culture with stromal cells [13–15] and embryoid body (EB) formation in serum or serum-free media [16], usually including specific morphogens that drive mesoderm and its subsequent patterning into blood progenitors [17, 18]. The current method drives hematopoietic differentiation on a PSC-derived stromal monolayer and offers several advantages: (1) Utilization of defined media and the absence of heterologous stromal cells allow for controlled differentiation conditions, consistency, and reproducibility. (2) A highly enriched population of hematopoietic progenitors (~90 % enriched) are released synchronously from the stromal layer, minimizing the need for subsequent purification steps. It is possible to generate primitive HPCs using a different protocol (adapted from Kennedy et al. [18]) in which PSCs are stimulated to form EBs in serum-free defined culture medium. The monolayer approach is advantageous due to the reasons listed above, but the EB-derived approach has other advantages including cellular maturation in a three-dimensional system that may more closely mimic the in vivo environment and the ability to culture EBs for longer periods of time, which may be important for maturation or switching to definitive-like lineages.

We have used the monolayer approach described here to study variation in hematopoietic potential between different normal iPSC lines [19], and we have used both EB and monolayer approaches to define disease phenotypes in iPSCs derived from patients with genetic disorders of hematopoiesis including Down syndrome [20], juvenile myelomonocytic leukemia (JMML) [21], and Diamond–Blackfan Anemia (DBA) [22].

2 Materials

2.1 Maintenance and Differentiation Base Media

1. hESC (HES) maintenance medium: Dulbecco's modified Eagle's medium/F12 (DMEM/F12) (Life Technologies 11330-057), supplemented with 20 % knockout serum replacement (KSR) (Life Technologies 10828-028), 100 μM nonessential amino acids (Life Technologies 11140050) solution, 50 U/ml penicillin and 50 g/ml streptomycin (Cellgro 30-002-CL), 2 mM glutamine (Cellgro 25-005-CL), 1 mM sodium pyruvate (Life Technologies 11360070), 0.075 % sodium bicarbonate (Life Technologies 25080094), 0.1 mM β-mercaptoethanol (Life Technologies 21985). Filter sterilize.

2. Differentiation base media and supplements: RPMI (Life Technologies 22400-089), StemPro-34 serum-free medium (SP34, Life Technologies 10639011), and serum-free differentiation (SFD) medium [23]: Iscove's Modified Dulbecco's Medium (IMDM, Life Technologies 12200069) (made according to the manufacturer's instructions), containing 25 % Ham's F12 (Cellgro 10-080-CV) supplemented with 0.5 % N2 (Life Technologies 17502-048), 1 % B27 without Vitamin A (Life Technologies 12587-010), and 0.05 % BSA diluted in PBS (Sigma A1470). The RPMI, SP34, and SFD base media are all supplemented with 2 mM glutamine (Cellgro 25-005-CL), 50 μg/ml ascorbic acid (AA, Sigma A4544), and 4×10^{-4} M monothioglycerol (MTG) (Sigma) (*see* **Note 1**).

2.2 Small Molecules, Morphogens, and Growth Factors

1. Small molecules: ROCK inhibitor (ROCKi, Y-27967) Tocris 1254), Chir99021 (CHIR) (Tocris 4423).

2. Growth factors and morphogens: Basic fibroblast growth factor (bFGF) (R&D systems 233-FB/CF), bone morphogenic protein-4 (BMP4) (R&D systems 314-BP/CF), vascular endothelial growth factor (VEGF) (R&D systems 293-VE/CF), wingless-type MMTV integration site-3a (Wnt3a) (R&D systems 1324-WN/CF), stem cell factor (SCF) (R&D systems PHC2113), interleukin-6 (IL-6) (R&D systems 206-IL/CF), fms-like tyrosine kinase 3 ligand (Flt3L) (Life Technologies PHC9413), erythropoietin (EPO) (Epogen NDC 55513-144-01), thrombopoietin (TPO) (R&D systems 288-TP/CF), interleukin-3 (IL-3) (R&D systems 203-IL/CF), and granulocyte/macrophage colony-stimulating factor (GM-CSF) (R&D systems 215-GM).

2.3 Cell Culture Reagents, Solutions, and Supplies

1. Wash buffer: IMDM containing 5 % KSR.

2. Dulbecco's Phosphate-buffered Saline (DPBS) (Corning 21-031-CM).

3. Bovine serum albumin (BSA) (Sigma A1470).

4. Growth Factor Reduced Matrigel (BD Biosciences 354230) diluted 1:3 in IMDM. Thaw the stock bottle (10 ml) overnight at 4 °C, mix with 20 ml cold IMDM, and keep on ice while aliquoting 3 ml into 5 ml snap-cap tissue culture tubes chilled in the –80 °C freezer overnight. Store aliquoted tubes at –20 °C.

5. MethoCult (Stem Cell Technologies, H4435): Thaw overnight at 4 °C, mix thoroughly, and aliquot 3.5 ml into 14 ml snap-cap tissue culture tubes. Store at –20 °C.

6. MegaCult (Stem Cell Technologies, 04962): Follow the manufacturer's instructions for storage of components.

7. Double Chamber Slide Kit (Stem Cell Technologies, 04963).

8. Gelatin (Sigma G-1890) diluted to 0.1 % in DPBS, autoclaved, aliquoted into 125 ml tissue culture bottles, and stored at 4 °C.

9. TrypLE dissociation reagent (Life Technologies 12605-010).

10. 25 cm, 2-Position blade cell scraper (Sarstedt 83.1830).

11. Fetal bovine serum (FBS) (tissue culture biological 101).

12. Dimethyl sulfoxide (DMSO) (Sigma D2650).

13. Low cluster 6-well tissue culture plates (Costar 2471).

14. Cell strainer (70 μm) (Corning Falcon 352350).

15. Cryovials (Sarstedt 2016-06).

16. Antibodies: SSEA3 AF488 (BioLegend 330306) at 1:400, SSEA4 AF647 (BioLegend 330408) at 1:400, Tra-1-60 AF488 (BD Biosciences 560173) at 1:20, Tra-1-81 AF647 (BioLegend

330706) at 1:50, CD41 PE (BD Biosciences 555467) at 1:40, CD42 FITC (BD Biosciences 558818) at 1:20, CD235a APC (BD Biosciences 551336) at 1:10,000, CD45 Pacific Blue (BioLegend 304022) at 1:100, CD18 APC (BD Biosciences 551060) at 1:20, hKDR PE (R&D Systems FAB357P) at 1:20, CD31 PE-Cy7 (BioLegend 303118) at 1:100.

3   Methods

An overview of the protocol illustrating sequential stages of differentiation, cytokine/morphogens added, and media used is shown in Fig. 1. Photomicrographs depicting appearance of the cultures at successive stages of differentiation are shown in Fig. 2. It is critical to monitor the kinetics of cell differentiation by using flow cytometry to interrogate cell surface marker expression. Figure 3 shows the expected cell surface profiles versus time during the differentiation process, but these kinetics of differentiation can vary by a day or two in different PSC lines.

3.1   PSC Expansion and Characterization Prior to Differentiation

It is essential to begin the differentiation process with healthy, subconfluent PSC cultures. Important factors in this regard include colony size, minimal spontaneous differentiation, use of high-quality irradiated MEFs, and a daily feeding regimen.

1. Plate $0.75–1.0 \times 10^6$ MEFs in a 6-well tissue culture plate precoated with 0.1 % gelatin and incubate at 37 °C, 5 % CO_2 and atmospheric O_2, for a maximum of 2–3 days.

2. When PSCs are ~85–90 % confluent (Fig. 2a), passage by adding 1 ml/well TrypLE and incubate at room temperature for 3–4 min until MEFs detach and PSC colonies appear loose. Aspirate TrypLE, rinse well with 2 ml wash medium, add 1 ml/well of HES media containing 10 µM ROCKi (inhibits dissociation-induced apoptosis of PSCs [24]), and scrape the colonies into small clumps. Add 2 ml/well of HES media,

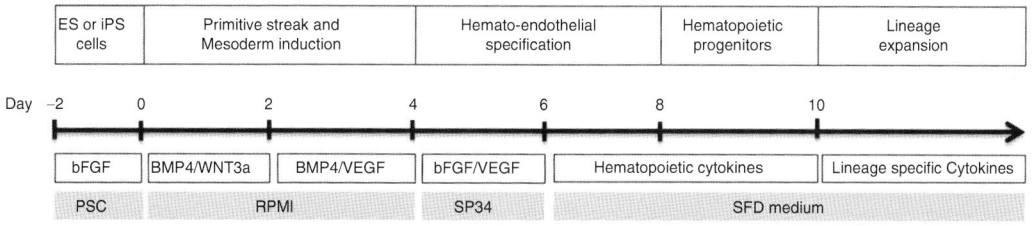

Fig. 1 A schematic of the adherent monolayer protocol showing sequential media and cytokine changes and emergence of differentiated cells

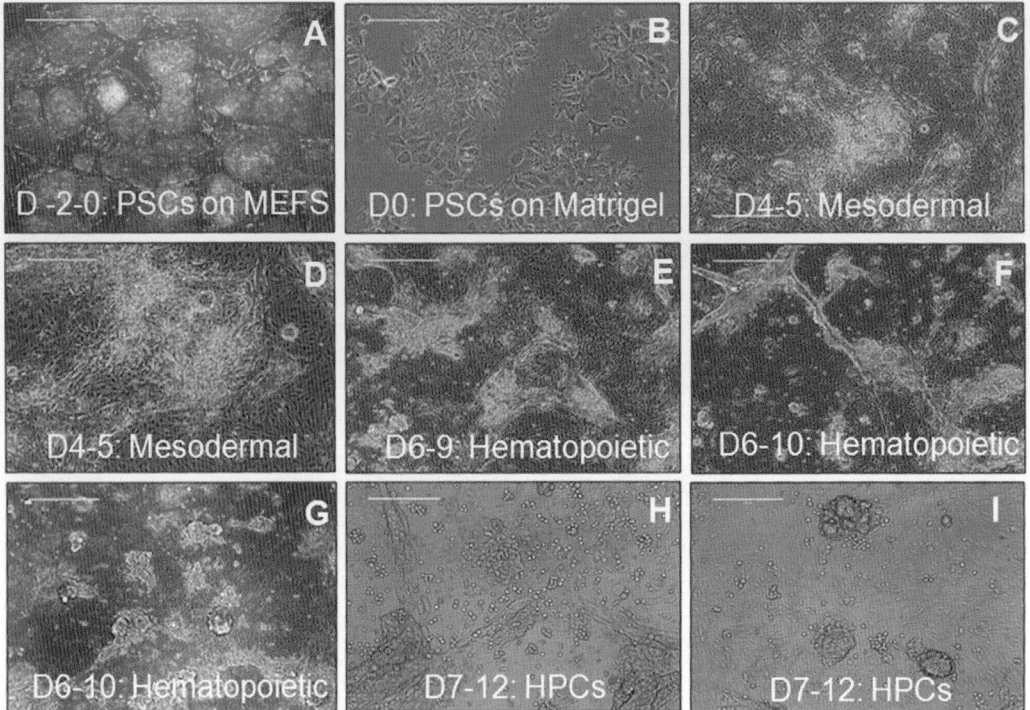

Fig. 2 Phase micrographs at various time points during differentiation. (**a**) Undifferentiated PSCs co-cultured with MEFs, ~80 % confluent; (**b**) density of PSCs on day 0 at ~50 % confluence; (**c, d**) cultures at differentiation days 4–5 showing stromal-like cells with small clusters of mesodermal derived prehematopoietic cells; (**e–g**) cells on days 6–10 showing increased density of stromal-like cells interspersed with larger clusters of hematopoietic cells; (**h, i**) HPCs shedding from the monolayer and cellular clusters

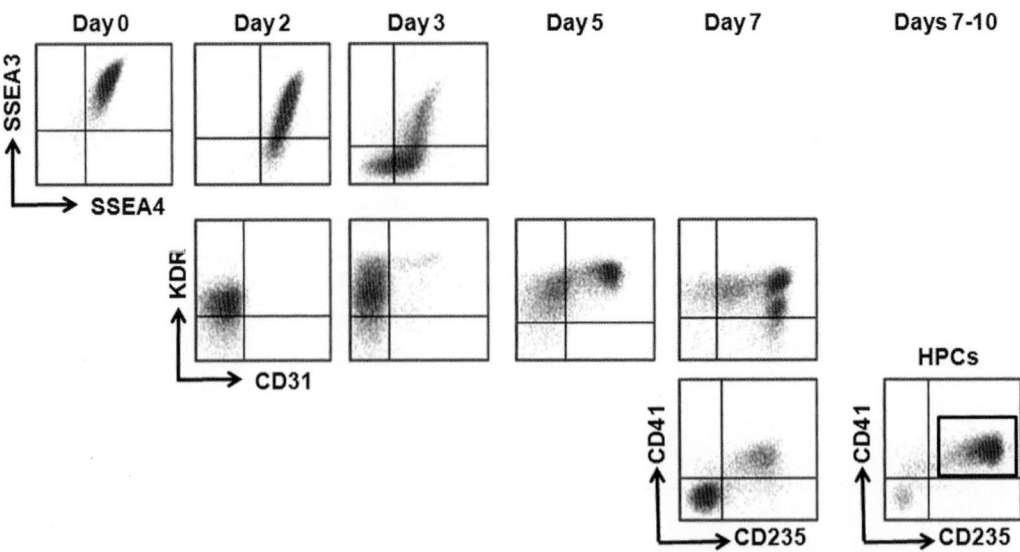

Fig. 3 Sequential appearance of cell surface markers used to track differentiation. The first row shows co-expression of the pluripotency markers, SSEA3 and SSEA4, at day 0 and the loss of these markers following 3 days of differentiation. The second row shows the emergence of patterned mesoderm by VEGFR2 or KDR expression followed by the emergence of hemato-endothelial cells that express PECAM or CD31. The third row shows emergence of single-cell HPCs that co-express the pan-hematopoietic marker CD41 and glycophorin A/CD235

pipet 3–5 times using a 5 ml pipet, add the cells to 10 ml of HES medium containing 5 ng/ml bFGF and 10 μM ROCKi, and plate 2 ml/well of cells in a 6-well tissue culture dish containing MEFs. Feed cells daily with fresh medium excluding ROCKi.

3. Sacrifice one well of cells to determine the pluripotent state of the cells before plating on Matrigel for MEF depletion. Greater than 90 % of the PSCs should express both pluripotency surface markers SSEA3 and SSEA4 (Fig. 3, day 0, and *see* **Note 2**).

3.2 MEF Depletion on Matrigel-Coated Tissue Culture Plates

This step maintains healthy PSCs while depleting the cultures of MEFs, which can inhibit subsequent differentiation steps. Matrigel is a mixture of extracellular-matrix-like proteins, secreted by Engelbreth–Holm–Swarm mouse sarcoma cells, used as a cell adhesion substrate. Matrigel must be kept cold as it is a solution at 4 °C and solidifies at 37 °C.

1. Thaw a tube of frozen Matrigel (1:3) overnight at 4 °C or 2–3 h on ice. Pre-chill 6-well tissue culture plates and pipets at −20 °C and pipet tips at 4 °C.

2. Chill tissue culture plates on ice, and fully coat each well by serial additions of Matrigel using a p1000 pipet. Aspirate residual Matrigel from each well, and incubate plates for ~15–60 min at 37 °C (*see* **Note 3**).

3. Harvest PSCs from MEF-containing plates, and count the PSCs using a hemocytometer for subsequent replating onto 6-well plates. The optimal density of starting cells varies for each PSC line depending on their growth rates and must be determined empirically (*see* **Note 4**). This number can range from 0.5 to 4×10^5 PSCs/well in a 6-well dish. We typically plate human H9 ES cells at 2–3×10^5/well. Add the designated number of PSCs to the Matrigel-coated tissue culture plates in a final volume of 2 ml HES medium containing 5 ng/ml bFGF and 10 μM ROCKi. Incubate for 24–48 h at 37 °C, 5 % CO_2, 5 % O_2, and 90 % N_2, at which time the cells should reach ~70 % confluency (Fig. 2b). Feed cells after 24 h and replace with fresh medium excluding ROCKi.

4. Sacrifice one well of cells before starting the differentiation for determining the number of cells to start the differentiation and to ensure that >90 % of the cells co-express the surface markers SSEA3 and SSEA4 (*see* **Note 2**).

3.3 Hematopoietic Differentiation

Table 1 outlines the differentiation process including sequential media and cytokine requirements.

1. Begin the differentiation when cells on Matrigel-coated plates are ~70 % confluent. Rinse two times with wash medium followed by addition of day-0 differentiation medium (Table 1)

Table 1
Timetable of differentiation showing medium and cytokine mixtures

Days	Medium	BMP4[a]	VEGF	Wnt3a	bFGF	SCF	Flt3L	TPO	IL-6
0–1	RPMI	5	50	25					
2	RPMI	5	50		20				
3	SP34	5	50		20				
4–5	SP34		15		5				
6	SFD		50		50	50	5		
7–10	SFD		50		50	50	5	50	10

[a]Cytokine concentrations in ng/ml

at 2 ml/well. Incubate cells for 24 h at 37 °C, 5 % CO_2, 5 % O_2, and 90 % N_2.

2. Change the cell media daily, according to Table 1 (*see* **Note 5**). During days 2–3, cell death occurs, followed by the formation of a stromal-like cell layer (Fig. 2c, d). Dead cells are removed during daily media changes. Between days 3 and 5, rounded mounds of cells begin to appear (Fig. 2e–g). Between days 6 and 7, bright round loosely adherent cells appear, which can be collected by carefully removing the supernatant (Fig. 2h, i). These cells, representing the multipotent HPC population, co-express CD41, a pan-hematopoietic marker, during embryonic hematopoiesis [25] and glycophorin A/CD235, which is typically found on mature erythrocytes (Fig. 3). Of note, CD235[+]/CD41[+] cells generated during similar PSC differentiation protocols also express CD43 and were reported to represent primitive megakaryocyte–erythroid progenitors [26, 27]. We find that these cells also have myeloid potential (Paluru et al. The Negative Impact of Wnt Signaling on Megakaryocyte and Primitive Erythroid Progenitors Derived From Human Embryonic Stem Cells, in press). These cells are continuously produced from the monolayer culture over days ~7–10 and can be analyzed in liquid expansion assays (*see* Subheading 3.3, **step 4**) or cryopreserved (*see* **Note 6**).

3. Perform flow cytometry to assess the differentiation state of cells during the entire protocol. Harvest adherent cells using TrypLE. At day 5, mesoderm-specified cells express the surface markers KDR and CD31 (Fig. 3, day 5). Typically, 20–50 % of the adherent cells express these markers depending on the PSC line. Lower percentages of this population indicate inefficient differentiation, predicting poor yield of HPCs. Between days 6 and 7, the HPCs begin to form and express the markers KDR[lo] and CD31 when analyzing all cells in the well (Fig. 3, day 7).

Table 2
Timetable of lineage expansion showing medium and cytokine mixtures

Lineage	Days	Medium	SCF[a]	EPO	TPO	IL-3	GM-CSF
Erythroid	7–14	SCF	50	2 U			
Megakaryocyte	5–7	SCF	50		50	10	
Myeloid	10–14	SCF	50			10	50

[a]Cytokine concentrations in ng/ml except EPO which is in units (U)

4. Collect the non-adherent HPCs by lightly pipetting the mono-layer cells using a 5 ml serological pipet (*see* **Note** 7). Non-adherent cells can be removed each day with the adherent monolayer being replenished with fresh medium. Centrifuge non-adherent cells at 335 × *g* for 3 min and resuspend in fresh medium for subsequent hematopoietic studies (*see* Subheading 3.4).

3.4 Liquid Expansion of HPCs to Erythroid, Megakaryocyte, and Myeloid Lineages

The HPCs that appear in the cultures between days 7 and 10 can be expanded in specific cytokine mixtures to generate erythroid, mega-karyocyte, and myeloid cells. The medium, cytokine mixtures, and timing to generate specific lineages are shown in Table 2.

1. *Primitive erythroblasts:* Plate HPCs at densities of ~1–10 × 10^5/ well in low-cluster 6-well tissue culture dishes in erythroid-specific culture medium and incubate at 37 °C, 5 % CO_2 and atmospheric O_2 (*see* Table 2). Collect cells every 2 days and resuspend in fresh medium. Within a week post-expansion, a homogenous population of CD235+ erythroid cells will be present (Fig. 4). Typically, approximately nine- to tenfold expansion of primitive erythroid cells occurs.

2. *Megakaryocytes:* Plate HPCs at densities of ~1–10 × 10^5/well in low-cluster 6-well tissue culture dishes in megakaryocyte-specific culture medium and incubate at 37 °C, 5 % CO_2 and atmospheric O_2 (Table 2). Collect cells every 2 days and resuspend in fresh medium. Within 4–5 days post-expansion, a population of CD41 + CD42+ cells will be present (Fig. 4). Depending on the cell line, this population can range from 20 to 80 % of the total cells with three- to fivefold expansion. Typically, cells can be maintained in culture for 7–10 days, but growth will decrease by ~5 days.

3. *Myeloid cells:* Collect HPCs at day 10 for myeloid expansion. Harvest HPCs on days 7–9 of differentiation, resuspend in day-7 differentiation medium, and put back into the wells of the adherent monolayer cultures (*see* Subheading 3.3, **step 4**). Harvest HPCs on day 10 and plate at densities of 1 × 10^5/well in low-cluster TC dishes in myeloid-specific culture media and

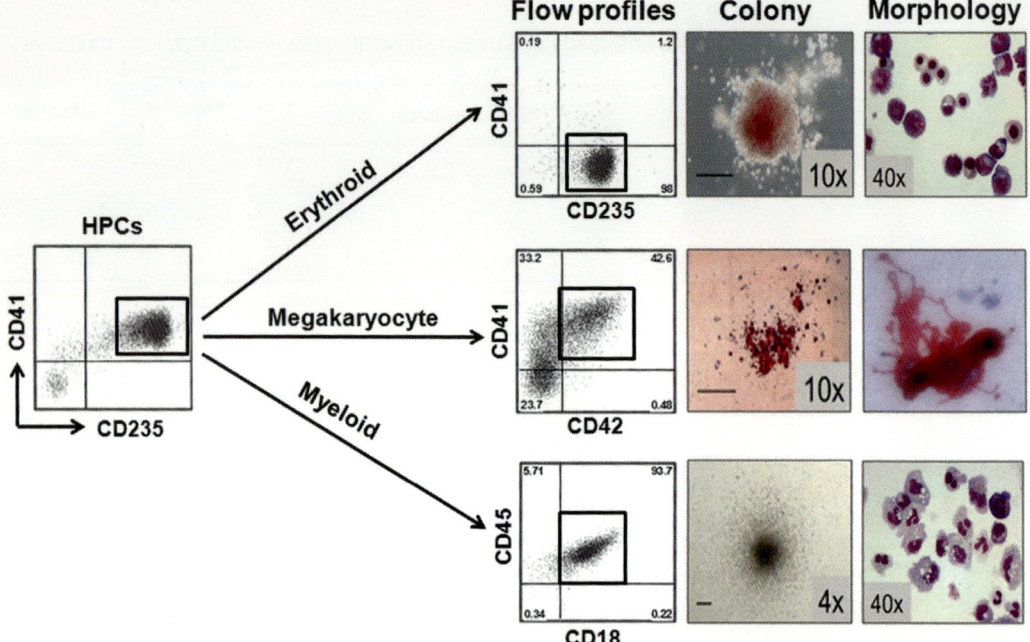

Fig. 4 Lineage-specific differentiation of HPCs. HPC-derived lineage expansion in liquid culture and progenitor potential in colony assays. The first column shows flow cytometry profiles of erythroid, megakaryocyte, and myeloid lineages generated in liquid expansion cultures. The second column shows erythroid and myeloid colonies generated in methylcellulose cultures and megakaryocyte colonies generated in a collagen-based assay. The third column shows morphologies of erythroid and myeloid cells stained with May-Grünwald Giemsa and megakaryocytes stained with anti-CD41 antibody. Note the proplatelet extensions emanating from the megakaryocytes

incubate at 37 °C, 5 % CO_2 and atmospheric O_2 (Table 2). Myeloid cells will expand four- to fivefold, giving rise to a population of cells expressing the markers CD45 and CD18 (Fig. 4). These cells are expanded to specific myeloid lineages, such as macrophages and granulocytes, as described by Choi et al. [28].

3.5 Colony Assays

The progenitor potential of HPCs is determined by quantitative colony assays in which the cells are plated in semisolid methylcellulose (erythroid, myeloid, mixed colonies) or collagen-based (megakaryocyte colonies) mixtures containing the appropriate cytokines (*see* **Note 8**). Figure 4 shows pictures of erythroid, megakaryocyte, and myeloid colonies. Cytospins of erythroid and myeloid colonies stained with May-Grünwald Giemsa and megakaryocyte colonies stained with an anti-CD41 antibody are shown.

1. *Methylcellulose:* Thaw one tube of MethoCult overnight at 4 °C or at room temperature for ~4 h. Add 1.5×10^4 (CD41+/ CD235+) HPCs/300uL IMDM per tube, vortex to mix, and put tube in hood for ~10 min to let the bubbles dissolve.

Dispense 1.0 ml/35 mm tissue culture in three dishes each using a 5 ml syringe and 16-gauge needle, rotate the dish to spread the methylcellulose evenly in the dish, and incubate in a humidified atmosphere at 37 °C, 5 % CO_2 and atmospheric oxygen. Incubate for 10–14 days, and count CFU-E (erythrocytes), CFU-GM (granulocytes/macrophage), and CFU-GEMM (granulocyte, erythrocyte, macrophage, megakaryocyte) (*see* the manufacturer's instructions for descriptions of colonies and images).

2. *MegaCult:* Follow the manufacturer's instructions for preparation of MegaCult including collagen and cytokine (TPO 50 ng/ml, IL-6 10 ng/ml, and IL-3 10 ng/ml) mixtures keeping all components on ice. Add 1.5×10^4 (CD41+/CD235+)/400 µl IMDM HPCs per tube, vortex to mix, and dispense into four chambers on two slides. Incubate in a humidified atmosphere at 37 °C, 5 % CO_2 and atmospheric oxygen, for ~10 days. Process and stain the slides according to the manufacturer's instructions.

4   Notes

1. The SFD and SP34 base media are stored for 2 weeks at 4 °C and kept in the dark by wrapping in aluminum foil. All cytokines, small molecules, ascorbic acid, and MTG are added fresh to the medium on the day of use.

2. To ensure that the majority of cells are undifferentiated, one well of cells is sacrificed to analyze for co-expression of the pluripotency markers SSEA3 and SSEA4 (Tra-1-60 and Tra-1-81 can also be monitored). A single-cell suspension is obtained by adding 1 ml/well TrypLE for ~5–7 min at room temperature and harvesting with a p1000 pipet.

3. Matrigel-coated tissue culture plates/dishes can be stored for 1 week at 4 °C prior to use. However, we find that coating the dish just prior to its use appears to improve the efficiency of differentiation.

4. The optimal PSC density at the start of differentiation varies with each ESC and iPSC line. For untested iPSC lines, we recommend starting with several different cell densities ranging from 0.5 to 3×10^5 cells/well. Stocks of PSCs should be expanded, aliquoted, and cryopreserved for future differentiations. Use after prolonged passage should be avoided. For continuously growing lines, we typically discard cultures and thaw fresh aliquots every 1–2 months.

5. Wnt3a is essential for hematopoietic differentiation, but it is an expensive cytokine. To reduce the cost, we found that the

GSK-3β inhibitor, CHIR99021, can be used instead at a concentration of 1 or 2 μM (determine empirically for each lot).

6. The single-cell HPCs can be cryopreserved in 90 % FBS + 10 % DMSO at a density of 10^6/vial. The viability after thawing ranges from 70 to 80 % of the original cell number. The ability to cryopreserve these cells permits storage of large numbers of aliquoted progenitors for multiple PSC lines, which can then be thawed and analyzed in parallel. This increases the convenience of comparative studies.

7. When harvesting the non-adherent HPCs, small clusters of nonhematopoietic cells can peel off of the monolayer. These cell clusters can be removed by passing the supernatant through a 70 μm (or 100 μm) cell strainer.

8. Depending on the iPSC line analyzed, the number of HPCs plated per colony assay may vary. Plating too many cells per dish or chamber can interfere with colony growth. HPCs from disease-specific iPSC lines may generate large colonies that will overgrow in the plate and be difficult to count accurately. In these situations, the numbers of HPCs plated will need to be determined empirically. We have encountered this issue while analyzing JMML iPSCs, which expand more rapidly than normal [21]. Unusually rapid expansion of a "normal" PSC line may also indicate acquired genetic abnormalities [19].

5 Conclusions

In this chapter, we have provided a useful protocol for the analysis of hematopoiesis from human PSCs. This "monolayer" system generates "primitive" HPCs that are easily isolated as single cells with multi-lineage potential. The HPCs are released from the monolayer as a homogeneous population that can be used in standard hematopoietic assays and be cryopreserved for distribution or analysis at later time points.

While human PSCs provide a useful model system for the study of developmental hematopoiesis and disease, there are several caveats to note for such experiments, particularly when examining the effects of disease-associated mutations through comparison of wild-type and mutant PSC lines:

First, different ESC and iPSC lines can vary greatly in their hematopoietic potential [29]. This is likely due to intrinsic genetic differences in the donor cells and/or mutations acquired before, during, or after iPSC generation [19]. Thus, it is important to characterize newly derived PSC lines and high-passage number stocks by karyotyping and SNP analysis to assess for acquired copy number variation (*see* also **Note 8**). It is also informative to examine multiple clones of PSCs from different patients before deriving general conclusions about a specific disease-associated mutation. Ideally, it is best to compare "isogenic" lines in which disease-related mutations of interest are created or corrected in the same line via transgenic or genome editing approaches [22].

Second, hematopoietic development during embryogenesis occurs in distinct waves producing different lineages with unique biological properties [6, 9]. Distinct hematopoietic lineages also arise in PSC cultures, presumably paralleling events that occur in the embryo. The extent to which PSC-derived hematopoiesis recapitulates in vivo events is not fully understood, and it has not yet been possible to generate definitive adult-type hematopoietic stem/progenitor cells from PSCs. The protocol described here provides an efficient, reproducible method to generate PSC-derived primitive-type HPCs that most closely resemble embryonic yolk sac progenitors. Using this approach, we showed that in vitro differentiation of patient-derived iPSCs recapitulate the pathophysiology of several human hematopoietic diseases [20–22]. However, disease-associated mutations are likely to manifest differently in primitive versus definitive hematopoietic lineages derived from PSCs. For example, primitive and definitive hematopoietic progenitors derived from trisomy 21 PSCs exhibit both overlapping and distinct abnormalities [20, 30]. In our experience, it has been difficult to generate definitive hematopoietic progenitors from PSCs quantitatively or consistently. Development of such protocols will enhance the utility of PSCs for disease modeling.

References

1. Yoshida Y, Yamanaka S (2010) Recent stem cell advances: induced pluripotent stem cells for disease modeling and stem cell-based regeneration. Circulation 122:80–87

2. Murry CE, Keller G (2008) Differentiation of embryonic stem cells to clinically relevant populations: lessons from embryonic development. Cell 132:661–680

3. Orkin SH, Zon LI (2008) Hematopoiesis: an evolving paradigm for stem cell biology. Cell 132:631–644

4. Migliaccio G, Migliaccio AR et al (1986) Human embryonic hemopoiesis. Kinetics of progenitors and precursors underlying the yolk sac—liver transition. J Clin Invest 78:51–60

5. Tavian M, Peault B (2005) Embryonic development of the human hematopoietic system. Int J Dev Biol 49:243–250

6. McGrath K, Palis J (2008) Ontogeny of erythropoiesis in the mammalian embryo. Curr Top Dev Biol 82:1–22

7. McGrath KE, Frame JM et al (2011) A transient definitive erythroid lineage with unique regulation of the beta-globin locus in the mammalian embryo. Blood 117:4600–4608

8. Schechter AN (2008) Hemoglobin research and the origins of molecular medicine. Blood 112:3927–3938

9. Dzierzak E, Speck NA (2008) Of lineage and legacy: the development of mammalian hematopoietic stem cells. Nat Immunol 9: 129–136

10. Choi KD, Vodyanik MA et al (2012) Identification of the hemogenic endothelial progenitor and its direct precursor in human pluripotent stem cell differentiation cultures. Cell Rep 2:553–567

11. Kennedy M, Awong G et al (2012) T lymphocyte potential marks the emergence of definitive hematopoietic progenitors in human pluripotent stem cell differentiation cultures. Cell Rep 2:1722–1735

12. Slukvin II (2013) Hematopoietic specification from human pluripotent stem cells: current advances and challenges toward de novo generation of hematopoietic stem cells. Blood 122:4035–4046

13. Kaufman DS, Hanson ET et al (2001) Hematopoietic colony-forming cells derived from human embryonic stem cells. Proc Natl Acad Sci U S A 98:10716–10721

14. Vodyanik MA, Bork JA et al (2005) Human embryonic stem cell-derived CD34+ cells: efficient production in the coculture with OP9 stromal cells and analysis of lymphohematopoietic potential. Blood 105:617–626

15. Takayama N, Nishikii H et al (2008) Generation of functional platelets from human embryonic stem cells in vitro via ES-sacs, VEGF-promoted structures that concentrate hematopoietic progenitors. Blood 111:5298–5306

16. Zambidis ET, Peault B et al (2005) Hematopoietic differentiation of human embryonic stem cells progresses through sequential hematoendothelial, primitive, and definitive stages resembling human yolk sac development. Blood 106:860–870

17. Ng ES, Davis RP et al (2005) Forced aggregation of defined numbers of human embryonic stem cells into embryoid bodies fosters robust, reproducible hematopoietic differentiation. Blood 106:1601–1603

18. Kennedy M, D'Souza SL et al (2007) Development of the hemangioblast defines the onset of hematopoiesis in human ES cell differentiation cultures. Blood 109:2679–2687

19. Mills JA, Wang K et al (2013) Clonal genetic and hematopoietic heterogeneity among human-induced pluripotent stem cell lines. Blood 122:2047–2051

20. Chou ST, Byrska-Bishop M et al (2012) Trisomy 21-associated defects in human primitive hematopoiesis revealed through induced pluripotent stem cells. Proc Natl Acad Sci U S A 109:17573–17578

21. Gandre-Babbe S, Paluru P et al (2013) Patient-derived induced pluripotent stem cells recapitulate hematopoietic abnormalities of juvenile myelomonocytic leukemia. Blood 121: 4925–4929

22. Garcon L, Ge J et al (2013) Ribosomal and hematopoietic defects in induced pluripotent stem cells derived from Diamond Blackfan anemia patients. Blood 122:912–921

23. Gadue P, Huber TL et al (2006) Wnt and TGF-beta signaling are required for the induction of an in vitro model of primitive streak formation using embryonic stem cells. Proc Natl Acad Sci U S A 103:16806–16811

24. Watanabe K, Ueno M et al (2007) A ROCK inhibitor permits survival of dissociated human embryonic stem cells. Nat Biotechnol 25:681–686

25. Mikkola HK, Fujiwara Y et al (2003) Expression of CD41 marks the initiation of definitive hematopoiesis in the mouse embryo. Blood 101:508–516

26. Klimchenko O, Mori M et al (2009) A common bipotent progenitor generates the erythroid and megakaryocyte lineages in embryonic stem cell-derived primitive hematopoiesis. Blood 114:1506–1517

27. Vodyanik MA, Thomson JA et al (2006) Leukosialin (CD43) defines hematopoietic progenitors in human embryonic stem cell differentiation cultures. Blood 108:2095–2105

28. Choi KD, Vodyanik MA et al (2009) Generation of mature human myelomonocytic cells through expansion and differentiation of pluripotent stem cell-derived lin-CD34+CD43+CD45+ progenitors. J Clin Invest 119:2818–2829

29. Choi KD, Yu J et al (2009) Hematopoietic and endothelial differentiation of human induced pluripotent stem cells. Stem Cells 27:559–567

30. Maclean GA, Menne TF et al (2012) Altered hematopoiesis in trisomy 21 as revealed through in vitro differentiation of isogenic human pluripotent cells. Proc Natl Acad Sci U S A 109:17567–17572

Ex Vivo Assays to Study Self-Renewal, Long-Term Expansion, and Leukemic Transformation of Genetically Modified Human Hematopoietic and Patient-Derived Leukemic Stem Cells

Pallavi Sontakke, Marco Carretta, Marta Capala, Hein Schepers, and Jan Jacob Schuringa

Abstract

With the emergence of the concept of the leukemic stem cell (LSC), assays to study them remain pivotal in understanding (leukemic) stem cell biology. Although the in vivo NOD-SCID or NSG xenotransplantation model is currently still the favored assay of choice in most cases, this system has some limitations as well such as its cost-effectiveness, duration, and lack of engraftability of cells from some acute myeloid leukemia (AML) patients. Here, we describe in vitro assays in which long-term expansion and self-renewal of LSCs isolated from AML patients can be evaluated. We have optimized lentiviral transduction procedures in order to stably express genes of interest or stably downmodulate genes using RNAi in primary AML cells, and these approaches are described in detail here. Also, we describe bone marrow stromal coculture systems in which cobblestone area-forming cell activity, self-renewal, long-term expansion, and in vitro myeloid or lymphoid transformation can be evaluated in human CD34+ cells of fetal or adult origin that are engineered to express oncogenes. Together, these tools should allow a further molecular elucidation of derailed signal transduction in LSCs.

Key words Leukemic stem cell (LSC), Acute myeloid leukemia (AML), Leukemic stem cell self-renewal, Ex vivo assay, Bone marrow stromal coculture, Lentiviral transduction of primary patient cells

1 Introduction

With the development of NOD-SCID and later the NOD-SCID-β2-microglobulin$^{-/-}$ and NSG (NOD-SCID-IL2Rγ$^{-/-}$) leukemia xenotransplantation models, it has become possible to firmly establish the concept of the leukemic stem cell (LSC) [1–4]. As in the normal hematopoietic system, it has been recognized that in leukemias the developing malignant clones comprise heterogeneous groups of cells that differ in their differentiation status (reviewed in [5]). Only a rare subset of the most immature cells termed

Kevin D. Bunting and Cheng-Kui Qu (eds.), *Hematopoietic Stem Cell Protocols*, Methods in Molecular Biology, vol. 1185, DOI 10.1007/978-1-4939-1133-2_13, © Springer Science+Business Media New York 2014

SCID-leukemia-initiating cells (SL-ICs), or LSCs as we refer to those cells here, are present in 0.2–100 per 10^6 mononuclear cells. It is exclusively these LSCs that are capable of initiating and sustaining leukemic growth in vivo in NOD-SCID or NSG mice [1, 2]. These SL-ICs have high self-renewal capacity as demonstrated in serial transplantation experiments [2, 6].

Although the NOD-SCID/NSG model has been used so far as the favoured model of choice in most cases, this system has some limitations such as its cost-effectiveness, duration, and lack of engraftability of cells from a large proportion of acute myeloid leukemia (AML) patients [7]. Moreover, these mice lack a human bone marrow environment which further hampers faithful recapitulation of human hematopoiesis or leukemogenesis [8]. Thus, in order to further understand the differences in molecular mechanisms that are involved in leukemic transformation of hematopoietic stem cells, accessible assays are required in which gene-function analyses in the LSC can be performed and in which phenotypes such as long-term expansion, self-renewal, and apoptosis can be monitored. It is clear that self-renewal of normal stem cells heavily depends on extrinsic cues they obtain from the bone marrow microenvironment. Yet, very little is known about how LSCs depend on the bone marrow microenvironment for self-renewal (reviewed in [9]), further stressing the need for assays in which interactions between stromal cells of the niche with HSCs/LSCs can be studied.

Based on assays that had been developed to study and enumerate normal human HSCs in vitro [10–14], it has been attempted to grow AML cells on bone marrow stromal layers and indeed a subset was capable of initiating long-term growth (leukemic long-term culture-initiating cells (L-LTC-IC) [15–17]. In these assays AML cells were plated onto stromal cell layers for a period of 5 weeks, after which methylcellulose was added to the wells for an additional 2 weeks to determine the L-LTC-IC frequencies. The L-LTC-IC represents a rare subpopulation within the AML clone(s), ranging from 1.6 to 37 in 10^5 mononuclear cells [17, 18], but this frequency is higher than the reported SL-IC frequencies suggesting that slightly different cell populations are being monitored in these assays. We have also utilized bone marrow stromal cocultures supplemented with human cytokines in which long-term leukemic expansion for 7 to >24 weeks could be established in about two-thirds of the studied cases ($n > 80$) [19–29]. Leukemic cobblestone area-forming cells (L-CAFCs) readily formed in these assays and self-renewal could be addressed by serial replating of cultures onto new stroma, whereby new L-CAFCs are generated. The in vitro assay also allows for the evaluation of LSC markers that identify populations with long-term self-renewal potential [19, 25, 29]. Functionally, we were able to introduce or downmodulate genes in the AML LTC-ICs by lentiviral (RNAi) approaches [19, 22–24, 27–29], and detailed information on our lentiviral transduction protocols is described in the second part of the chapter.

In addition to using patient samples to gain further insight into LSCs, model systems have been generated in which oncogenes such as MLL-AF9 [30–32], BCR-ABL [22, 33], FLT3-ITDs [34], KRAS [35], STAT5 [36], and NUP98-HOXA9 [37] were introduced into healthy human HSCs. For many of these studies fetal cord blood (CB) CD34+ cells have been used; however we have observed that distinct phenotypes can be obtained when oncogenes are introduced into fetal versus adult human CD34+ cells [32]. In the third part of this chapter we discuss in vitro stromal cocultures in which myeloid and lymphoid transformation of transduced human CD34+ can be evaluated.

2 Materials

2.1 Isolation of Human Stem/Progenitor Cells from CB, BM, PB, or AML Patient Samples

1. Lymphocyte Separation Medium (LSM1077; PAA Laboratories GmbH, Cölbe, Germany).

2. Minimum Essential Medium (MEM) Alpha media (αMEM; Cambrex, Verviers, Belgium), phosphate-buffered saline (PBS).

3. Fetal Calf Serum (FCS) and Fetal Horse Serum are obtained from Sigma (Zwijndrecht, The Netherlands).

4. Stem/progenitor cells are isolated by MoFlo cell sorting (DakoCytomation, Carpinteria, CA, USA) or by utilizing the MiniMACS system from Miltenyi to isolate CD34+ cells from CB, BM, PB, or AML samples (130-056-701, Miltenyi Biotec, Amsterdam, The Netherlands). For further purification of LSCs from AML patients we use antibodies against CD34, CD38, CD90, CD123, CD45RA, CD135, CD47, or ITGA6 obtained from Becton Dickinson (BD, Alphen a/d Rijn, The Netherlands) (*see* also **Note 1**).

In case further purification of normal healthy HSCs and progenitor subpopulations is required we use the following antibodies (all from BD unless otherwise indicated): CD34-APC (581), CD38-PE (HB7), CD123-Pecy7 (6H6), and CD45RA-BV421 (HI100, Biolegend). Our lineage cocktail consists of the following Pecy5-conjugated antibodies (from eBiosciences unless otherwise indicated): anti-CD20 (2H7), -CD56 (B159, BD), -CD7 (124-D1, BD), -CD10 (MEM78), -CD235 (HIR2, BioLegend), -CD8 (RPA-T8), -CD19 (HIB19), -CD4 (RPA-T4), -CD2 (RPA 2.10), -CD3 (UCHT1), -CD11b (ICRF44, BioLegend), and -CD14 Tricolor (TUK4, Invitrogen).

2.2 Long-Term Coculture of AML Cells on Bone Marrow Stroma

1. MS5 murine bone marrow stromal cells (ACC 441) can be obtained from the Deutsche Sammlung von Mikroorganismen und Zellkulturen (DSMZ GmbH, Braunschweig, Germany).

2. MS5 growth medium for propagation: αMEM supplemented with heat-inactivated 10 % FCS, penicillin and streptomycin (Sigma), and 2 mM Glutamine (Sigma).

3. Long-term culture (LTC) medium: αMEM supplemented with heat-inactivated 12.5 % FCS, heat-inactivated 12.5 % horse serum, penicillin and streptomycin, 2 mM glutamine, 57.2 μM β-mercaptoethanol (Sigma), and 1 μM hydrocortisone (Sigma). Furthermore, IL-3, G-CSF, and TPO (20 ng/ml each) are added to the LTC medium.

4. Gelatin (Sigma) was prepared as 0.1 % stock solutions in PBS which is used to coat flasks or plates prior to plating MS5 stromal cells. AML LTC are typically performed in 12-well plates or T25 flasks. Prior to plating of MS5, plates or flasks are precoated with 0.1 % gelatin in PBS for 2 h at room temperature (in order to firmly attach the MS5 to the plates or the flasks). Then, gelatin is removed and MS5 cells are plated in MEM (10 % FCS) such that confluency is reached within 24 h. We do not routinely irradiate the MS5 stroma, although this is optional to further prevent overgrowth of stromal cells.

5. To isolate human cells from trypsinized MS5 bone marrow stroma before replating, we use anti-human CD45 antibodies from Becton Dickinson.

2.3 Lentiviral Transductions: Preparation of Particles

1. HEK 293T cells (ATCC number CRL-11268).

2. Dulbecco's Modified Eagle Medium (DMEM + Glutamax and 4.5 g/l D-Glucose, Gibco, Breda, The Netherlands).

3. Hematopoietic Progenitor Growth Medium (HPGM) (Cambrex).

4. Fugene 6 or HD (Roche, Almere, The Netherlands).

5. We use a third-generation lentiviral packaging system with the following vectors: packaging construct (pCMV Δ8.91), a vector encoding the VSV-G glycoprotein envelop (pMD2.G), and various lentiviral vectors with the gene of interest (GOI) or an shRNA-expressing cassette, and a marker gene such as GFP, mCherry, YFP, or the truncated NGF receptor.

6. Millex HV low protein binding filters (SLHV013SL Millipore).

7. Leica DM-IL fluorescence microscope (Leica Microsystems, Rijswijk, The Netherlands) with a 20×/0.30 or 40×/0.60 objective.

2.4 Lentiviral Transductions: Transduction of CB/ BM/PB/AML CD34+ Cells

1. Retronectin (Takara, Tokyo, Japan) is dissolved in H_2O at a stock concentration of 1 mg/ml which was stored at −20 °C. Prior to use, the retronectin stock is diluted in PBS to a final concentration of 25–50 μg/ml. Plates are coated with retronectin for 2 h at room temperature, and the diluted stock was stored at 4 °C and reused for about four times within a period of 2 weeks.

2. Polybrene (Hexadimethrine Bromide, Sigma) is prepared as a stock concentration of 4 mg/ml in PBS which was used 1:1,000.

2.5 Long-Term Coculture of Transduced Human CD34⁺ Cells on Bone Marrow Stroma Under Myeloid Permissive Conditions

1. Conditions are comparable to AML stromal cocultures, but instead of IL3, G-CSF, and TPO we use other cytokines in the LTC-IC medium. For transduced CB CD34⁺ cells we usually do not add additional cytokines to cocultures, except for CB CD34⁺ cells transduced with MLL-AF9 oncogenes, whereby 10 ng/ml FLT3-L, IL-3, and SCF are added. For adult BM and PB cocultures we add 20 ng/ml IL3 and TPO.

2. To enumerate progenitors in the suspension of MS5 cocultures we perform assays in methylcellulose (H4230, Stem Cell Technologies) supplemented with 20 ng/ml IL-3, IL-6, SCF, G-CSF, Flt-3L, 10 ng/ml GM-CSF, and 1 U/ml Epo.

2.6 Long-Term Coculture of Transduced Human CD34⁺ Cells on Bone Marrow Stroma Under Lymphoid Permissive Conditions

1. Lymphoid cultures contain the same components as the myeloid cultures with the exception of hydrocortisone, horse serum, and IL-3 and the addition of 50 μg/ml ascorbic acid (Sigma) and 10 ng/ml IL7 (R&D Systems).

3 Methods

3.1 Isolation of Human Stem/Progenitor Cells from CB, BM, PB, or AML Patient Samples

1. Healthy CB, BM, PB, or PB/BM cells from AML patients are studied after informed consent. Mononuclear cells (MNCs) are harvested by density-gradient centrifugation over lymphocyte separation medium according to the manufacturer's instructions. MNCs are routinely cryopreserved in aliquots of 50–100 × 10⁶ cells (AML samples) in 80 % MEM, 10 % dimethylsulfoxide (DMSO), and 10 % FCS in liquid nitrogen. The recovery rate is typically in the range of 50–80 %. MNCs from CB, BM, or PB are usually not frozen but immediately used for isolation of stem/progenitor cells.

2. Upon thawing, DMSO is removed by resuspending cells in 6 ml serum in the presence of 200 μl of 1 mg/ml DNAse and 200 μl of 25 mM MgCl₂, followed by centrifugation for 5 min at $800 \times g$.

3. The LSC phenotype in AML is still under investigation and is most likely rather heterogeneous, but for many experiments we start with sorted AML CD34⁺ and CD34⁻ cells (*see* **Note 1**). 50–100 × 10⁶ AML cells are incubated with 10 μl anti-CD34 antibodies in 200 μl PBS for 30 min at room temperature.

Cells are washed once with PBS followed by cell sorting on a MoFlo. In some cases, we have utilized the MiniMACS system from Miltenyi to isolate CD34$^+$ AML cells. Then, 50–100×10^6 AML cells were incubated with 100 μl MultiSort CD34 MicroBeads and isolation was performed according to the manufacturer's instructions (*see* **Note 2**).

4. CD34$^+$ stem/progenitor cells from CB, BM, or PB are typically isolated by MiniMACS. In case we require further purification of HSCs and progenitor we use the following sorting strategies: HSCs are defined as CD34$^+$/CD38$^-$/Lin$^-$, LMPPs are defined as CD34$^+$/CD38$^-$/CD45RA$^+$, CMPs are defined as CD34$^+$/CD38$^+$/CD123$^+$/CD45RA$^-$, MEPS are defined as CD34$^+$/CD38$^+$/CD123$^-$/CD45RA$^-$, and GMPs are defined as CD34$^+$/CD38$^+$/CD123$^+$/CD45RA$^+$. After washing, cells were resuspended in 1 μg/ml propidium iodide to exclude dead cells.

3.1.1 Long-Term Coculture of AML Cells on Bone Marrow Stroma

1. MS5 murine bone marrow stromal cells are routinely propagated in MEM (10 % FCS and pen/strep). Cells are subcultured three times a week and should never be grown to full confluency during passaging as MS5 overgrowth or detachment might occur later during LTC.

2. Typically, 4×10^4 AML CD34$^+$ cells are plated per well in gelatin-coated 12-well plates, while 2×10^5 AML CD34$^+$ cells are plated in gelatin-coated T25 flasks. AML cells are plated in LTC medium supplemented with IL-3, G-CSF, and TPO (20 ng/ml each). Although we usually use this cytokine cocktail, it is likely that differences exist in cytokine dependency between individual AML samples (see also [21]). For instance, for MLL-AF9 AML cocultures we routinely add 10 ng/ml FLT3-L, IL-3, and SCF, but it is well plausible that optimal cytokine conditions need to be determined for each individual AML sample. As normal healthy cells can also expand under these conditions, it will be important to validate that expanded cells indeed belong to the leukemic clone(s). Therefore, it is important to confirm mutation status by PCR or by karyotyping after 4–5 weeks of culture.

3. Instead of using 12-well plates, we are also using 96-well plates in order to determine the LSC frequency in vitro (*see* **Note 3**). Furthermore, various markers can be used to enrich for the LSC fraction (*see* **Note 4**).

4. Cultures are kept at 37 °C and 5 % CO_2 and are weekly demidepopulated. Plates or flasks are swirled gently, and half of the medium is collected and is set aside for analysis. Fresh LTC medium including cytokines are added back to the cultures. The harvested cells are spun down and counted and can be

analyzed by, e.g., FACS, Western blotting, or q-PCR. Our experience so far indicates that in about two-thirds of the investigated cases ($n > 80$) the AML CD34$^+$ fraction contains cells that can give rise to expanding LTC on MS5.

5. While normal CD34$^+$ cells derived from cord blood or bone marrow give rise to phase-dark cobblestone area-forming cells (CAFCs) only after 5 weeks of plating onto MS5, we typically observe leukemic cobblestone areas (L-CAs) earlier in the cultures, ranging from 1 to 5 weeks when the first L-CAs arise.

6. Incubate for 5 weeks at 37 °C, 5 % CO_2. Cobblestone areas can be counted at week 5 to determine LTC-IC frequencies in bulk (*see* **Note 3**). Cultures can also be serially replated; *see* Subheading 3.1.2.

3.1.2 Addressing Self-Renewal: Replating of Long-Term AML Cocultures onto New Bone Marrow Stroma

1. Self-renewal is addressed by serial replating of cocultures. Typically, cocultures are replated at week 5, but in case high numbers of L-CAs arise early on in the cocultures or when a lot of suspension cells are produced and the MS5 stroma no longer seems capable of sustaining the cultures we have replated cocultures earlier than week 5, at week 3 or 4.

2. Harvest suspension cells, wash adherent layer twice with PBS, and collect all fractions.

3. Trypsinize adherent cells for 5 min at 37 °C.

4. Collect adherent fraction in PBS.

5. Stain human cells with an antibody recognizing human CD45, and sort cells on a MoFlo (*see* **Note 5**).

6. Combine the sorted CD45$^+$ human AML cells with the suspension cells, and replate 1/20 to 1/3 of the total cells onto a new 12-well plate or T25 flask that was precoated with new MS5 the day before. We have also replated adherent and suspension cells separately onto new stroma. We find that the majority of replating activity resides in the adherent L-CA population, but some replating activity is also present in the suspension cells in some cases.

7. At weeks 10 and 15 (or at earlier time points when MS5 stromal cells deteriorate due to high numbers of leukemic cells) **steps 2–6** can be repeated to initiate third and fourth cultures (*see* **Notes 6** and **7**).

3.2 Lentiviral Transductions: Preparation of Particles

Detailed molecular analysis of signal transduction pathways in human LSCs depends upon efficient delivery of gene targeting vectors, by which loss-of-function and gain-of-function analyses can be performed. The efficient delivery involves optimal preparation of lentiviral particles, which takes about 4 days. Each day is described below in detail.

3.2.1   *Day 1*

1. Coat 10 cm dishes or T75 flasks for 2 h with 0.1 % gelatin at room temperature.

2. Remove gelatin, plate $2–4 \times 10^6$ 293T cells in 10 ml DMEM plus 10 % FCS per group, and incubate overnight at 37 °C, 5 % CO_2. Multiple plates can be seeded at the same time and later harvested into one large batch of virus, but the detailed protocol below describes the procedure for one plate. Obtaining a larger batch of virus can be useful from the point of experiment to experiment consistency as well as reducing the overall amount of work, since fewer preparations need to be done.

3.2.2   *Day 2*

1. Transient transfection of 293T cells using Fugene. Change medium to 5 ml DMEM + 10% FCS, pen-strep 2–6 h prior to transfection. Before transfection, prepare two tubes with the following reagents. Tube 1: DMEM without FCS and penicillin/streptomycin (100 μl), packaging construct (pCMV Δ8.91) (3 μg), glycoprotein envelop plasmid (pMD2.G) (0.7 μg), vector construct containing GOI, and GFP (3 μg). Tube 2: DMEM without FCS and penicillin/streptomycin (400 μl) and Fugene 6 or HD (21 μl). The amounts of DNA have been carefully optimized, but further optimization might be required for other plasmid combinations (*see* **Note 8**).

2. Add tube 1 to tube 2, flick gently, and allow complex formation for 20 min at room temperature.

3. After 20 min add mixture dropwise to 293T cells, swirl gently, and incubate cells overnight at 37 °C, 5 % CO_2.

3.2.3   *Day 3*

Check the transfection efficiency of the 293T cells. If the 293T cells are still solidly attached and a bright GFP signal is observed through the fluorescent microscope, the medium on the 293T cells can be changed to 4.5–6 ml HPGM and cells can be incubated overnight at 37 °C, 5 % CO_2. If 293T cells are detaching or a weak GFP signal is observed, an optimization of the growth and transfection procedure is necessary to obtain higher viral titers, mandatory for efficient AML transduction (*see* **Notes 7–9**).

3.2.4   *Day 4*

After an approximate 12 h of virus production into HPGM medium, virus can be harvested by removing 4.5 ml medium from 293T cells and filtering over low-protein-binding filters, Millex HV filters, to remove residual 293T cells. As we have observed that some 293T cells might be detaching during the harvest period, this step is necessary. After filtering, use virus directly for infection of AML cells or freeze virus-containing supernatant in aliquots of 500 μl in cryotubes in –80 °C and thaw upon use.

3.3 Lentiviral Transductions: Transduction of AML CD34⁺ Cells

The procedure for AML CD34$^+$ transduction takes 3 days to complete and can be started on day 4 of the lentiviral preparation. A typical AML transduction involves three rounds of infection of approximately 8–12 h, and we have reached transduction efficiencies ranging form 25 to 80 % depending upon AML sample, viral preparations, and the vector of interest.

3.3.1 Day 1

1. Typically, 1.5×10^6 AML CD34$^+$ cells are isolated with MACS columns or sorted by MoFlo and incubated at a cell density of 0.5×10^6 cell/ml in LTC medium supplemented with IL-3, G-CSF, and TPO (each 20 ng/ml) for 4 h at 37 °C, 5 % CO$_2$.

2. During this incubation period, wells from a 12-well plate are coated with 0.5 ml of retronectin (50 µg/ml in PBS) at room temperature (1 well per group). After 2 h retronectin is removed and the wells are blocked immediately with 2 % BSA/PBS for 30 min. Wash the plate twice with PBS and keep at 4 °C until use.

3. After 4 h, the preincubated AML CD34$^+$ cells are split into various groups that will be transduced with lentiviral batches of interest. Use at least 1.5×10^5 cells per group, but not more than 5×10^5 in 500 µl per well in 12-well plates. Make sure to include both an empty vector control group that only expresses your marker gene as well as a no-virus control group which will not be transduced but will be used for setting FACS gates and also allow a comparison of the growth of non-transduced AMLs with empty vector-transduced AMLs.

4. Per transduction group, plate 500 µl AML cell suspension per well in the retronectin-coated 12-well plates. Add 500 µl lentivirus supernatant to each well. In the no-virus control group, add 500 µl HPGM. Furthermore, add the following to each group: 110 µl FCS (final concentration 10 %), 20 ng/ml hIL-3, 20 ng/ml TPO, 20 ng/ml GCSF, and 4 µg/ml polybrene. Incubate overnight at 37 °C, 5 % CO$_2$. This will be the first round of transduction.

5. Meanwhile, MS5 stromal cells need to be cultured as described above in α-MEM supplemented with 10 % FCS, so that, e.g., a T75 reaches 80 % confluency at day 2, which should be sufficient for 3T25 flasks (in case you want to plate a no-virus control group, an empty vector control group, and an experimental group).

3.3.2 Day 2

1. Repeat transduction procedure in the morning (round 2). Add 500 µl new viral supernatants (or HPGM) to the wells as well as FCS, growth factors, and polybrene at concentrations indicated above. Washing of cells or removal of the first 500 µl

of viral supernatant that was added in round 1 is not necessary and will only lead to loss of cells. Incubate for 8 h at 37 °C, 5 % CO_2.

2. Transduction round 3 should be started in the evening by adding another 500 μl of viral supernatant as well as FCS, growth factors, and polybrene at concentrations indicated above.

3. Furthermore, coat three T25 flasks with 0.1 % gelatin for 2 h at room temperature. When done, trypsinize an 80 % confluent T75 flask with MS5, resuspend in 15 ml α-MEM with 10 % FCS, and plate 5 ml per gelatin-coated per T25 (these MS5 cultures will reach confluency on day 3, optimal for seeding with AML cells). When more than three T25 are required, make sure to prepare multiple T75 flasks with MS5 that reach 80 % confluency at day 2.

3.3.3   Day 3

1. The transduced AML CD34+ cells from each group are washed 3–5 times with PBS and are resuspended in 1.5 ml of LTC medium supplemented with IL-3, G-CSF, and TPO (each 20 ng/ml).

2. Take a 50 μl aliquot for FACS to assess transduction efficiency (use cells from the no-virus group to set the gates).

3. Plate equal amounts of AML CD34+ cells on MS5 in LTC medium (at least 1.5×10^5 cells per T25, no more than 5×10^5). MoFlo sorting of transduced cells can be performed, but is not strictly necessary as the untransduced cells within each culture can serve as an internal control. However, note that the, e.g., GFP/YFP expression usually does not reach steady-state levels until 2 days after the infection procedure, so analyzing a small sample after 2 days of expansion on MS5 stromal layers will give a more reliable transduction efficiency (*see* **Note 10**).

3.4 Loss- and Gain-of-Function Studies with Long-Term Coculture of Transduced AMLs on MS5

1. LTC of transduced AML cells are essentially the same as long-term cocultures of non-transduced AML cells as described in Subheading 3.2. Remove half of the suspension (2.5 ml) from the cultures weekly, and add 2.5 ml fresh LTC medium supplemented with IL-3, G-CSF, and TPO (each 20 ng/ml). The removed half can be used for FACS analysis, cell count, cytospins, etc. Determining the percentage of marker genes such as GFP or YFP as well as the use of differentiation makers in FACS analyses on these weekly demipopulations give a rapid insight into growth and differentiation characteristics of transduced cells. We typically compare GFP/YFP-expressing cells with the non-transduced cells within each group as negative controls as well as with the control group that contains AML cells that were not transduced at all (examples can be found in [19, 22–24, 27–29]).

3.5 Long-Term Coculture of Transduced Human CD34+ Cells on Bone Marrow Stroma Under Myeloid or Lymphoid Permissive Conditions

1. Long-term coculture and serial replating procedures for transduced human CD34+ cells on bone marrow stroma under myeloid or lymphoid permissive conditions are essentially the same as described under Subheadings 3.1.1 and 3.1.2 for AML cocultures. Examples can be found in [22, 30–37]. Both the huCD45+ adherent fraction and the suspension fraction are typically combined for replating onto new stroma, and the timing of replating depends heavily on the expansion kinetics of the transformed cells, which can vary from once every 2 weeks to once every 5 weeks. While often CB CD34+ cells are used due to their easy availability, it is important to note that important differences exist in terms of transformation potential between fetal CB cells and adult BM or PB cells ([32] *see* **Note 11**).

2. We evaluate the presence of myeloid progenitors by plating cells in methylcellulose. 10,000 Cells obtained after demidepopulation of cultures are plated in 1 ml methylcellulose supplemented with 20 ng/ml IL-3, IL-6, SCF, G-CSF, Flt-3L, 10 ng/ml GM-CSF, and 1 U/ml Epo at 37 °C, 5 % CO_2. After 2 weeks, CFCs are scored. Self-renewal of progenitors can be evaluated by serial replating of CFCs. Pipet 1 ml PBS to the content of one dish containing methylcellulose colonies, place in 50 ml tube, and wash cells three times with 45 ml PBS. Cells can, e.g., be used for FACS or cytospins. For second CFCs, plate 50,000–200,000 into 1 ml new methylcellulose supplemented with 20 ng/ml IL-3, IL-6, SCF, G-CSF, Flt-3L, 10 ng/ml GM-CSF, and 1 U/ml Epo at 37 °C and 5 % CO_2. After 2 weeks, second CFCs are scored.

4 Notes

1. The FACS phenotype of LSCs is most likely rather heterogeneous and differs between individual patients, and the CD34+/CD38− HSC phenotype is clearly not valid for all AML patients [25, 38]. In most cases, the AML LSCs are contained within the CD34+ fraction [19, 38], but exceptions exist as well, e.g., for NPMcyt AML [39] or in MLL-AF9-rearranged samples (our unpublished observations). Besides CD34, various other LSC markers such as ITGA6, FLT3, CD47, CD33, CLL1, and CD96 have been suggested and can be included in the MoFlo sorting procedures, but functional studies are required to first elucidate in which population the LSCs reside.

2. Instead of sorting CD34+ AML cells on the MoFlo, we have been utilizing the MiniMACS columns using anti-CD34 magnetic beads as well (Miltenyi Biotec, Amsterdam, The Netherlands). It must be taken into account that the maximum binding capacity of the columns is (in our experience) around

$20–30 \times 10^6$ cells, so when the CD34 percentage of an AML is high, the flow through cannot be considered as CD34-.

3. Instead of using 12-well plates, we have been able to initiate long-term AML cocultures on 96-well plates coated with MS5 to determine the in vitro LSC frequency and to allow a more high-throughput analysis. Depending on the sorting procedures to enrich for LSCs (see **Note 4**), we have initiated LTC with 1–5 cells per well, but large variations in frequency can be observed between different AML samples and depending on sorting schemes prior to starting the assay. Week-5 leukemic LTC-IC frequencies can be read out by scoring CAFC-containing wells at week 5 as positive after which L-LTC-IC frequencies are determined using L-Calc. As primary AML cells usually do not efficiently form CFCs in methylcellulose we routinely do not add methylcellulose to each well at week 5 as we would do for normal LTC-IC assays.

4. The culture system obviously also allows for sorting on the basis of different markers, and we have been able to use CD34, CD38, CD123, D117, and HLA-DR as potential markers to identify LSCs. While the AML cells that give rise to LTC almost always reside in the CD34+, and are absent from the CD34- fraction [19], we have observed in our analyses of various AML samples a much more heterogeneous expression pattern of the other markers on cells that can initiate LTC on MS5 stroma.

5. Upon replating of cocultures, we have been able to use human CD45 MicroBeads (130-045-801) from Miltenyi instead of sorting by MoFlo in order to isolate human AML cells from the adherent fraction. However, it must be taken into account that some MS5 cells will contaminate your CD45 isolation as it is difficult to deplete all MS5 cells from the column during the wash steps. Alternatively, we have been able to remove the majority of MS5 stromal cells from AML cells by preplating of trypsinized adherent fractions on T75 flasks in 15 ml MEM (10 % FCS) for 5–10 min. The majority of MS5 cells will attach to the plastic within this period, while the AML cells do not and can be harvested from the non-adherent suspension fraction.

6. An important issue is how to deal with leukemic versus normal stem cells that might be present in the CD34 isolation steps as well. We establish the leukemic origin of the expanding cultures by performing PCR analysis for the presence of genetic markers such as the Flt3-ITD and by the fact that cultures generate second, third, and fourth L-CAs, a feature of self-renewing cells that we do not observe with normal CB stem cells. Also, L-CAs derived from PB should represent leukemic cells as the PB of healthy donors typically does not contain significant

numbers of stem/progenitor cells that give rise to cobblestone areas on stroma.

7. We have observed that around two-thirds of all investigated AMLs (>80 until date of publication) can give rise to long-term expanding cultures on MS5 stroma. The growth characteristics were categorized into the risk groups according to the new World Health Organization classification, but no significant differences were observed. It was intriguing however that out of the AMLs under investigation that belonged to the good risk group (containing either AML1-ETO or INV [16] translocations), no AML LTC-ICs could be established, suggesting that good risk AMLs do not perform well in our ex vivo assay.

8. When transducing AML CD34$^+$ cells, efficient transduction can be achieved when using viral supernatants that contain a titer of ~10^7 viral particles per ml (TU/ml). By infecting 293T cells using a standard viral titration protocol this number can be easily calculated. If titers are low, this might have been due to low transient transfection efficiencies of 293T cells. Ensure that the producing 293T cells are proliferating optimally, cells are attached, and correct Fugene 6(HD)/DNA complexes are used. Use clean DNA to transfect the 293T cells with an OD 260/280 of >1.8. Furthermore, the ratio of the various plasmids in the DNA/Fugene 6(HD) complex as described above may vary when slightly different plasmids are used. The protocol described above has been optimized for various plasmids, but we have observed (unpublished results) that different promoters, different plasmid sizes, and different sizes of the GOI can also influence transfection and transduction efficiencies. In such cases, transfections need to be optimized with different ratios of plasmid versus Fugene 6/HD.

9. Another possibility when low viral titers are obtained is to concentrate the virus particles. Standard concentration of virus involves ultracentrifugation for several hours, but an easier and faster method can be employed when using Centriprep columns (YM-50, Amicon, Millipore). Following the manufacturer's instructions, 15 ml of virus supernatant can be concentrated 15–20× in two rounds of 20 min of centrifugation using any normal centrifuge. This method is therefore ideally suited when an expensive ultracentrifuge is not at hand and ensures that virus can be frozen or used as fast as possible without the risk of losing virus due to a possible short half-life.

10. Since the transductions are performed in the absence of stromal cells and the efficient culture of AML cells is dependent upon interactions with stroma, an important point of interest is how many AML cells undergo cell death during the transduction procedure. During most of our transductions we have observed

that the number of AML cells after transductions is ~20–50 % lower than what was plated upon the start of the transduction procedure, while only in some exceptional cases we have observed a slight expansion during the transduction procedures. We have therefore always counted viable cells after the transduction procedure, before subjecting them to long-term stromal cocultures, ensuring that we had equal numbers of AML cells at the start of the culture. We typically start with a minimum of 1.5×10^5 cells per T25. In the first 1 or 2 weeks, an initial drop in cell number is usually observed, but in all long-term experiments with transduced AML CD34$^+$ cells we have performed till date rapidly expanding cultures could be established within a few weeks after plating.

11. When we transduced MLL-AF9 into human fetal CB CD34$^+$ cells and human adult BM or PB cells we observed that transformation was more difficult to achieve in adult cells and that differentiation was biased towards the myeloid lineage, while both lymphoid and myeloid transformation could efficiently be induced in fetal CB cells [32]. This is in line with what is observed in pediatric MLL-AF9 patients that can develop both AML and ALL. While adult MLL-AF9 patients typically develop only AML and not ALL, it exemplifies that the cell of origin has great impact on the phenotypes that can be generated in these model systems.

Acknowledgements

We would like to acknowledge all members of the Experimental Hematology lab for helpful discussions. This work was supported by grants from the NWO (VENI 91611105, VIDI 91796312), KWF (2009-4411), and EU (FP7 EuroCSC ITN).

References

1. Lapidot T, Sirard C, Vormoor J, Murdoch B, Hoang T, Caceres-Cortes J, Minden M, Paterson B, Caligiuri MA, Dick JE (1994) A cell initiating human acute myeloid leukaemia after transplantation into SCID mice. Nature 6464:645–648

2. Bonnet D, Dick JE (1997) Human acute myeloid leukemia is organized as a hierarchy that originates from a primitive hematopoietic cell. Nat Med 7:730–737

3. Kollet O, Peled A, Byk T, Ben Hur H, Greiner D, Shultz L, Lapidot T (2000) beta2 microglobulin-deficient (B2m(null)) NOD/SCID mice are excellent recipients for studying human stem cell function. Blood 10:3102–3105

4. Vargaftig J, Taussig DC, Griessinger E, Anjos-Afonso F, Lister TA, Cavenagh J, Oakervee H, Gribben J, Bonnet D (2012) Frequency of leukemic initiating cells does not depend on the xenotransplantation model used. Leukemia 26:858–860

5. Valent P, Bonnet D, De MR, Lapidot T, Copland M, Melo JV, Chomienne C, Ishikawa F, Schuringa JJ, Stassi G, Huntly B, Herrmann H, Soulier J, Roesch A, Schuurhuis GJ, Wohrer S, Arock M, Zuber J, Cerny-Reiterer S, Johnsen HE, Andreeff M, Eaves C (2012) Cancer stem cell definitions and terminology: the devil is in the details. Nat Rev Cancer 11:767–775

6. Hope KJ, Jin L, Dick JE (2004) Acute myeloid leukemia originates from a hierarchy of leukemic

stem cell classes that differ in self-renewal capacity. Nat Immunol 7:738–743

7. Pearce DJ, Taussig D, Zibara K, Smith LL, Ridler CM, Preudhomme C, Young BD, Rohatiner AZ, Lister TA, Bonnet D (2006) AML engraftment in the NOD/SCID assay reflects the outcome of AML: implications for our understanding of the heterogeneity of AML. Blood 3:1166–1173

8. Groen RW, Noort WA, Raymakers RA, Prins HJ, Aalders L, Hofhuis FM, Moerer P, van Velzen JF, Bloem AC, van Kessel B, Rozemuller H, van Binsbergen E, Buijs A, Yuan H, de Bruijn JD, de Weers M, Parren PW, Schuringa JJ, Lokhorst HM, Mutis T, Martens AC (2012) Reconstructing the human hematopoietic niche in immunodeficient mice: opportunities for studying primary multiple myeloma. Blood 3:e9–e16

9. Rizo A, Vellenga E, de Haan G, Schuringa JJ (2006) Signaling pathways in self-renewing hematopoietic and leukemic stem cells: do all stem cells need a niche? Hum Mol Genet 15(2):R210–R219

10. Gartner S, Kaplan HS (1980) Long-term culture of human bone marrow cells. Proc Natl Acad Sci U S A 8:4756–4759

11. Eaves CJ, Casman JD, Eaves AC (1991) Methodology of long-term culture of human hemopoietic cells. J Tiss Cult Meth 13:55–62

12. Sutherland HJ, Eaves CJ, Eaves AC, Dragowska W, Lansdorp PM (1989) Characterization and partial purification of human marrow cells capable of initiating long-term hematopoiesis in vitro. Blood 5:1563–1570

13. Coulombel L, Eaves AC, Eaves CJ (1983) Enzymatic treatment of long-term human marrow cultures reveals the preferential location of primitive hemopoietic progenitors in the adherent layer. Blood 2:291–297

14. Sutherland HJ, Lansdorp PM, Henkelman DH, Eaves AC, Eaves CJ (1990) Functional characterization of individual human hematopoietic stem cells cultured at limiting dilution on supportive marrow stromal layers. Proc Natl Acad Sci U S A 9:3584–3588

15. Gartner S, Kaplan HS (1981) Long-term culture of normal and leukemic human bone marrow. Haematol Blood Transfus 26:276–288

16. Scholzel C, Lowenberg B (1985) Stimulation of proliferation and differentiation of acute myeloid leukemia cells on a bone marrow stroma in culture. Exp Hematol 7:664–669

17. Ailles LE, Gerhard B, Hogge DE (1997) Detection and characterization of primitive malignant and normal progenitors in patients with acute myelogenous leukemia using long-term coculture with supportive feeder layers and cytokines. Blood 7:2555–2564

18. Sutherland HJ, Blair A, Zapf RW (1996) Characterization of a hierarchy in human acute myeloid leukemia progenitor cells. Blood 11:4754–4761

19. van Gosliga D, Schepers H, Rizo A, van der Kolk D, Vellenga E, Schuringa JJ (2007) Establishing long-term cultures with self-renewing acute myeloid leukemia stem/progenitor cells. Exp Hematol 10:1538–1549

20. Schuringa JJ, Schepers H (2009) Ex vivo assays to study self-renewal and long-term expansion of genetically modified primary human acute myeloid leukemia stem cells. Methods Mol Biol 287–300

21. Han L, Wierenga AT, Rozenveld-Geugien M, van de Lande K, Vellenga E, Schuringa JJ (2009) Single-cell STAT5 signal transduction profiling in normal and leukemic stem and progenitor cell populations reveals highly distinct cytokine responses. PLoS One 11:e7989

22. Rizo A, Horton SJ, Olthof S, Dontje B, Ausema A, van Os R, van den Boom V, Vellenga E, de Haan G, Schuringa JJ (2010) BMI1 collaborates with BCR-ABL in leukemic transformation of human CD34+ cells. Blood 22: 4621–4630

23. Rizo A, Dontje B, Vellenga E, de Haan G, Schuringa JJ (2008) Long-term maintenance of human hematopoietic stem/progenitor cells by expression of BMI1. Blood 5:2621–2630

24. Rizo A, Olthof S, Han L, Vellenga E, de Haan G, Schuringa JJ (2009) Repression of BMI1 in normal and leukemic human CD34(+) cells impairs self-renewal and induces apoptosis. Blood 8:1498–1505

25. Bonardi F, Fusetti F, Deelen P, van Gosliga D, Vellenga E, Schuringa JJ (2013) A proteomics and transcriptomics approach to identify leukemic stem cell (LSC) markers. Mol Cell Proteomics 3:626–637

26. Rozenveld-Geugien M, Baas IO, van Gosliga D, Vellenga E, Schuringa JJ (2007) Expansion of normal and leukemic human hematopoietic stem/progenitor cells requires rac-mediated interaction with stromal cells. Exp Hematol 5:782–792

27. Schepers H, Wierenga AT, van Gosliga D, Eggen BJ, Vellenga E, Schuringa JJ (2007) Reintroduction of C/EBPalpha in leukemic CD34+ stem/progenitor cells impairs self-renewal and partially restores myelopoiesis. Blood 4:1317–1325

28. Schepers H, van Gosliga D, Wierenga AT, Eggen BJ, Schuringa JJ, Vellenga E (2007) STAT5 is required for long-term maintenance of normal and leukemic human stem/progenitor cells. Blood 8:2880–2888

29. Woolthuis CM, Han L, Verkaik-Schakel RN, van Gosliga D, Kluin PM, Vellenga E,

Schuringa JJ, Huls G (2012) Downregulation of MEIS1 impairs long-term expansion of CD34(+) NPM1-mutated acute myeloid leukemia cells. Leukemia 26:848–853

30. Barabe F, Kennedy JA, Hope KJ, Dick JE (2007) Modeling the initiation and progression of human acute leukemia in mice. Science 5824:600–604

31. Wei J, Wunderlich M, Fox C, Alvarez S, Cigudosa JC, Wilhelm JS, Zheng Y, Cancelas JA, Gu Y, Jansen M, Dimartino JF, Mulloy JC (2008) Microenvironment determines lineage fate in a human model of MLL-AF9 leukemia. Cancer Cell 6:483–495

32. Horton SJ, Jaques J, Woolthuis C, van Dijk J, Mesuraca M, Huls G, Morrone G, Vellenga E, Schuringa JJ (2013) MLL-AF9-mediated immortalization of human hematopoietic cells along different lineages changes during ontogeny. Leukemia 27:1116–1126

33. Chalandon Y, Jiang X, Christ O, Loutet S, Thanopoulou E, Eaves A, Eaves C (2005) BCR-ABL-transduced human cord blood cells produce abnormal populations in immunodeficient mice. Leukemia 3: 442–448

34. Chung KY, Morrone G, Schuringa JJ, Wong B, Dorn DC, Moore MA (2005) Enforced expression of an Flt3 internal tandem duplication in human CD34+ cells confers properties of self-renewal and enhanced erythropoiesis. Blood 1:77–84

35. Fatrai S, van Gosliga D, Han L, Daenen SM, Vellenga E, Schuringa JJ (2011) KRAS(G12V) enhances proliferation and initiates myelomonocytic differentiation in human stem/progenitor cells via intrinsic and extrinsic pathways. J Biol Chem 8:6061–6070

36. Schuringa JJ, Wu K, Morrone G, Moore MA (2004) Enforced activation of STAT5A facilitates the generation of embryonic stem-derived hematopoietic stem cells that contribute to hematopoiesis in vivo. Stem Cells 7:1191–1204

37. Chung KY, Morrone G, Schuringa JJ, Plasilova M, Shieh JH, Zhang Y, Zhou P, Moore MA (2006) Enforced expression of NUP98-HOXA9 in human CD34(+) cells enhances stem cell proliferation. Cancer Res 24:11781–11791

38. Eppert K, Takenaka K, Lechman ER, Waldron L, Nilsson B, van Galen P, Metzeler KH, Poeppl A, Ling V, Beyene J, Canty AJ, Danska JS, Bohlander SK, Buske C, Minden MD, Golub TR, Jurisica I, Ebert BL, Dick JE (2011) Stem cell gene expression programs influence clinical outcome in human leukemia. Nat Med 9:1086–1093

39. Taussig DC, Vargaftig J, Miraki-Moud F, Griessinger E, Sharrock K, Luke T, Lillington D, Oakervee H, Cavenagh J, Agrawal SG, Lister TA, Gribben JG, Bonnet D (2010) Leukemia-initiating cells from some acute myeloid leukemia patients with mutated nucleophosmin reside in the CD34(−) fraction. Blood 10:1976–1984

Chapter 14

Ex Vivo Expansion of Murine and Human Hematopoietic Stem Cells

Phuong L. Doan and John P. Chute

Abstract

Hematopoietic stem cells have the capacity to self-renew and give rise to the entirety of the mature blood and immune system throughout the lifespan of an organism. Here, we describe methods to isolate and culture murine bone marrow (BM) CD34⁻ckit⁺Sca1⁺Lineage⁻ (CD34⁻KSL) hematopoietic stem cells (HSCs). We also describe a method to measure functional HSC content via the competitive repopulation assay. Furthermore, we summarize methods to isolate and culture human CD34⁺CD38⁻Lineage⁻ cells which are enriched for human hematopoietic stem and progenitor cells.

Key words Hematopoietic stem cell, Self-renewal, Cell expansion, Regeneration, Reconstitution, Competitive repopulation assay, Cord blood

1 Introduction

Cell surface markers have been utilized for isolation of HSCs. These isolation methods employ a combination of magnetic column separation to enrich for lineage-depleted cells, followed by fluorescence-activated cell sorting (FACS), which facilitates the collection of hematopoietic cells based on multiple surface markers that are tagged with fluorescent probes. The ability to isolate live HSCs provides the capability to measure the effects of specific treatments (i.e., cytokines, genetic modulation) on HSC and progenitor populations [1, 2]. HSCs can be characterized functionally via the competitive repopulation assay, in which limiting dilutions of HSCs are transplanted via tail vein injection into lethally irradiated, congenic mice which also receive host BM competitor cells [1–4]. Over time, donor hematopoietic cell engraftment can be measured via flow cytometric analysis of recipient peripheral blood (PB) or BM cells by distinguishing surface CD 45.1⁺ or CD 45.2⁺ expression. The levels of donor chimerism can be tracked in the peripheral blood through 20 weeks to determine the kinetics of donor cell engraftment and to estimate donor long-term HSC content.

Kevin D. Bunting and Cheng-Kui Qu (eds.), *Hematopoietic Stem Cell Protocols*, Methods in Molecular Biology, vol. 1185, DOI 10.1007/978-1-4939-1133-2_14, © Springer Science+Business Media New York 2014

Limiting dilution analysis can be performed using 3–5 donor hematopoietic cell doses, such that a subset of recipient mice will demonstrate non-engraftment of donor cells. This approach allows for Poisson statistical analysis to estimate the frequency of donor HSCs in comparative donor sources [1, 2]. Our laboratory has utilized these same methods to be employed to estimate the residual frequency of HSCs in the BM of mice following exposure to genotoxic stressors such as ionizing radiation [1, 2].

Like murine HSCs, human hematopoietic cell populations can be enriched via fluorescence activated cell sorting, and then placed in culture for genetic modification or in order to expand subpopulations ex vivo. Cord blood units are a rich source of human HSCs and have been used in HSC transplantation in adults [5–7]. Here, we describe methods to isolate and expand human CB HSCs ex vivo, which has potential implications to improve donor CB engraftment in patients following transplantation.

2 Materials

2.1 Isolation and Culture of Murine CD34⁻ ckit⁺Sca1⁺Lineage⁻ Bone Marrow Cells

1. Dissecting tools: scissors and forceps. Autoclaved and are sterile.

2. Bone marrow collection media: Iscove's DMEM (IMDM, Cellgro, Manassas, VA), 10 % Fetal Bovine Serum (FBS), 1 % penicillin–streptomycin. Prepare 500 ml. Sterilize with 0.2 μm filter. Store at 4 °C (*see* **Note 1**).

3. Buffer for Lineage Depletion: 10 % FBS, 1 % penicillin–streptomycin in phosphate buffered saline (PBS), pH 7.2.

4. Staining buffer: 1 % FBS in PBS.

5. Plasticware: 28.5 gauge insulin syringes, 40 μm mesh nylon strainer, 15 and 50 ml conical tubes, tubes for FACS, 96-well U-bottom clear polystyrene plate (*see* **Note 2**).

6. ACK Lysing Buffer (Lonza, Walkersville, MD). Store at room temperature.

7. Trypan Blue (Lonza, Walkersville, MD).

8. Magnetic cell sorting: Lineage Depletion Kit, MACS LS columns, and MidiMACS Separator (Miltenyi Biotec, Auburn, CA).

9. Antibodies and reagents for FACS: CD34, ckit, and Sca1 antibodies conjugated with fluorophores, 7-aminoactinomycin D (7-AAD, *see* **Note 3**).

10. Thrombopoietin, Stem Cell Factor, Flt3 Ligand (TSF) Mouse Cytokine media: 20 ng/ml Thrombopoietin, 125 ng/ml Stem Cell Factor, 50 ng/ml Flt3 Ligand in bone marrow collection media (*see* **Note 4**).

2.2 Isolation and Culture of Human CD34⁺CD38⁻Lineage⁻ Cord Blood Hematopoietic Cells

1. Wash Buffer: 10 % FBS, 1 % penicillin–streptomycin in PBS, pH 7.2.

2. Priming Media: PBS without Ca^{+2} and Mg^{+2} at room temperature or degassed (*see* **Note 5**). No serum or protein additives.

3. Separation Media: 2 % FBS in PBS.

4. Cord blood units (*see* **Note 6**).

5. Human Hematopoietic Progenitor Cell Enrichment Kit: Human hematopoietic progenitor cell enrichment cocktail, magnetic colloid, pH 7.0–7.5 (StemCell Technologies, Vancouver, BC, Canada) (*see* **Note 7**).

6. StemSep Magnet, Negative selection columns, and Peristaltic Pump with Pump Tubing (StemCell Technologies, Vancouver, BC, Canada) (*see* **Note 8**).

7. Thrombopoietin, Stem Cell Factor, Flt3 Ligand (TSF) Human Cytokine media: 20 ng/ml Thrombopoietin, 125 ng/ml Stem Cell Factor, 50 ng/ml Flt3 Ligand in 10 % FBS with IMDM.

3 Methods

Carry out all procedures under sterile conditions. Procedures should be done expeditiously to improve cell viability.

3.1 Isolation of Murine CD34⁻ ckit⁺Sca1⁺Lineage⁻ Bone Marrow Cells

1. Euthanize mice using protocols approved by the Institutional Animal Care and Use Committee. Dissect bilateral femurs and tibias. Clip both ends of femurs. Insert insulin syringe filled with bone marrow collection media. Flush bone marrow into 50 ml conical tube filled with collection media. Repeat several times on both ends of femurs until femur appears white. Repeat with tibias.

2. Pellet cells in centrifuge at $350 \times g$, 5 min.

3. Aspirate supernatant.

4. Lyse red blood cells with 3 ml of ACK Lysing Buffer and vortex. Incubate 3 min at room temperature (*see* **Note 9**).

5. Neutralize lysis buffer with greater than 3 volumes of buffer for lineage depletion.

6. Pass cells through 40 μm mesh nylon strainer (*see* **Note 10**).

7. Pellet cells in centrifuge at $350 \times g$, 5 min.

8. Aspirate supernatant.

9. Resuspend cells in 40 μl of buffer for lineage depletion per 10^7 cells (*see* **Note 11**). Volume is dependent on total number of mice in collection.

10. Obtain cell count by trypan blue exclusion.

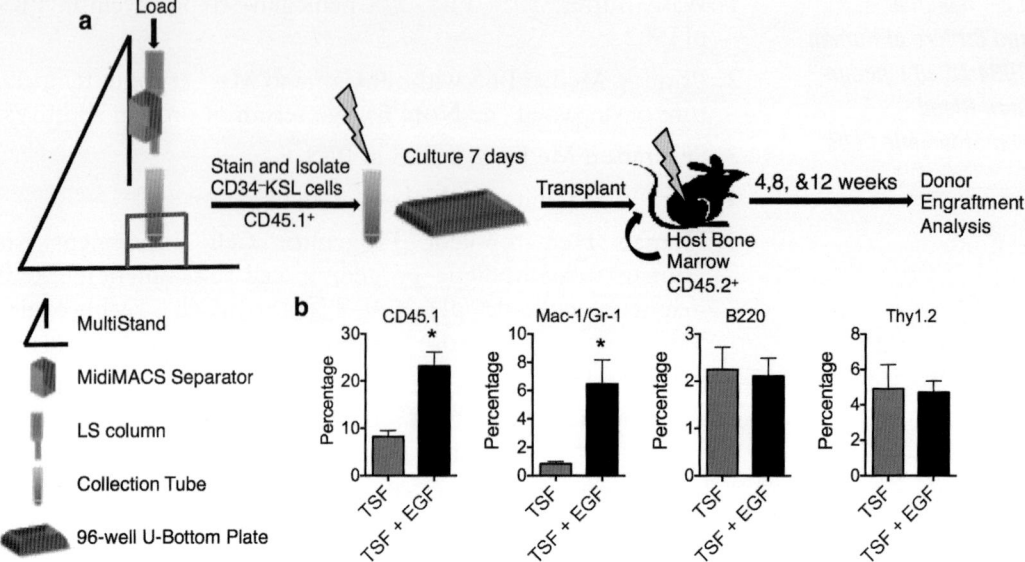

Fig. 1 Isolation of murine CD34⁻KSL cells followed by irradiation, ex vivo culture, and competitive repopulation assay. (**a**) Schematic diagram of method to isolate BM CD34⁻KSL cells. Red blood cell-depleted bone marrow from CD 45.1⁺ mice is labeled for magnetic enrichment of lineage-negative cells. Labeled sample is loaded into a primed LS column. Lineage-negative cells are collected, stained, and FACS-sorted for CD34⁻, ckit⁺, and Sca1⁺ (CD34⁻ KSL) cells. To measure the regenerative capacity of these cells, cells were exposed to 300cGy ionizing radiation (Cs137 source). Cells were cultured with recombinant cytokines (thrombopoietin, stem cell factor, and flt3 ligand = TSF) or TSF supplemented with epidermal growth factor (TSF + EGF). Following 7 days in culture, CD34⁻KSL cells and their progeny were collected and transplanted by tail vein injection into lethally irradiated (950cGy) CD45.2⁺ recipients along with 2×10^5 host BM cells as competitors. Total and multi-lineage donor engraftment in the blood were measured by flow cytometric analysis of the PB between 4 and 20 weeks following transplantation. (**b**) Peripheral blood donor engraftment at 8 weeks of CD45.2⁺ recipient mice following transplantation of 100 CD34⁻KSL cells exposed to 300cGy followed by 7-days in culture with TSF + EGF or TSF alone. Total CD 45.1⁺ donor engraftment, myeloid (Mac-1/Gr-1), B cell (B220), and T cell (Thy1.2) engraftment are shown. *$p < 0.05$. (*see* ref. 2, Fig. 1g)

11. Add 10 μl of Biotin-antibody cocktail per 10^7 cells. Mix well and incubate on ice or at 4 °C for 10 min (*see* **Note 12**).

12. Add 30 μl of buffer per 10^7 cells.

13. Add 20 μl of Anti-biotin Microbeads per 10^7 cells (*see* **Note 13**). Mix well and incubate on ice or at 4 °C for 15 min.

14. Wash cells by adding at least 2 ml per 10^7 cells of buffer for lineage depletion and centrifuge at $300 \times g$ for 10 min.

15. Aspirate supernatant completely.

16. Resuspend up to 10^8 cells in 500 μl of buffer for lineage depletion (*see* **Note 14**).

17. Place LS column(s) in the magnetic field of MACS Separator (*see* **Note 15**, *see* Fig. 1).

18. Prime columns with 3 ml of buffer for lineage depletion per column. Discard effluent. Place clean 15 ml conical tubes underneath columns.

19. Add cell suspension to column(s) and collect effluent in 15 ml conical tubes.

20. Wash columns with 3 ml of buffer for lineage depletion three times. Collect all effluent (*see* **Note 16**).

21. Pellet cells in centrifuge at $350 \times g$, 5 min.

22. Obtain cell count by trypan blue exclusion (*see* **Note 17**).

23. Label cells with CD34, ckit, and Sca1 antibodies in staining buffer (*see* **Note 18**). Incubate 30 min at 4 °C. Rinse cells and resuspend cells at a density of approximately 300 µl per 10^7 cells.

24. Add 15 µl of 7-AAD and proceed to cell sorting (*see* **Note 19**).

25. Pellet cells and resuspend cells in TSF cytokine media (*see* **Note 20**).

26. Seed cells into a 96-well U-bottom plate. Incubate plate at 37 °C, 5 % humidity for the time point of interest (*see* **Note 21**).

27. Collect cells and pellet. Perform cell counts by trypan blue exclusion.

28. Prepare host bone marrow competing cells using **steps 1–8**.

29. Perform cell counts by trypan blue exclusion of host bone marrow cells.

30. Aliquot competing cell dose into sterile eppendorf tubes (*see* **Note 22**).

31. Add donor cells from culture to each eppendorf tube (*see* **Note 23**).

32. Perform tail vein injections of donor and competing host cells into recipient mice that have been lethally irradiated (*see* **Note 24**, *see* Fig. 1).

33. Starting 4 weeks following transplantation and at 4-week intervals, collect peripheral blood by maxillary vein puncture and label for total donor engraftment and multi-lineage engraftment for flow cytometric analysis (*see* **Note 25**).

3.2 Isolation and Culture of Human CD34⁺CD38⁻Lineage⁻ Cord Blood and Bone Marrow Hematopoietic Cells

Cord blood and bone marrow cells arrive in the laboratory with all patient information de-identified. Methods will describe procedure for cord blood stem cell isolation. Identical methods can be applied for human bone marrow stem cell isolation.

1. Transfer cord blood into sterile container(s) (*see* **Note 26**).

2. Aliquot 15 ml of ficoll into 50 ml conical tubes.

3. Overlay ficoll with 30 ml of cord blood per tube (*see* **Note 27**).

4. Centrifuge at $350 \times g$, 35 min at 25 °C (*see* **Note 28**).

5. Carefully aspirate upper layer of serum without disturbing buffy coat layer, which contains mononuclear cells.

6. Collect mononuclear cell layer and transfer to new 50 ml conical tubes (*see* **Note 29**).

7. Bring total volume in tubes to 50 ml with wash buffer.

8. Centrifuge at $400 \times g$, 10 min. Discard supernatant.

9. Add 20 ml ACK Lysing Buffer. Vortex (*see* **Note 30**). Incubate cells at 37 °C in water bath for 15–30 min.

10. Bring volume to 50 ml with wash buffer. Centrifuge at $400 \times g$, 10 min. Discard supernatant. Repeat.

11. Resuspend cells in wash buffer and perform cell counts with trypan blue exclusion (*see* **Note 31**).

12. Add StemSep enrichment cocktail at 100 µl/ml. Mix and incubate on ice for 30 min or at room temperature for 15 min (*see* **Note 32**).

13. Add magnetic colloid at 60 µl/ml. Mix and incubate on ice for 30 min or at room temperature for 15 min (*see* **Note 33**).

14. Assemble column into magnet. Assemble pump tubing and prime column with priming media (*see* **Note 34**, *see* Fig. 2).

15. Wash from the top down with the appropriate volume of separation media (*see* **Note 35**).

16. Load sample and repeat **step 15**. Collect sample volume and wash volume (*see* **Note 36**).

17. Pellet cells by centrifugation $350 \times g$, 5 min. Perform cell count by trypan blue exclusion (*see* **Note 37**).

18. Label cells with CD34 and CD38 antibodies in staining buffer (*see* **Note 38**). Incubate 30 min at 4 °C. Rinse cells and resuspend cells at a density of approximately 300 µl per 10^7 cells.

19. Add 15 µl of 7-AAD and proceed to cell sorting (*see* Fig. 2).

20. Pellet cells and resuspend in human TSF cytokine media. Aliquot cells into 96-well U-bottom plate for culture.

21. Perform tail vein injections of cultured cells into NOD-SCID or NOD-scid *Il2rg* recipient mice that have been conditioned with radiation.

22. Starting 4 weeks following transplantation and at 4-week intervals, collect peripheral blood by maxillary vein puncture and label for total donor engraftment and multi-lineage engraftment for flow cytometric analysis (*see* **Note 39**, *see* Fig. 2).

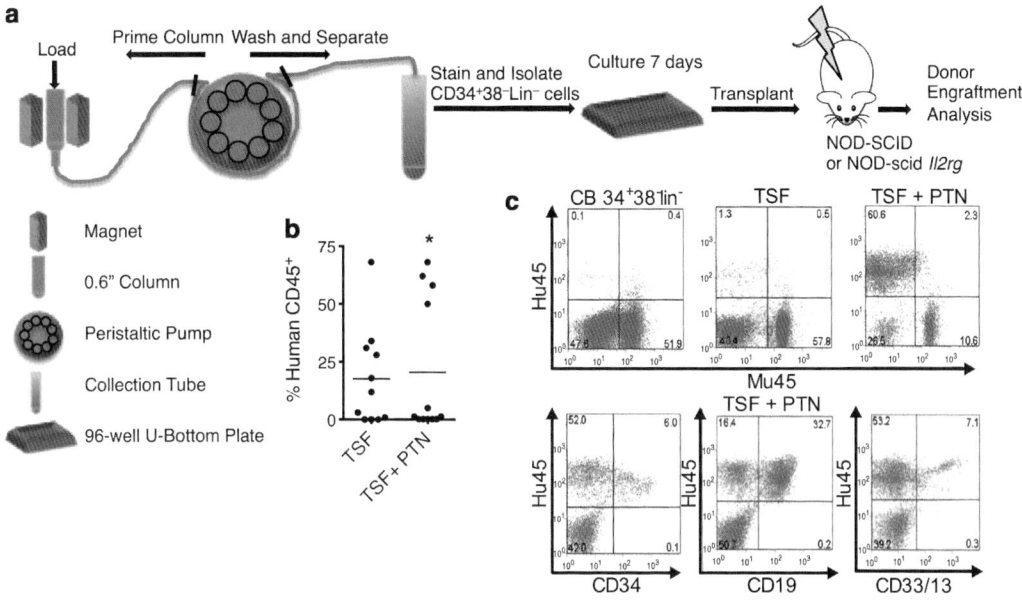

Fig. 2 Isolation and ex vivo culture of human CB CD34+CD38−Lin− cells followed by tail vein injection into NOD/SCID mice. (**a**) StemSep TAC Magnetically labeled cord blood was loaded into a primed and washed 0.6″ magnetic column. The action of the peristaltic pump separated lineage-negative cells into a collection tube. Lineage-negative cells were stained and FACS-sorted for CD34+CD38−Lin− cells. Following 7 days in culture with 500 ng/ml pleiotrophin and human TSF cytokine media (TSF + PTN) or TSF alone, CD34+CD38−Lin− cells and progeny were transplanted via tail vein injection into NOD-SCID or NOD-scid *Il2rg* mice that were conditioned with 300cGy total body irradiation. (**b**) Bone marrow engraftment at 8 weeks is shown following transplantation of the progeny of 2,500 CD34+CD38−Lin− cells following 7 days of culture with TSF or TSF + PTN. *$p < 0.05$. (**c**) Top, representative FACS plots of total human CD45+ hematopoietic cell engraftment at 8 weeks following transplantation of CB CD34+CD38−Lin− cells or their progeny following culture with TSF or TSF + PTN; at bottom, myeloid (CD 33/13) and B cell (CD 19) engraftment in NOD-SCID mice at 8 weeks after transplantation with the progeny of 2,500 cord blood CD34+CD38−Lin- cells following culture with TSF + PTN (*see* ref. 1, Fig. 3d, e)

4   Notes

1. Reagents are made on the day of experiment.

2. All plasticware should be sterile.

3. We routinely use CD34 FITC, ckit PE, and Sca1 APC-cy7.

4. Cytokines are lyophilized and require resuspension in IMDM, 1 % FBS, 1 % penicillin–streptomycin and are filter-sterilized through a 0.2 μm Nalgene bottle. Store at 100× concentrations at −80 °C.

5. Degassed media reduces the introduction of air bubbles into columns, which can cause channeling and a decrease in the cell capacity of the column. If cell density exceeds the cell capacity of the column, then cell purity can be compromised.

6. Note the cord unit cell count, volume, identification number, and expiration date and time. We process cord units within

48 h of collection to improve cell viability. Wear personal protective equipment (i.e., lab coat, gloves, goggles) when handling human specimens. Working with human specimens may require approval from the Institutional Review Board.

7. Enrichment kit comes with enough reagents to label 5×10^9 cells.

8. Columns come in a variety of sizes. We use 0.6″ columns, which has a column capacity to lineage deplete 10^8–1.5×10^9 cells. While cell separation could be performed with gravity alone, we use a peristaltic pump to both prime and wash cells during separation.

9. To improve cell yield, it is critical to ensure cell pellet is fully dispersed through vigorous vortexing or aggressive pipetting.

10. This filtration step is critical to remove bone fragments following flushing of bone marrow.

11. After red blood cell lysis, we routinely obtain about 25 million whole bone marrow cells per mouse. For fewer than 10^7 cells, use same volume as indicated for 10^7 cells. When working with larger number of cells, increase volumes accordingly. For example, lineage depletion of 2×10^7 cells requires 80 µl of buffer.

12. Working on ice may increase incubation times. Longer incubation times may increase nonspecific cell labeling. Biotin-antibody cocktail includes monoclonal antibodies to CD5, CD45R (B220), CD11b, Anti-Gr-1, anti-7-4, and Ter119.

13. Vortex Anti-biotin Microbeads immediately prior to adding to labeled cells since beads may settle in container. Microbeads are conjugated to a monoclonal anti-biotin antibody (clone: Bio3-18E7.2; mouse IgG1).

14. Scale volumes up according for greater than 10^8 cells. It is critical that cells are in single-cell suspensions.

15. LS columns should fit snugly into MidiMACS Separator. Each LS column has the capacity to isolate a maximum of 10^8 labeled cells and a maximum of 2×10^9 total cells. They are packaged sterilely and come individually wrapped. Columns are "flow stop" and will not run dry. They have a void volume of 400 µl and a reservoir volume of 8 ml. They are intended for single use only.

16. Add buffer when reservoir is empty. This effluent is the enriched lineage-negative cells. The enrichment rate is between 50- and 1,000-fold, depending on the specificity of magnetic labeling. Lineage-positive cells are retained in columns within the magnetic field. If lineage-positive cells are required, remove the LS column from the magnet and place into a new 15 ml conical tube. Add 5 ml of buffer for lineage depletion to reservoir and immediately apply firm pressure to plunger supplied with column to flush out the lineage-positive cell fraction.

17. We typically obtain between 8×10^5 and 10^6 lineage negative cells per mouse.

18. The concentrations of antibodies used for staining require titration based on the specific antibody clone and cell density. In general, we use 1 μg/ml per 10^6 cells.

19. 7-AAD allows for flow cytometric exclusion of dead cells, which will be 7-AAD positive. Note the number of 34⁻KSL cells collected from the sorting procedure. If the anticipated number of 34⁻KSL cells to be collected is less than 5×10^5 cells, then we collect in 1.5 ml eppendorf tubes containing 300 μl of TSF cytokine media. If collecting greater than 5×10^5 cells, then we collect into flow tubes containing 1 ml of TSF cytokine media.

20. Cultures of 34⁻KSL cells in TSF cytokine media serve as controls for testing hematopoietic stem cell expansion with other cytokines (i.e., Pleiotrophin, PTN, or epidermal growth factor, EGF) (*see* ref. [1, 2]). To measure the regenerative capacity of 34⁻KSL cells following injury such as radiation, cells are irradiated immediately following cell sorting prior to seeding into the 96-well U-bottom plates.

21. We typically seed between 30 and 300 cells per well in 200 μl of TSF cytokine media. For culture of KSL cells, we have maintained these cultures at 5,000 cells per well for 7 days without exhausting the TSF cytokine media. Most of our in vitro hematopoietic stem cell assays are analyzed at 72 h and 7 days.

22. Competing cell dose is usually 2×10^5 whole bone marrow cells (*see* ref. 4). We tend to use CD45.1⁺ donor cells and CD45.2⁺ recipient mice.

23. We determine both the cell dose(s) and number of replicates for competitive transplantation prior to setting up cultures. For example, if we intend to transplant 100 34⁻KSL cells and progeny donor cells, we seed each well with 100 cells. If we want to have 15 mice per group, then 15 wells per group are seeded with 34⁻KSL cells. At time of transplantation, the entire contents of each well are transplanted into one mouse. Injection media should contain 2–10 % FBS in IMDM. Total volume of injection may range from 100 to 250 μl per mouse.

24. Mice are irradiated 6–24 h prior to transplantation. We maintain all mice on water with trimethoprim-sulfamethoxazole following total body irradiation.

25. We collect 100–200 μl of peripheral blood and perform a RBC lysis as described in Subheading 3.1, **steps 4, 5**. Cells are labeled for flow cytometric analysis as follows: total engraftment (CD 45.1⁺), myeloid (Mac-1/Gr-1), B cell (B220), and T cell (Thy1.2).

26. We use sterile T175 flasks. Autoclaved, sterile 1 l bottles would work as well.

27. We use 2 volumes of cells to 1 volume of ficoll.

28. Be careful to not disturb interface between cells and ficoll when handling tubes. We disable the brake mechanisms on the centrifuge for this step.

29. Adjust hand pipet to the slowest speed for collection to provide better control. We combine mononuclear cells from several tubes (i.e., 3–4) into new conical tubes for ease of handling fewer tubes.

30. As in **Note 9**, aggressive vortexing to disrupt cell pellet will improve cell yield.

31. Cell density should be about 5×10^7 cells per ml, or within the range of $2–8 \times 10^7$ cells per ml. If cells are still clumped, we recommend that cell suspension be passed through a 40 μm mesh nylon strainer.

32. Enrichment cocktail contains mouse monoclonal IgG_1 antibodies bound to a bispecific tetrameric antibody complex directed against both dextran and the following human hematopoietic surface markers: CD2, CD3, CD14, CD16, CD19, CD24, CD56, CD66b, and glycophorin A. Cocktail is in PBS and is stable between 2 and 8 °C for 2 years. Longer incubation times may lead to nonspecific binding.

33. Colloid is stable at 2–8 °C for 6 weeks or at −20 °C for 1 year. Colloid should be vortexed prior to refreezing.

34. Column should be inserted from above down into the gap of the magnet. Do not insert column from front of magnet. Connections between columns and pump tubing should be checked to ensure no leaks are present. Priming speed depends on column size. For 0.6″ column, column should be primed at 0.6 ml/min with a pump setting of 3.0. Priming is complete when priming media is visible above the magnetic layers of the column.

35. For 0.6″ columns, 25 ml of separation media is added to the columns. The direction of flow for the peristaltic pump should be reversed and changed to speed of 2 ml/min at a setting of 10. At no point should the column run dry or air bubbles will be introduced.

36. These cells are lineage-depleted cells and are ready to use. Sample purity could be measured by flow cytometric analysis with anti-CD34 antibody. We routinely achieve greater than 90 % purity with these methods.

37. Lineage-negative cells could be stored in liquid nitrogen until ready to use. Alternatively, they may be stored at 4 °C overnight with excellent cell viability.

38. Volume of antibodies for staining will need to be titrated depending on the antibody clone and fluorophore used.

39. We collect 100–200 µl of peripheral blood and perform a RBC lysis as described in Subheading 3.1, **steps 4, 5**. At the end of the study, usually between 8 and 16 weeks, bone marrow cells may be collected and analyzed for donor engraftment. Cells are labeled for flow cytometric analysis as follows: total engraftment (human CD45), myeloid (human CD 33/13), B cell (human CD 19), and T cell (human CD 3). We also stain for a phenotypic hematopoietic stem cell population with CD 34 and CD 38.

References

1. Himburg HA, Muramoto GG, Daher P et al (2010) Pleiotrophin regulates the expansion and regeneration of hematopoietic stem cells. Nat Med 16:475–482

2. Doan PL, Himburg HA, Helms KA et al (2013) Epidermal growth factor regulates hematopoietic regeneration after radiation injury. Nat Med 19:295–304

3. Ema H, Sodu K, Seita J et al (2005) Quantification of self-renewal capacity in single hematopoietic stem cells from normal and Lnk-deficient mice. Dev Cell 8:907–914

4. Purton LE, Scadden DT (2007) Limiting factors in murine hematopoietic stem cell assays. Cell Stem Cell 1:263–270

5. Koh L-P, Chao NJ (2004) Umbilical transplantation in adults using myeloablative and nonmyeloablative preparative regimens. Biol Blood Marrow Transplant 10:1–22

6. Laughlin MJ, Eapen M, Rubinstein P et al (2004) Outcomes after transplantation of cord blood or bone marrow from unrelated donors in adults with leukemia. N Engl J Med 351:2265–2275

7. Majhail MS, Brunstein CG, Tomblyn M et al (2008) Reduced-intensity allogeneic transplant in patients older than 55 years: unrelated umbilical cord blood transplant is safe and effective for patients without a matched related donor. Biol Blood Marrow Transplant 14:282–289

Chapter 15

Using Microfluidics to Investigate Hematopoietic Stem Cell and Microniche Interactions at the Single Cell Level

Byungwook Ahn, Zhengqi Wang, David R. Archer, and Wilbur A. Lam

Abstract

In recent years, microfluidic devices have become widely used in biology, and with the advantage of requiring low sample volumes, enables previously technologically infeasible experiments in hematopoietic stem cell (HSC) research. Here, we introduce a microfluidic device to investigate dynamic interactions between HSC and model niches in vitro. The device comprises a pneumatic valve which enables the culturing of different types of niche cells in different parts of the same device. Single HSCs can then be injected into the microfluidic device, manipulated, and placed onto different niches within the same device as controlled by the user. Here, we describe the device fabrication method, the HSC collection methodology, and the operational procedure for the device.

Key words Microfluidics, Pneumatic valve, Endothelial cell, Niche, Microenvironment, Hematopoietic stem cell

1 Introduction

The first applications for microfluidics were focused on the capability of these devices to handle small amounts of liquids in the nanoliter and picoliter range and led to the advent of the current ink-jet technology [1]. In late 1990s, "soft lithography" techniques enabled microfluidic devices to be fabricated with the silicone elastomer polydimethylsiloxane (PDMS) [2], which brought forth the advantages of low cost, simple fabrication steps, and bio-compatibility. As such, microfluidics devices are now used in biology and biochemistry as a tool that leverage those capabilities and have enabled enumerable "lab-on-a-chip" applications. Indeed, almost 10,000 papers about microfluidics have been published in last 10 years alone [3].

Microfluidic devices are comprised of glass, thermal-plastic, or as mentioned above, PDMS. Glass or silicon wafer based microfluidics require processes to "etch" channels into the material and a thermal bonding process to seal the channels. Glass-based

Kevin D. Bunting and Cheng-Kui Qu (eds.), *Hematopoietic Stem Cell Protocols*, Methods in Molecular Biology, vol. 1185, DOI 10.1007/978-1-4939-1133-2_15, © Springer Science+Business Media New York 2014

microfluidics can yield sophisticated three-dimensional structure, but cannot be reused. On the other hand, PDMS based microfluidics has relatively simple fabrication steps via soft-lithography, and the process yields a silicon master mold, in which large numbers of PDMS devices can be fabricated. PDMS is porous and flexible, enabling features such as pneumatic valves to be incorporated into the device [4].

More recently, researchers have used other materials to produce microfluidic device. Polystyrene microfluidics can be fabricated by the hot embossing technique [5]. As polystyrene is the same material as the tissue culture flasks, the familiar material properties may be advantageous for biologists. Like PDMS, polystyrene is also optically transparent, which enables live cell imaging. Recently, paper based microfluidics has been developed for disease diagnostic purpose [6]. Unlike the other types of the microfluidics, the paper based microfluidics leverages the wettability of the material. The small amount of liquid is absorbed and transferred by a highly wettabile material (unpatterned area), but it is blocked or guided by a low wettability material (patterned area with wax or photoresist).

Microfluidic devices are suitable for biology and biochemistry area with the advantages of low volume consumption of reagent and samples, and PDMS microfluidics has key properties—optical transparency, high gas permeability and biocompatibility—that enable them to be amenable for biological and biochemical research [7]. In addition, microfluidic components such as valves, mixers, pumps have been incorporated into those microfluidic systems. Beside the basic components, one can embed optical or electrical parts into a device for sensing or actuating. Finally, the small size of microfluidics enables control and manipulation of single cells, which we describe here.

Microfluidics has recently been utilized for hematopoietic stem cell research and techniques such as purifying, analyzing, or culturing. Wu et al. [8] introduces an integrated microfluidic system capable of sorting of HSCs from cord blood that sorting efficiency is as high as 88 %. The total sorting time of the device takes only 40 min which is much faster than with the traditional method. A filter embedded PDMS device has also been shown to purify HSCs from the bone marrow with an efficiency that is comparable with flow cytometry [9]. Furthermore, the micro-well array [10, 11], cell trapping [12, 13], or valved chamber structures [14] have all recently been adaptable to microfluidic devices that enable to single HSC analysis and manipulation under dynamic conditions.

Here, we describe a microfluidic device that enables investigating the dynamic interactions between HSCs and the niche microenvironments of the bone marrow in vitro. The device consists of three layers made of PDMS: the valve channel, the thin membrane, and the fluidic channel (Fig. 1a). The sandwiched PDMS membrane is

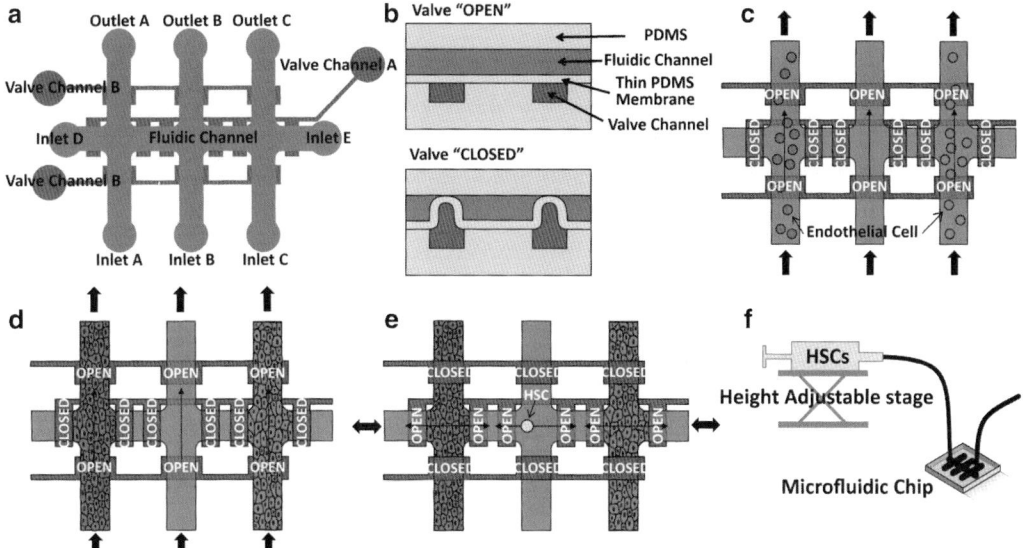

Fig. 1 A schematic of PDMS device and its operation process. (**a**) Top view (**b**) cross-sectional view of PDMS device. (**c**) Cell seeding process. The fluidic channels are separated by controlling the valve (the valve **a**: Closed, the valve **b**: Open). (**d**) Cell culture for 2 days until get confluence in an incubator. (**e**) Valve modification. The valve pressure is controlled to connect the fluidic channels each other (the valve **a**: Open, the valve **b**: Closed). (**f**) HSCs injecting and transporting. The injected HSC is transported by the pressure from the syringe placing over the height adjustable stage

flexible, so it can be deflected to the fluidic channel and blocking it depends on the pressure applied through the valve channel, pneumatically (Fig. 1b). A set of the valve channel A is utilized to seed and culture the different niches in the chambers. Once, valve channel B is actuated, the injected HSC is guided along the fluidic channel to expose it to the different niches.

2 Materials

2.1 Microfluidic Chip Fabrication Process Components

1. Cleanroom facility (including spin coater, hotplate, and aligner).

2. SU-8 photoresist and developer (Microchem, Newton, MA, USA).

3. Silicon wafer (NOVA electronic materials, Flower Mound, TX, USA).

4. Photomask.

5. HMDS (Hexamethyldisilazane) (Sigma-Aldrich, St. Louis, MO, USA).

6. Sylgard 184 (PDMS) (Dow Corning, Midland, MI, USA).

7. Micro-punch (Syneo, West Palm Beach, FL, USA).

2.2 HUVEC Culturing and Seeding Components

1. PBS.
2. EGM-2 cell culture media (Lonza).
3. Dextran from Leuconostoc spp. (Sigma-Aldrich, St. Louis, MO, USA).
4. Fibronectin solution: 5 % fibronectin stock solution in PBS (final fibronectin concentration, 50 µg/ml).
5. Syringe pump (Harvard Apparatus, Holliston, MA, USA).
6. Blunt needle (VWR, Suwanee, GA, USA).
7. Small tubing (30 gauge) (Cole-Parmer, Vernon Hills, IL, USA).
8. Large tubing (14 gauge) (Cole-Parmer, Vernon Hills, IL, USA).
9. Lab jack (Fisher Scientific, Pittsburgh, PA, USA).

2.3 HSC Component

1. Sterile 100×15 mm polystyrene petri dishes.
2. Sterile scissors to remove femur and tibias.
3. Sterile 10 ml syringe, 21 G and 25 G needle to expel bone marrow from medullary cavities.
4. BD Falcon cell strainer, 40 µm (#352340) and BD Falcon 5 ml polystyrene round-bottom tube with cell strainer cap (#352235).
5. MACS Separation LS column (Miltenyi Biotec Inc, Auburn, CA).
6. StemSep Magnet (STEMCELL Technologies Inc., Vancouver, BC, Canada) or Miltenyi AutoMACS Pro separator (Miltenyi Biotec Inc, Auburn, CA).
7. Mouse Lineage Cell Depletion Kit (Miltenyi Biotec Inc, Auburn, CA).
8. Blocking buffer: 5 % normal mouse serum in PBS.
9. Washing and staining buffer: PBS supplemented with 2 % heat-inactivated fetal bovine serum.
10. Monoclonal antibodies (eBioscience, San Diago, CA): lineage antibodies include Ly-6G (Gr1)~FITC (clone RB6-8C5), CD11b~FITC (clone M1/70), Ter119(Ly-76)~FITC (Clone TER119), CD45R(B220)~FITC (Clone RA3-6B2), CD8a~FITC (Clone 53-6.7) and CD4 (Clone GK1.5). CD150~PE (Clone mShad150), Ly-6A/E (Sca-1)~PE-Cy7 (Clone D7), CD117 (c-Kit)~APC (Clone 2B8), and CD48~pacific blue (clone HM48-1, Biolegend, San Diago, CA).
11. BD FACS Aria II Cell Sorter (BD Biosciences, San Jose, CA).

3 Methods

3.1 SU-8 Master Mold Process (Fig. 2a–d)

This process should be done in the cleanroom facility with proper personal protective equipment. For the best results of the SU-8 process, the users should follow the specific protocols (spin coater speed, UV exposure dose, bake temperature and time, development time, etc.) that have been offered by the commercial supplier of the photoresist or the cleanroom administrator.

1. Design photomask patterns using CAD software. The designed patterns can be transferred to the chrome photomask from commercial supplier.

2. Clean the silicon wafer (*see* **Note 1**).

3. Spin-coat the SU-8 2050 photoresist over the silicon wafer with the spin speed of $500 \times g$ for 30 s to get the final film thickness to 50 μm.

4. Pre-bake the spin-coated wafer over the hotplate set to 95 °C for 10 min (*see* **Note 2**).

5. Exposure the UV light using aligner tool.

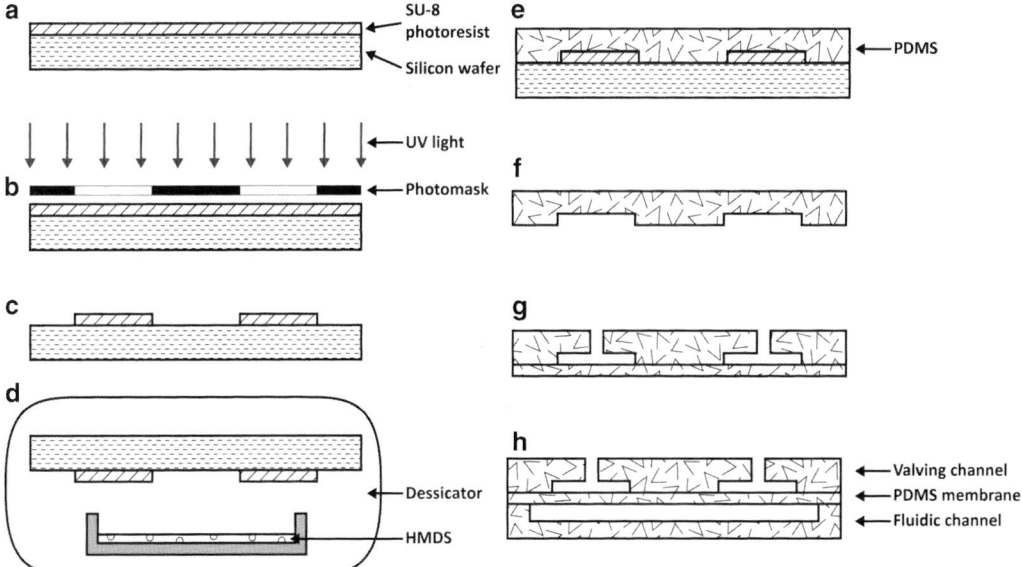

Fig. 2 Device fabrication process. A mold for PDMS device was fabricated by conventional photo-lithography process. (**a**) Spin coating of PR over a silicon wafer. (**b**) Exposing UV light through the photomask. (**c**) Developing the PR by immersing the UV exposed wafer. (**d**) Silanizing the developed wafer to prevent the PDMS adhesion during a curing process. (**e**) Pouring the mixed PDMS over a mold and cure. (**f**) Peeling off the cured PDMS. (**g, h**) Oxygen plasma bonding with another PDMS slab

6. Post-exposure bake the UV exposed wafer over the hotplate set to set to 95 °C for 8 min.

7. Immerse the baked wafer into the developer and stir the developer gently until all the unexposed photoresist dissolve in the developer (~5 min).

8. Wash the residue of the developer with IPA.

9. Place the patterned master mold with petri-dish containing 1 ml of HDMS (Hexamethyldisilazane) in the desiccator. Connect the desiccator to the vacuum line of the fume hood, and then let HMDS to evaporate at least 1 h. During the process, the surface of the master mold is coated with HMDS vapor which prevents the cured PDMS from sticking over the master mold (*see* **Note 3**).

3.2 PDMS Fabrication Process (Fig. 2e–h)

1. Mix the PDMS base and the PDMS curing agent thoroughly with 10:1 ratio by weight. Place the mixed PDMS in a vacuum desiccator until the air bubbles disappear.

2. Pour the PDMS over the HMDS treated master mold. Place the PDMS poured master mold in the convection oven which temperature is set to 65 °C overnight.

3. Using a knife or scalpel, cut out around the cured PDMS device and peel off the cured PDMS device from the master mold gently. Cut the PDMS slap into individual devices if necessary (*see* **Note 4**).

4. Spin-coat ($500 \times g$, 5 min) the PDMS mixture (10:1 ratio mixture) over the HMDS treated glass slide to make a thin PDMS membrane (~10 μm thickness). And cure the spin-coated glass slide over the 110 °C hotplate for 20 min.

5. Expose the surface of the PDMS device (valve channel) and thin PDMS coated glass slide to oxygen plasma for 30 s using a plasma cleaner. Then, place the PDMS device (valve channel side) over the thin PDMS coated glass slide to bond each other.

6. Place the bonded device over the 85 °C hotplate overnight to increase the bonding strength.

7. Peel off the plasma bonded device from the slide glass.

8. Punch the inlets and outlets of the PDMS devices with 0.75 mm diameter punch.

9. Do the plasma bond the fluidic channel PDMS device with the PDMS membrane bonded device to make three PDMS layers (valve channel, membrane and fluidic channel).

3.3 Collect the HSC

3.3.1 Bone Marrow Harvesting

1. Euthanize the mouse (*see* **Note 5**) and then wet the pelt thoroughly with 70 % ethanol.

2. Strip skin from hind limbs, remove excess tissue; use sterile sharp scissors, cut off the legs at the hip joints, separate femurs and tibias at the knee joints.

3. Flush the marrow using 21G (femurs) and 25G (tibias) needles attached to a syringe. Collect the bone marrow cells in PBS containing 2 % FBS.

4. Make a single cell suspension using 21G needle and pass cells through 40 µM cell strainer to remove cell clumps.

5. Count the cell number using hemocytometer after dilution in 2 % acetic acid (1:1) and Trypan blue (1:1) (*see* **Note 6**). Set aside about 5 million cells for control staining. Take the remaining cells for depletion of cells positive for lineage.

3.3.2 Magnetic Separation and Cell Sorting for HSC

1. Isolation of lineage negative cells from bone marrow cells is carried out essentially as described in the protocol of Mouse Lineage Cell Depletion Kit (Miltenyi Biotec). After magnetic labeling, magnetic separation is performed with either LS column (Miltenyi Biotec) or Miltenyi AutoMACS Pro separator (Miltenyi Biotec).

2. Count the enriched lineage negative fraction cells on a hemo-cytometer and pellet cells at $300 \times g$ for 5 min at 4 °C. Resuspend cells into 100–200 µl of blocking buffer for 15–20 min at 4 °C. Then wash with >4 volume of washing buffer, mix, and pellet cells at $300 \times g$ for 5 min at 4 °C.

3. Prepare lineage-FITC antibody cocktail (Anti-Gr1, Mac1, B220, Ter119, CD4 and CD8) along with Anti-c-Kit-APC, Anti-Sca1-PE-Cy7, Anti-CD150-PE, Anti-CD48-Pacific blue in 100–200 µl staining buffer.

4. Resuspend cell pellet in antibody cocktail and incubate at 4 °C for 15–20 min.

5. Meanwhile set control staining with the bone marrow cells set aside. Use 0.5–1 million bone marrow cell per control stain-ing. No stain control and single color staining control for each fluorochrome used (FITC, PE, APC, PE-CY7, Pacific Blue). Anti-B220 of each fluorochrome could be used for single color staining control.

6. Add >4 volume of washing buffer, mix, and pellet cells at $300 \times g$ for 5 min at 4 °C.

7. Wash cells once more and resuspend cell pellet in 0.5–1.0 ml of washing buffer. Get single cell suspension by passing through cell-strainer cap on BD Falcon 5 ml polystyrene tube.

8. Sort HSC with BD FACS Aria II Cell Sorter by gating on Lineage⁻c-Kit⁺Sca-1⁺CD150⁺CD48⁻ (Fig. 3) (*see* **Note 7**).

9. Pellet the sorted HSC at $300 \times g$ for 5 min at 4 °C. Resuspend into HSC culture medium (Iscove's modified Dulbecco's medium (IMDM), 15 % FBS (HyClone), 2 % penicillin/strep-tomycin B supplemented with 50 ng/ml recombinant murine stem cell factor, 20 ng/ml recombinant murine interleukin-3,

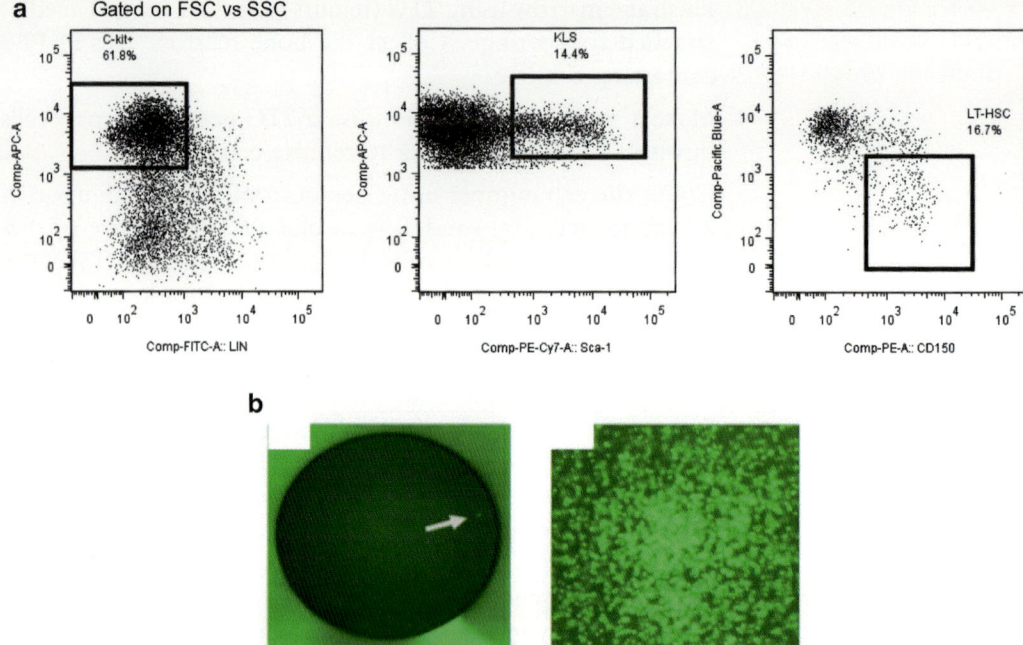

Fig. 3 Gating scheme for sorting hematopoietic stem cells (HSC). (**a**) Each dot plot represents the events contained in the previous electronic gate. The percentage shown in each dot plot represents the fraction of events that fall within the indicated gate. The flow cytometry analysis program FlowJo 7.0 was used to generate the dot plots shown. (**b**) (*Left*) Example of single GFP expressing HSC (*arrow*) sorted into a single well of a plate for in vitro studies and (*right*) a hematopoietic colony formed from a single HSC following 12 days of culture in semisolid medium

50 ng/ml recombinant human interleukin-6 (R&D system) and ready to culture HSC in microfluidic device (*see* **Note 8**).

10. Direct cell sorting of single HSC into the wells of the microfluidic device can be achieved by using the automated cell deposition unit (ACDU) on the FACSAria. The ACDU allows cells to be deposited at specified locations in a pre-or self-defined plate format. The coordinates of the wells are programmed into the ACDU using standard procedures and a precise number of cells are deposited into each well. Of note, the software FACSDiva v8 on the FACSAria is capable of indexed cell sorting allowing for the identification of the phenotype of an individual single cell selected by the well in which it was deposited. Functional evaluation of HSC can be performed by sorting single cells into each well of a plate and following survival and growth for 3–5 days. Optimal conditions for this sort are using cells from a eGFP expressing mouse and sorting into a Terasaki plate that has 20 µl wells. When HSC are plated this way, we often achieve identification and differentiation of single cells in 90–95 % of wells.

3.4 Microfluidic Device Operation and Endothelial Cell Seeding

1. Prepare three 1 ml syringes filled with PBS.

2. Connect the smaller tubing (~1 m long) to the larger tubing (~2 cm long), which is connected to a syringe with a blunt needle.

3. Load the syringes to the syringe pump and connect to the microfluidic chip (valve channel A, B).

4. Operate the syringe pumps with the slow flow rate setup (~1 µl/min) until all the air pushed away in the valve channel through the PDMS membrane. Release the pressure after removing all the air from the valve channels (*see* **Note 9**).

5. Inject the fibronectin solution (50 µg/ml) to the fluidic channel, and create a small (~100 µl) fibronectin droplet at the both inlet and outlet to ensure that the channel stays wet. And incubate the microfluidic chip in the temperature (37 °C), CO_2 (5 %) and humidity (90 %) controlled incubator for 1 h.

6. Operate the syringe pump which connected the valve channel A to add extra the PBS (~20 µl) into the valve channel A. Stop the syringe pump when the PDMS membrane is deflected and blocked the fluidic channel (Fig. 1c).

7. Prepare 5,000,000 cells/ml of HUVECs in 100 µl of endothelial growth media (EGM-2) with 8 % dextran.

8. Inject the prepared HUVECs using a pipette into the chambers through the inlet A and C (Fig. 1d).

9. Incubate the HUVECs seeded device in an incubator for 1 h.

10. Load three 10 ml syringe filled with EGM-2 at the syringe pump and connect the syringes to the microfluidic device through the inlet A to C.

11. Perfuse EGM-2 into the device with the flow rate of 2 µl/min. A monolayer of endothelial cell is formed on the inside of the device within 2–3 days (Fig. 1d).

12. Once HUVECs become confluent, release the pressure from the "valve A" by disconnecting the tubing from the device. Then, apply the pressure for the "valve B" by adding extra PBS (~20 µl) through the valve channel B using the syringe pump. Now, all the chambers are connected each other for transporting a HSC (Fig. 1e).

13. Prepare the HSCs in 1 ml syringe and connect to the device through inlet D.

14. Place the 1 ml syringe on the lab jack and adjust the height of the lab jack to control the position of the HSC (Fig. 1f) (*see* **Note 10**).

4 Notes

1. Remove thin oxide or natural oxide over the silicon wafer by dipping into buffered oxide etchant (BOE) in 5 min, then washed the wafer thoroughly with DI water. Clean the silicon wafer by general wafer cleaning step (acetone, methanol, and DI water). Then, dehydrate the wafer in a convection oven whose temperature is set to 150 °C in 30 min.

2. The baked wafer should be cool down slowly to the room temperature before proceeding to the next procedure.

3. HDMS is toxic and corrosive liquid. All the HMDS process should be done in the fume hood.

4. Clean the PDMS surface with tape whenever necessary.

5. The use of isoflurane is the preferred method of anesthesia because of simple administration (breathing vapors) without any noticeable adverse effects on normal function.

6. The use of acetic acid causes additional lysis of red blood cells and provides more accurate white blood cells count.

7. The purity of sorted cells could be verified by running a small fraction of cells again on BD FACS Aria II Cell Sorter.

8. For better cell tracking, HSC could be purified from EGFP transgenic mice instead of normal wild type mice.

9. PDMS membrane is gas permeable, so the air can pass through the membrane, but not liquid.

10. We can control the position of the HSC by adjusting the height of the lab jack. The pressure difference due to the height difference between the 1 ml syringe with HSCs and device induces the small amount of flow from syringe to device or vice versa.

References

1. Le HP (1998) Progress and trends in ink-jet printing technology. J Imaging Sci Tech 42:49–62

2. Xia YN, Whitesides GM (1998) Soft lithography. Annu Rev Mater Sci 28:153–184

3. Mark D, Haeberle S, Roth G et al (2010) Microfluidic lab-on-a-chip platforms: requirements, characteristics and applications. Chem Soc Rev 39:1153–1182

4. Unger MA, Chou HP, Thorsen T et al (2000) Monolithic microfabricated valves and pumps by multilayer soft lithography. Science 288:113–116

5. Young EW, Berthier E, Guckenberger DJ et al (2011) Rapid prototyping of arrayed microfluidic systems in polystyrene for cell-based assays. Anal Chem 83:1408–1417

6. Martinez AW, Phillips ST, Whitesides GM et al (2010) Diagnostics for the developing world: microfluidic paper-based analytical devices. Anal Chem 82:3–10

7. Whitesides GM (2006) The origins and the future of microfluidics. Nature 442:368–373

8. Wu HW, Hsu RC, Lin CC, Hwang SM, Lee GB (2010) An integrated microfluidic system for isolation, counting, and sorting of hematopoietic stem cells. Biomicrofluidics 4(2), pii:024112

9. Schirhagl R, Fuereder I, Hall EW et al (2011) Microfluidic purification and analysis of hematopoietic stem cells from bone marrow. Lab Chip 11:3130–3135

10. Lecault V, Vaninsberghe M, Sekulovic S et al (2011) High-throughput analysis of single hematopoietic stem cell proliferation in

microfluidic cell culture arrays. Nat Methods 8:581–586

11. Dykstra B, Ramunas J, Kent D et al (2006) High-resolution video monitoring of hematopoietic stem cells cultured in single-cell arrays identifies new features of self-renewal. Proc Natl Acad Sci U S A 103:8185–8190

12. Kobel SA, Burri O, Griffa A et al (2012) Automated analysis of single stem cells in microfluidic traps. Lab Chip 12:2843–2849

13. Faley SL, Copland M, Wlodkowic D et al (2009) Microfluidic single cell arrays to interrogate signalling dynamics of individual, patient-derived hematopoietic stem cells. Lab Chip 9:2659–2664

14. Glotzbach JP, Januszyk M, Vial IN et al (2011) An information theoretic, microfluidic-based single cell analysis permits identification of subpopulations among putatively homogeneous stem cells. Plos One 6:e21211

Part V

Transplantation Assays and Imaging Engraftment

Chapter 16

Five-Lineage Clonal Analysis of Hematopoietic Stem/Progenitor Cells

Ryo Yamamoto, Yohei Morita, and Hiromitsu Nakauchi

Abstract

Hematopoietic stem cells (HSCs) have self-renewal activity and multipotency. Clonal analysis and determination of HSC differentiation potential into platelets and erythrocytes as well as leukocytes are essential for the study of self-renewal and lineage commitment in HSC. However, due to technical limitations, platelet and erythrocyte differentiation potentials have not been assessed. This chapter describes principles and methods for single-cell sorting, single-cell transplantation, and identification and quantitative analysis of cell contribution to platelets and erythrocytes in addition to leukocytes in mouse chimeras.

Key words Hematopoietic stem cell, Hematopoietic progenitor cell, Single-cell sorting, Single-cell transplantation, Erythrocyte, Platelet, Granulocyte, B lymphocyte, T lymphocyte

1 Introduction

When bone marrow cells are transplanted into lethally irradiated myeloablated mice, the entire hematopoietic system is durably reconstituted with transplanted cells. Based on this observation, transplanted bone marrow is inferred to contain cells that home to bone marrow and clonally differentiate into cells of multiple lineages while self-renewing. Hematopoietic stem cells (HSCs) can be isolated by flow cytometry based on surface-marker expression. Use of a combination of cell surface markers has proved valuable for HSC purification.

CD34$^{-/low}$, c-Kit$^+$, Sca-1$^+$ lineage-marker$^-$ (CD34$^-$KSL) cells, CD150$^+$CD48$^-$KSL cells, and Thy-1^{low}Flt-3$^-$KSL cells are all significantly enriched in HSCs [1–3]. Clonal assays are crucial for analysis of HSCs since HSC differentiation and self-renewal potentials must be evaluated at the single-cell level. In our laboratory, we began to apply fluorescence-activated cell-sorting (FACS) clone-sorting technology to studies of hematopoiesis 25 years ago.

With flow cytometry, monoclonal antibodies against the allelic CD45 marker, which is expressed on all hematopoietic cells except

Kevin D. Bunting and Cheng-Kui Qu (eds.), *Hematopoietic Stem Cell Protocols*, Methods in Molecular Biology, vol. 1185, DOI 10.1007/978-1-4939-1133-2_16, © Springer Science+Business Media New York 2014

mature erythrocytes and platelets, can be used to distinguish host and donor cells in multiple lineages. However, platelet and erythrocyte differentiation potentials cannot be assessed at a clonal level using CD45 status. To address this issue, we have developed a transgenic mouse line in which all blood cells, including platelets and erythrocytes, express Kusabira-Orange (KuO) fluorescent protein [4], enabling in vivo tracing of five mature blood lineages (granulocytes, B lymphocytes, T lymphocytes, platelets, and erythrocytes). Furthermore, single-cell analysis is essential to assess differentiation potentials of transplanted cells precisely [5].

In this chapter, we describe detailed methods to sort single hematopoietic stem/progenitor cells and to perform single-cell transplantation.

2 Materials

2.1 Isolation of Bone Marrow

1. 8–12-Week-old KuO transgenic mice.
2. 6 cm Tissue culture dishes.
3. 2.5 mL Syringes with 25 G needle.
4. Mortar and muddler (*see* **Note 1**).
5. 45 μm Nylon mesh (Sansho, Tokyo, Japan) to filter the bone marrow cells after isolation.
6. 15 mL Tubes to stain bone marrow cells.
7. Hemocytometer.
8. Ice-cold staining medium: Phosphate-buffered saline (PBS, without calcium and magnesium) supplemented with 5 % heat-activated fetal bovine serum (FBS).

2.2 Staining of Bone Marrow (See Note 2)

1. Allophycocyanin (APC)-conjugated anti-c-Kit (CD117) antibody (2B8, eBioscience, San Diego, CA).
2. Alexa Fluor 700-conjugated anti-CD34 (RAM34, eBioscience).
3. Brilliant Violet 421-conjugated CD150 (TC15-12F12.2, BioLegend, San Diego, CA).
4. FITC-conjugated anti-CD41 (MWReg30, eBioscience).
5. PE-Cy7-conjugated anti-Sca-1 (D7, eBioscience).
6. Streptavidin conjugated with APC-Cy7 (BioLegend).
7. Anti-APC MicroBeads (Miltenyi Biotec, Auburn, CA).
8. Lineage-marker antibody cocktail.
9. 100 μL Biotin-conjugated anti-mouse Gr-1 antibody (RB6-8C5, eBioscience).
10. 50 μL Biotin-conjugated anti-mouse B220 (CD45RA) antibody (RA3-6B2, eBioscience).

11. 25 μL Biotin-conjugated anti-mouse CD4 antibody (RM4-5, eBioscience).

12. 25 μL Biotin-conjugated anti-mouse CD8 antibody (53-6.7, eBioscience).

13. 100 μL Biotin-conjugated anti-mouse TER-119 antibody (TER-119, eBioscience).

14. 50 μL Biotin-conjugated anti-mouse IL-7R antibody (A7R34, eBioscience).

15. 350 μL Staining medium.

16. LS columns (Miltenyi Biotec).

17. Propidium iodide (PI; Sigma-Aldrich, St. Louis, MO) in PBS at 200 μg/mL: A stock solution (10 mg/mL) is dissolved in water and stored at –20 °C. The final concentration of PI in sample should be 1 μg/mL.

2.3 Single-Cell Sorting

1. A FACS machine with at least seven-color capability is required.

2. We use BD FACS Aria II (BD Bioscience) equipped with a green laser.

2.4 Single-Cell Transplantation and Peripheral Blood Analysis

1. C57BL/6-Ly5.2 as recipient mice (8–12 weeks old).

2. X-ray irradiator.

3. Myjector insulin syringes with attached 27 G needles (TERUMO, Tokyo, Japan).

4. 96-Well round-bottom plate.

5. Glass-type capillary tubes (plain) for blood collection.

6. 0.1 M EDTA (Sigma-Aldrich).

7. Eppendorf tubes.

8. Isoflurane for animal (Mylan, Canonsburg, PA).

9. All-in-one anesthetizer for small animals (Muromachi, Tokyo, Japan).

10. Red blood cell (RBC) lysis buffer, 140 mM ammonium chloride (Wako, Osaka, Japan).

11. Peripheral blood analysis is performed on a Gallios (Beckman Coulter, Fullerton, CA) and FACS Cant II (BD Bioscience). Collected data are analyzed with FlowJo software (Tree Star, Ashland, OR).

12. Plt/RBC staining cocktail:
2 μL APC-conjugated anti-Ter-119 (TER-119, eBioscience).
4 μL eFluor450-conjugated anti-CD41 (MWReg30, eBioscience).
6 μL FITC-conjugated anti-CD42d (Gon.C2, Emfret, Eibelstadt, Germany).
4 mL Staining medium (for 20 mice).

13. WBC staining cocktail:

4 μL PE-Cy7-conjugated anti-CD45.1/Ly5.1 (A20, eBioscience).

4 μL Pacific Blue-conjugated anti-CD45.2/Ly5.2 (104, BioLegend).

2 μL FITC-conjugated anti-Gr-1 (RB6-8C5, eBioscience).

2 μL FITC-conjugated anti-Mac-1 (M1/70, eBioscience).

2 μL APC-eFluor780-conjugated anti-B220 (RA3-6B2, eBioscience).

3 μL APC-conjugated anti-CD3e (145-2C11, eBioscience).

1 mL Staining medium (for 20 mice).

3 Methods

3.1 Isolation of Bone Marrow

Obtain bone marrow from 8- to 12-week-old mouse.

1. Euthanize the mouse according to the method of choice (*see* **Note 3**).

2. Dissect the femurs, tibias, and coxal bones by small scissors and forceps, taking care to dissect away as much as muscle tissue as possible (*see* **Note 4**).

3. Place them in a 60 mm tissue culture dish containing 6 mL ice-cold PBS.

4. Crash all bones with a muddler (*see* **Note 5**).

5. Disaggregate bone marrow tissues by repeated aspirations using 2.5 mL syringe with 25 G needle.

6. Filter cells through 45 μm nylon mesh into a 15 mL tube (*see* **Note 6**).

7. Count nucleated cells with a hemocytometer.

8. Centrifuge cells for 5 min at $440 \times g$ at 4 °C.

3.2 Staining of Bone Marrow

1. Aspirate the supernatant, and suspend bone marrow cells in antibodies at a density of 5×10^7 cells per mL.

2. Add 0.5 μL APC-conjugated anti-c-Kit antibody per 10^7 cells.

3. Incubate cells for 30 min in the refrigerator (2–8 °C).

4. Centrifuge cells with 14 mL staining medium for 5 min at $440 \times g$ at 4 °C. Aspirate supernatant completely.

5. Resuspend cell pellet in 80 μL of staining medium per 10^7 cells.

6. Add 0.2–0.5 μL of Anti-APC MicroBeads per 10^7 total cells (*see* **Note 7**).

7. Mix well, and incubate the cells for 15 min in the refrigerator.

8. Centrifuge cells with 14 mL staining medium for 5 min at $440 \times g$ at 4 °C. Aspirate supernatant completely.

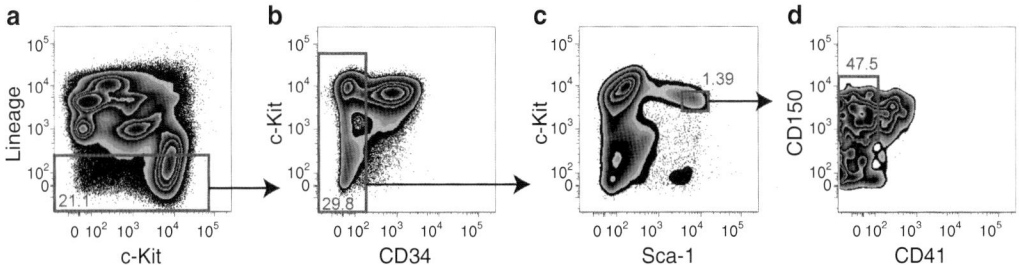

Fig. 1 Sorting gate for CD150⁺CD41⁻CD34⁻KSL cells. Flow cytometry for analysis and sorting is shown. (**a**) c-Kit+ enriched cells (doublets and dead cells are excluded) are displayed for c-Kit and lineage expression with the sorting gate for Lin− cells. (**b**) Lineage⁻ cells are displayed for CD34 and c-Kit expression with the sorting gate CD34⁻ cells. (**c**) CD34⁻Lineage⁻ cells are displayed for c-Kit and Sca-1 expression with the gating for Sca-1⁺ and c-Kit⁺ cells. (**d**) CD34⁻KSL cells are displayed for CD41 and CD150 expression with the gating for CD150⁺ and CD41⁻ cells

9. Add 3 mL degassed staining medium (*see* **Note 8**).

10. Apply cell suspension onto an LS column (*see* **Note 9**).

11. Collect unlabeled cells that pass through, and wash column twice with 4 mL degassed staining medium.

12. Count nucleated cells with a hemocytometer (*see* **Note 10**).

13. Spin down cells for 5 min at $440 \times g$ at 4 °C. Aspirate supernatant completely.

14. Resuspend 10^7 cells per 100 μL staining medium.

15. Add 2.5 μL lineage marker Ab cocktail per 10^7 cells.

16. Incubate the cells for 30 min in the refrigerator.

17. Centrifuge cells with 14 mL staining medium for 5 min at $440 \times g$ at 4 °C. Aspirate supernatant completely.

18. Resuspend 10^7 cells per 50 μL staining medium.

19. Add 4 μL Alexa 700-anti-CD34, 2 μL Brilliant Violet 421-anti-CD150, 1 μL FITC-anti-CD41, 1 μL APC-anti-c-Kit, 0.5 μL PE-Cy7-Sca-1, and 1 μL streptavidin APC-Cy7 per 10^7 cells.

20. Incubate the cells for 120 min in the refrigerator (*see* **Note 11**).

21. Wash the cells with 14 mL staining medium.

22. Aspirate the supernatant. Filter the cells through nylon mesh, and suspend cells at a density of 10^7 cells per mL to sort the cells by FACSAria II.

3.3 Single-Cell Sorting and Single-Cell Transplantation

The following protocols describe single-cell transplantation of bone marrow stem cells isolated from KuO mice as well as assessment of engraftment and reconstitution activity in recipient hosts.

1. Using a flow cytometer, sort KuO⁺CD150⁺CD34⁻CD41⁻KSL cells from KuO bone marrow (Fig. 1), at one cell per well, into a round-bottomed 96-well microtiter plate of which each well contains 100 μL staining medium (*see* **Note 12**).

2. Plate the plate on ice for 2–3 h so that cells settle in the well bottoms, permitting easier identification under an inverted microscope (*see* **Note 13**).

3. Visually verify under an inverted microscope that one cell is present per well.

4. To each well where one cell is found add 100 µL PBS containing 2×10^5 whole bone marrow cells from 8- to 12-week B6-Ly5.1/Ly5.2-F1 mice (*see* **Note 14**).

5. Irradiate B6-Ly5.2 mice at a total dose of 9.8Gy (4.9Gy, each of the two doses, 4 h apart) (*see* **Note 15**).

6. Aspirate all medium from each well of the plate into an insulin syringe (*see* **Note 16**).

7. Inject cells into mice via tail vein (*see* **Note 17**).

3.4 Peripheral Blood Analysis

Peripheral blood analysis is performed periodically (*see* **Notes 18** and **19**; Fig. 2).

1. Take up a small amount (10–20 µL) of 0.1 M EDTA into a glass capillary tube by capillary attraction (*see* **Note 20**).

2. Obtain 50–100 µL of peripheral blood from the retro-orbital plexus of recipient mice under isoflurane anesthesia into a capillary tube containing EDTA.

3. Transfer blood into an Eppendorf tube.

4. For analysis of platelets and erythrocytes, add 1 µL of blood sample to prepared tube containing 200 µL Plt/RBC staining cocktail.

5. Incubate the tube at 4 °C for 30 min in the refrigerator. After the incubation, chimerism of platelets and erythrocytes is analyzed on a Gallios.

6. For analysis of WBC, add 800 µL red blood cell lysis buffer to each blood sample and incubate at room temperature for 10 min.

7. Spin the tube at $440 \times g$ for 5 min at 4 °C.

8. Discard the supernatant.

9. Add 50 µL of WBC staining cocktail, and incubate cells for 30 min in the refrigerator.

10. Wash cells with staining medium. Discard the supernatant.

11. Resuspend cells in 100–200 µL staining medium containing PI at a final concentration of 1 µg/mL.

12. Keep cells on ice and in the dark until cells are analyzed by FACS Canto II.

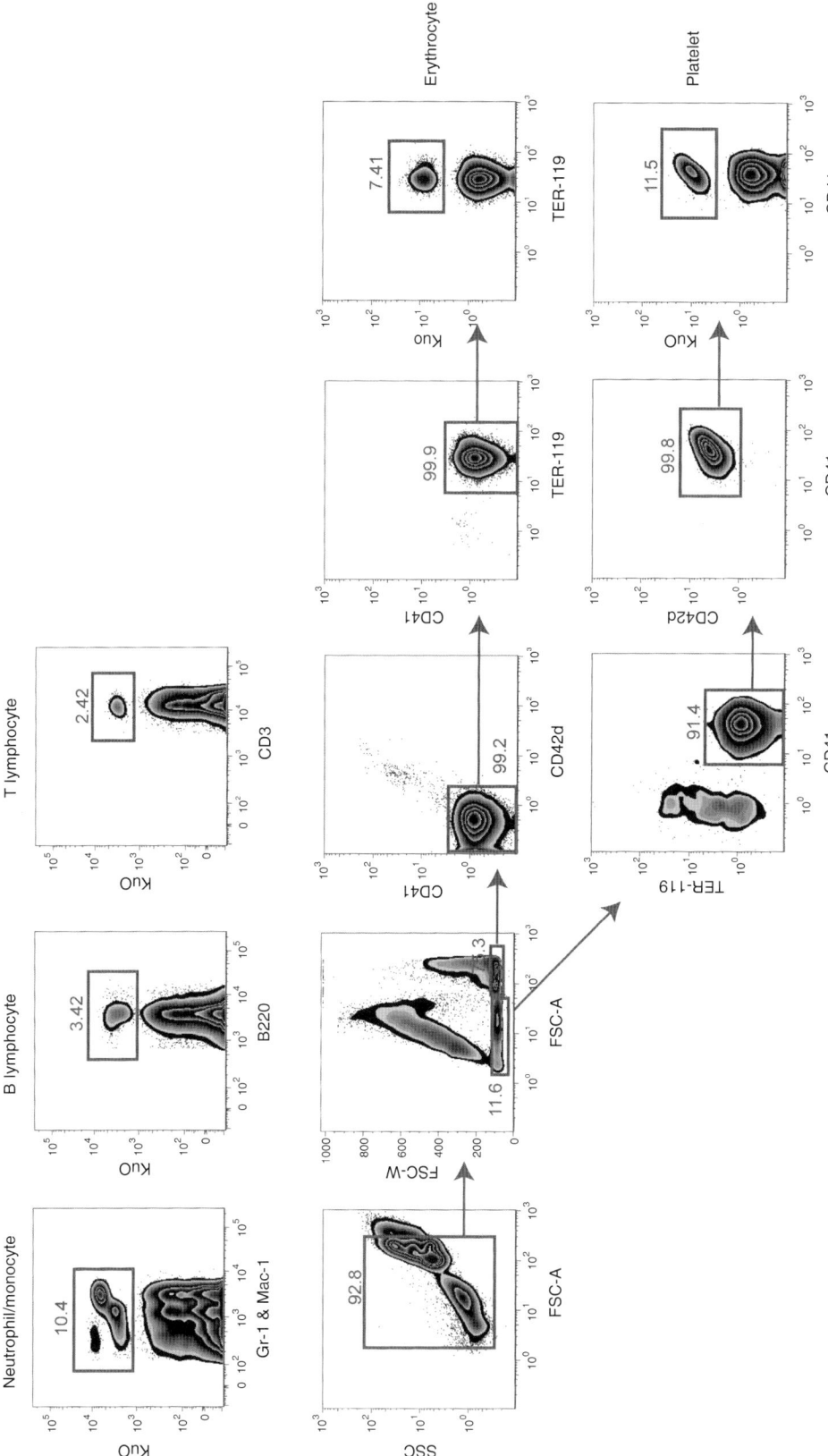

Fig. 2 The percentage of chimerism of neutrophil/monocytes, B cells, T cells, erythrocytes, or platelets was defined as the percentage of KuO[+] cells among CD45.1[+]CD45.2[-]B220[-]CD3[-]Gr-1[+]Mac-1[+] cells, CD45.1[+]CD45.2[-]Gr-1[-]Mac-1[-]CD3[-]B220[+] cells, CD45.1[+]CD45.2[-]Gr-1[-]Mac-1[-]B220[-]CD3[+] cells, CD45.1[-]CD45.2[-]Gr-1[-]Mac-1[-]B220[-]CD3[-] cells, CD41[-] CD42d[-]TER-119[+] cells, or TER-119[-]CD41[+] CD42d[+] cells. The FSC[high] gate and FSC[low] gate were used for analysis of erythrocytes and platelets, respectively

4   Notes

1. When you do not dissect coxal bones, prepare 3 mL syringes with 25 G needles to flush marrow out of femurs and tibias.

2. All antibodies should be titrated before use and should be used at dilutions that brightly stain antigen-positive cells without nonspecifically staining antigen-negative cells.

3. Animal experiments are to be performed in accordance with relevant authorities' guidelines and regulations.

4. If sterility is required, the procedure may be executed in a biological safety cabinet.

5. Be careful not to crush too much and damage cells.

6. Prepare a single-cell suspension by drawing the marrow and staining medium through the needle into the syringe. The resulting cell suspension must be filtered, as it is expelled into the tube, by placing a nylon screen over the mouse of the 15 mL tube.

7. Efficiency of recovery varies depending on the amount of monoclonal antibody and magnetic beads. Therefore, a series of pretests is required to determine efficient recovery rates.

8. Degassed medium is required for efficient recovery of c-Kit-positive cells.

9. Prepare a MACS Column and MACS Separator according to the manufacturer's protocol.

10. The typical recovery rate of c-Kit-positive cells is 2–5 %.

11. A minimum of 90 min is required for good staining of cells with Alexa 700-anti-CD34 Ab. 120-min staining is desired. We have shown that nonspecific binding by anti-CD34 Ab remains low with this protocol.

12. It is important to verify the accuracy of single-cell deposition by the cell sorter used. We adjusted our cell sorter so that single-cell sorting and good recovery are achieved. Positive lines should be determined via fluorescence minus one (FMO) controls.

13. One can centrifuge the plate; however, single cells do not drop at the center of wells. It becomes difficult to verify the presence of cells by microscopy.

14. We recommend the use of 2×10^5 bone marrow cells as competitor cells because these cells contain minimal numbers of HSCs and progenitors to support survival of lethally irradiated mice under most circumstances. However, the larger the number of competitor cells, the lower the sensitivity of detection of single-HSC engraftment.

15. The dose of irradiation needs to be verified empirically. Mice can also be irradiated the day before transplantation.

16. First, push out air from the Myjector. Before all medium is aspirated from a well, repeat aspiration and discharge medium several times while avoiding formation of air bubbles. In practice approximately 95 % of the volume of medium is injectable into mice.

17. Total volume from 200 to 300 µL can be transferred into mice via tail vein. Before injection, mice should be warmed, which makes it easier to inject bone marrow cells into tail veins completely. Additional hypodermic injection of PBS is used for arrest of bleeding.

18. The timing of PB analyses depends on the kinds of transplanted cells (e.g., CD34⁻KSL cells, CD34⁺KSL cells, or more mature cells) and on experimental aims. When transplanting single CD34⁻KSL cells we usually begin analyses at 10–14 days after transplant because platelets appear first, around days 10–12, among granulocytes, B lymphocytes, T lymphocytes, erythrocytes, and platelets. In contrast, when single CD34⁺KSL cells are transplanted, we begin PB analyses at 1 week after transplantation because granulocytes appear around days 6–8.

19. The analysis should extend until at least 16 weeks after transplant. We usually analyze PB up to 24 weeks after transplant. Secondary transplantation should be performed for determination of self-renewal activity. We usually analyze PB from 4 to 20 (24) weeks after transplant every 4 weeks. For secondary transplantation, $0.5–1 \times 10^7$ whole bone marrow cells are transplanted into lethally irradiated mice.

20. Heparinized capillary tubes should not be used since they can reduce platelet numbers.

References

1. Kiel MJ, Yilmaz OH, Iwashita T et al (2005) SLAM family receptors distinguish hematopoietic stem and progenitor cells and reveal endothelial niches for stem cells. Cell 121: 1109–1121

2. Osawa M, Hanada K, Hamada H et al (1996) Long-term lymphohematopoietic reconstitution by a single CD34-low/negative hematopoietic stem cell. Science 273:242–245

3. Spangrude GJ, Aihara Y, Weissman IL et al (1988) The stem cell antigens Sca-1 and Sca-2 subdivide thymic and peripheral T lymphocytes into unique subsets. J Immunol 141: 3697–3707

4. Hamanaka S, Ooehara J, Morita Y et al (2013) Generation of transgenic mouse line expressing Kusabira Orange throughout body, including erythrocytes, by random segregation of provirus method. Biochem Biophys Res Commun 435:586–591

5. Yamamoto R, Morita Y, Ooehara J et al (2013) Clonal analysis unveils self-renewing lineage-restricted progenitors generated directly from hematopoietic stem cells. Cell 154:1112–1126

Chapter 17

Intravital Imaging of Hematopoietic Stem Cells in the Mouse Skull

Juwell W. Wu, Judith M. Runnels, and Charles P. Lin

Abstract

Over the past 50 years, much insight has been gained into the biology of hematopoietic stem cells (HSCs). Much of this information has been gained though isolation of specific bone marrow populations, and transplantation into irradiated recipients followed by characterization of chimeras months later. These studies have yielded insights into the function of HSCs, but have shed little light on the interactions of individual stem cells with their environment. Characterization of the behavior of single HSCs awaited the use of relatively noninvasive intravital microscopy, which allows one to identify rare cells in real time and follow them in multiple imaging sessions. Here we describe techniques used to image transplanted HSCs in the mouse calvarium using hybrid confocal/multi-photon microscopy and second harmonic imaging. For detection, fluorescently tagged HSCs are transplanted into a recipient mouse. The architecture of the bone marrow can be delineated using a combination of fluorescent probes and vascular dyes, second harmonic generation to detect the collagen signal from bone, and transgenic recipient mice containing specific fluorescent support cell populations.

Key words Intravital imaging, Multi-photon imaging, Confocal imaging, Second harmonic generation signal, Bone marrow, Hematopoietic stem cell transplantation

1 Introduction

The study of HSCs in vitro and in vivo has relied heavily on that population's ability to regenerate the hematopoietic compartment. In vitro, HSCs have been verified by their multilineage colony forming ability [1]. Likewise, in vivo, demonstration of multilineage reconstitution following bone marrow (BM) cell transplant identifies the presence of transplanted HSCs [2]. Both of these methods require significant time periods from the initiation of the assays to data collection, and yield no information on early HSC engraftment or HSC interaction with the in vivo environment. Some information regarding HSC location in the BM environment has come from histology [3], and functional studies have pointed to the importance of the interaction of HSCs with other BM cell types, e.g., endothelial cells [4–7], perivascular cells [7–10],

Kevin D. Bunting and Cheng-Kui Qu (eds.), *Hematopoietic Stem Cell Protocols*, Methods in Molecular Biology, vol. 1185, DOI 10.1007/978-1-4939-1133-2_17, © Springer Science+Business Media New York 2014

osteoblasts [11–14], and regulatory T cells [15]. However, real-time in vivo studies have required the introduction of intravital imaging techniques [16–19]. Bioluminescence imaging has demonstrated the regeneration of precursor populations after HSC transplant [20–22], but does not have the sensitivity or resolution to detect single cells. In order to visualize the interactions of single HSCs with its surrounding environment single cell resolution is required. Through the use of in vivo confocal and multi-photon microscopy to image fluorescently tagged BM structures and cells, early events and interactions of transplanted HSCs have been uncovered [16]. The relatively thin, flat mouse skull is accessible as the imaging site with minimal manipulation [23], and hence, perturbation of the bone marrow. We describe here the procedures we use to image HSCs in the mouse calvarial bone marrow.

2   Materials

2.1   Recipient and Donor Mice

Any strain of mice can be used, provided appropriate genetically compatible donors and recipients are used (*see* **Note 1**). Strains differ somewhat in their sensitivities to radiation [24] and ease of intravenous injection post irradiation. We standardly use C57BL/6 (Charles River Laboratories, Wilmington, MA or Jackson Laboratory, Bar Harbor, ME) and strains derived from them. There are many transgenic fluorescent mice that are available in his background (Jackson Laboratory, Bar Harbor, ME).

2.2   Reagents

1. Fluorescently conjugated antibodies for HSC isolation (eBioscience, San Diego, CA; BioLegend, San Diego, CA; BD Biosciences, San Jose, CA. *See* ref. 25 for a list of potential candidates).

2. PBS: Phosphate-buffered saline without calcium and magnesium.

3. Saline: 0.9 % sodium chloride solution (Sigma Aldrich, St Louis, MO).

4. FBS: Fetal bovine serum (Life Technologies Corp., Carlsbad, CA).

5. BSA: Bovine serum albumin (Sigma Aldrich, St Louis, MO).

6. PBS–BSA: Phosphate-buffered saline without calcium and magnesium, 0.1 % BSA.

7. PBS–FBS: Phosphate-buffered saline without calcium and magnesium, 2 % FBS.

8. Lipophilic carbocyanine membrane dyes (optional, depending upon experiment): Vybrant DiO/DiI/DiD/DiR (Life Technologies Corp., Carlsbad, CA).

9. Fluorescent labels for in vivo labeling (optional, depending upon experiment):

- Vascular: AngioSense (Perkin Elmer, Waltham, MA), Rhodamine B–or Tetramethylrhodamine–dextran 70k-2E6 MW (Life Technologies Corp., Carlsbad, CA; Sigma Aldrich, St Louis, MO), or Qtracker vascular labels (Life Technologies Corp., Carlsbad, CA).

- Cell nuclei: Hoechst 33342 (Life Technologies Corp., Carlsbad, CA).

- New bone (hydroxylapatite): OsteoSense (Perkin Elmer, Waltham, MA).

- Cell markers, adhesion molecules, chemokines: appropriate antibodies, fluorescent probes for antibody conjugation (Cy3, Cy5, Cy5.5, Cy7 (GE Healthcare, Piscataway, NJ); Alexa Fluor 488, Alexa Fluor 647 (Life Technologies Corp., Carlsbad, CA)), antibody conjugation and clean-up kits (Thermo-Scientific, Rockford, IL).

10. Anesthesia, such as Isoflurane (Forane; Baxter, Deerfield, IL) or Ketamine–Xylazine mixture (Ketamine Hydrochloride Injection, Bioniche Pharma USA LLC, Lake Forest, IL; Xylazine: AnaSed Injection, Lloyd Laboratories, Shenandoah, IA).

11. Methocel: 2 % solution (Methocel MC, Sigma Aldrich, St. Louis, MO) in PBS or Saline.

2.3 Supplies and Equipment

1. Gamma Irradiator: In this work, we have used a Gammacell 40 Exactor (MDS Nordion, Ottawa, ON, Canada).

2. Fluorescence Activated Cell Sorter such as the BD FACS Aria series (BD Biosciences, San Jose, CA).

3. Dissection tools: forceps, scissors, suture-holding hemostat (World Precision Instruments, Saratosa, FL).

4. Water bath or cell incubator (if using lipophilic carbocyanine membrane dyes for labeling HSCs).

5. Centrifuge and/or microcentrifuge (depending upon need to label with membrane dye and volume of cells recovered).

6. General lab supplies: sterile centrifuge and microcentrifuge tubes; pipettes and micropipettes; foam tip applicators (Catalog# 15-960-3J, Fisher Brand, Waltham, MA); alcohol wipes; Kimwipes.

7. Disposable syringes (1 ml) and needles (30 G, ½ in.) (Becton Dickinson, Franklin Park, NJ), or insulin syringes with permanently attached needle (29 G, ½ in.; EXEL, Los Angeles, CA).

8. Gas anesthesia vaporizer (if using gas anesthesia; SurgiVet, Dublin, OH).

9. Home-built or commercial confocal/multi-photon scanning microscope, customized for live mouse imaging. Appropriate accessories for the microscope, such as glass slides and No. 1 coverslips (Fisher Brand, Waltham, MA).

10. Nylon black monofilament suture, 6-0, PC-1 needle (Ethilon 1956G, Ethicon, L.L.C., San Lorenzo, Puerto Rico).

11. Bacitracin + neomycin + polymyxin B triple antibiotic topical ointment (for multiple imaging sessions; CVS Corporation, Woonsocket, RI).

12. Acetaminophen (for multiple imaging sessions; Tylenol or Children's Tylenol, McNeil Consumer Healthcare, Fort Washington, PA).

3 Methods

3.1 Bone Marrow Transplant

3.1.1 Preparation of BM Recipients

Because the HSC niche is normally occupied in healthy animals under homeostatic conditions, the recipient animals are irradiated with 6.5 Gy (650 rads), a sublethal dose, or 9.5 Gy (950 rads), a lethal irradiation dose, approximately 24 h before transplant using a Cs-137 source in order to facilitate engraftment of donor HSCs.

3.1.2 Preparation of the Donor HSCs

Standard protocols may be followed to isolate either the Lin⁻cKit⁺ Sca-1⁺ (LKS) population or long-term stem cells (LT-HSC) [26] from the donor animal. One LT-HSC can be found in approximately 100,000 whole bone marrow (BM) cells, while LKS cells are present at approximately ten times that frequency. In order to image the cells in vivo, however, they need to be fluorescently tagged. This can be accomplished by starting with donors that carry fluorescent proteins (FP) in their stem cells: for example, universal GFP or DsRed mice (expressed under the control of ubiquitin or actin promoters), available from Jackson Laboratory, Bar Harbor, ME (*see* **Note 2**). Using mice carrying FP as donors has the advantage of allowing one to follow repopulation of the niche and measure chimerism over time as all the progeny of the FP-HSC will also be marked.

Alternatively, if follow-up of donor HSC behavior is short-term (within a week), nonfluorescent mice can be used as donors. Following HSC purification, the HSCs can then be labeled using lipophilic carbocyanine membrane dyes. Vybrant DiO, DiI, DiD and DiR (Life Technologies Corp., Carlsbad, CA) are recommended for this purpose (*see* **Note 3**). Labeling the HSC population post FACS isolation has the advantage in that the fluorescence of the membrane dye does not have to be considered in designing the HSC isolation strategy. These dyes do not appear to interfere with homing or other cellular processes and can be detected on dividing cells in vivo for 5–7 cell divisions [15–17, 27].

The following protocol is based on DiD labeling. Other Vybrant dyes behave similarly.

1. In order to label isolated HSCs, they need to be harvested from the FACS collection media by centrifugation at $500 \times g$ for 5 min (*see* **Note 4**).

2. Resuspend the HSCs at a concentration of 1×10^6 cells/ml in PBS–BSA. For ease of working, the cells can be transferred to a microcentrifuge tube at this point.

3. Add the membrane dye at a final concentration of 5 μM. Mix well, but do not vortex.

4. Place the tube with cells and dye in a pre-warmed 37 °C bath, and incubate for 20 min.

5. After the incubation period, add at least 1 ml of PBS–FBS (*see* **Note 5**). Mix well (FBS concentration should be at least 2 %.).

6. Wash the cells twice with 1 ml of PBS–FBS, centrifuging at $500 \times g$ for 5 min.

7. Resuspend the cells in PBS–FBS at the proper concentration for injection (*see* **Note 6**).

Recipient mice can be injected with donor bone marrow cells either through the tail vein or retro-orbitally. Tail veins can be injected with the help of a commercially available restrainer, or by placing the mouse under a beaker under a weight [28] to restrict the mouse's movement (*see* **Note 7**).

3.2 In Vivo Imaging   HSCs reside in the bone marrow, which is found in both long bones and flat bones. We choose the calvarium (skull) bone marrow for in vivo imaging as its cortical bone layer is sufficiently thin to allow for light penetration without the need for bone thinning [23], which can easily disrupt the intricate microarchitecture of the bone marrow environment. It has been reported that calvarial HSCs are comparable in frequencies, immunophenotype and repopulating ability compared to long bones [16, 29].

Our lab uses video rate laser scanning hybrid confocal/two-photon microscopes specifically designed for intravital bone marrow imaging in live mice [15, 16, 19, 30, 31]. Polygon-based scanning allows multi-channel (RGB) image acquisition at video rate (30 frame per second). The microscopes are equipped with multiple diode laser light sources (491/532/561/633 nm) for one-photon, confocal imaging and a Ti:Sapphire femtosecond pulse laser (700–1,000 nm; Mai Tai, Spectra Physics, Santa Clara, CA) for two photon excitation; detection wavelengths for each optical channel are fully customizable. For calvarial imaging, we use 30–60× water immersion objective lenses (0.9–1.0 NA); at 30× magnification, each field of view measures ~330 × 330 μm. The microscopes are also equipped with a XYZ motorized stage control at micrometer-precision (MP385, Sutter Instrument, Novato, CA), a PID temperature-feedback heating system (TCAT-2AC controller, Physitemp, Clifton, NJ) and a custom-made holder and mount for mouse placement. Isoflurane anesthesia is delivered from a stand-alone system to the animal via a fitted nose cone.

In vivo mouse calvarial imaging can also be performed with commercial microscopes [27, 32–34]. As microscopes vary in setup and imaging software, we will focus on the approach and potential pitfalls of the imaging process in subsequent sections. There are no restrictions on the choice of imaging systems as long as animal wellbeing is considered.

3.2.1 Microscope Requirements for In Vivo Imaging

Considerations for the Mouse

Microscopes for in vivo bone marrow imaging should, first of all, include a stage and holder that secures the animal in a steady but comfortable position. Even under anesthesia, mice movement can be significant during imaging from breathing and heartbeats. The microscope should also have a means of keeping the mouse warm (34 °C surface body temperature). All parts accessible to the animal should be easy to dissemble for cleaning and disinfection.

Considerations for Optimal Fluorescence Detection

In vivo images of HSCs are rarely meaningful without including reference structures from the bone marrow space. These images require multichannel setup during imaging, which captures information from multiple fluorescent probes that each labels a different structure. It is important to plan ahead and make sure that the imaging system is capable of observing and recording these signals prior to HSC and animal preparation.

In the bone marrow cavities, the structures that have been to our interest in our HSC studies include the surrounding bone (collagen), blood vessels, osteoblasts, and hematopoietic and stromal cells in the cavities. Fluorescent labeling of these structures should be specific and exclusive. Additional guidelines for probe selection are as follows:

- Endogenous labels, if available, are preferred.
- Overlaps between the fluorescence spectra of participating probes should be kept to minimum (*see* **Note 8**).
- Literature on in vivo imaging offers the best guidance on in vivo probe choices. Fluorescent probes designed for in vitro use may not be biocompatible or require higher dosages for in vivo use. Certain probes may be cleared too quickly from the animals for effective labeling.
- Fluorescent probes used on HSCs (such as for isolation) prior to imaging should be checked for in vivo signal strength before proper experiments.

We have been able to image the bone by the endogenous second harmonic generation (SHG) signal from bone collagen. HSCs,

Table 1
Fluorescent probes suitable for in vivo bone marrow imaging

Target	Fluorescent probe	Notes
Vasculature (*see* **Note 9**)	Rhodamine-Dextran, 70k-2E6 MW Qtracker Vascular Labels Angiosense	+ Low cost – Relatively fast leakage from vasculature + Narrow fluorescence spectra with large Stokes shift – Relatively fast clearance (2 hrs) + Little vascular permeability, slow clearance (>24 h)
Cells	Hoechst 33342	Labels most non-HSC cell nuclei (*see* **Note 10**)
Bone (hydroxylapatite)	OsteoSense	On newly formed bone only
Osteoblasts	Col2.3-eGFP	Fluorescent reporter mouse [35]
Cell markers, adhesion molecules, and chemokines (e.g., CD31, CD62E, SDF-1)	Antibodies conjugated with: • Cy3, Cy5, Cy5.5, Cy7 • Alexa Fluor 488, Alexa Fluor 647	+ Versatile; can identify specific cell populations, such as endothelial and stromal cells – 10^1 µg antibodies/mouse needed for in vivo signal; antigen may still fail detection due to low expression

as discussed above, are tagged with lipophilic carbocyanine dyes and fluorescent proteins. Fluorescent probes we have used successfully for intravital labeling are listed in the Table 1 by target. Spectra of many probes in this list overlap significantly, especially towards the red and longer wavelengths. Examples include Vybrant DiI/Cy3, Vybrant DiD/Angiosense680/Osteosense680/Cy5/Cy5.5, Vybrant DiR/Cy7. This is often due to similar fluorophore structures (cyanines). In some cases, probes with similar fluorescence emission wavelengths can be distinguished by their excitation spectra, particularly in two-photon imaging mode; in such cases, the probes can be used simultaneously by choosing appropriate laser wavelengths. Minor overlaps, on the other hand, do not preclude the use of a probe pair as long as the "tail" of the leaked fluorescent signal remains below the detection limit of the affected channel.

Sample setups we have used for in vivo HSC imaging are shown in Table 2. An RGB image derived using these setups would include information on HSC and two other structures. Additional fluorescently tagged structures may be included and recorded into a separate RGB image. For example, "Setup B" can include vascular imaging with Rhodamine B–Dextran (561 nm excitation, 573–613 nm detection). "Setup C" uses exclusively two-photon imaging mode with a single excitation wavelength. Compared to confocal imaging, two-photon imaging offers improved optical sectioning and is especially suited for observing the microarchitecture of the bone marrow environment.

Table 2
Sample optical channel setups for in vivo HSC imaging

	Setup A	Setup B	Setup C
HSC	eGFP 491 nm excitation 509–547 nm detection	Vybrant DiD 638 nm excitation 667–722 nm detection	–
Bone	SHG 840 nm excitation 415–455 nm detection	SHG 840 nm excitation 415–455 nm detection	SHG 920 nm excitation 420–500 nm detection
Blood vessels	Rhodamine B–Dextran 561 nm excitation 573–613 nm detection	–	Qdot655 920 nm excitation 625–672 nm detection
Osteoblasts	–	Col2.3-eGFP 491 nm excitation 509–547 nm detection	Col2.3-eGFP 920 nm excitation 505–590 nm detection

Autofluorescence

Autofluorescence from bone marrow cells may cause "false-positives" when identifying fluorescently labeled HSCs in vivo. Its excitation and emission spectra are wide, spanning the visible wavelengths—strongest in blue and green—and becoming negligible only in the near infrared (>700 nm). Intensity varies from cell to cell. In response to this issue, we often include an "autofluorescence channel" during in vivo imaging, especially if the HSCs fluoresce <600 nm (Vybrant DiI, eGFP, DsRed). This channel should detect in the autofluorescence spectral range, without signal contribution from any fluorescently labeled bone marrow cells (HSCs and others). In "Setup A" (Table 2), for example, the autofluorescence channel can share the same detection setup as the blood vessel channel, but with 492 nm excitation turned off to make certain no residual eGFP fluorescence from the HSCs may leak into that channel. If the suspect HSC remains bright compared to neighboring cells without 492 nm excitation, it is considered a non-HSC with bright autofluorescence, otherwise it is considered an HSC.

3.2.2 Preparing the Mouse for Imaging

1. The remaining steps will be done while the animal is under anesthesia. Anesthesia can be accomplished using gas [isoflurane (4 % for induction, 2.5 % maintenance) mixed with oxygen in a vaporizer] and delivered through a nose cone, or though an intraperitoneal injection of 80 mg/kg ketamine + 12 mg/kg xylazine cocktail.

2. Additional fluorescent probes for intravital bone marrow labeling (Table 1) are injected either through the tail vein or the retro-orbital plexus (*see* **Note 6**).

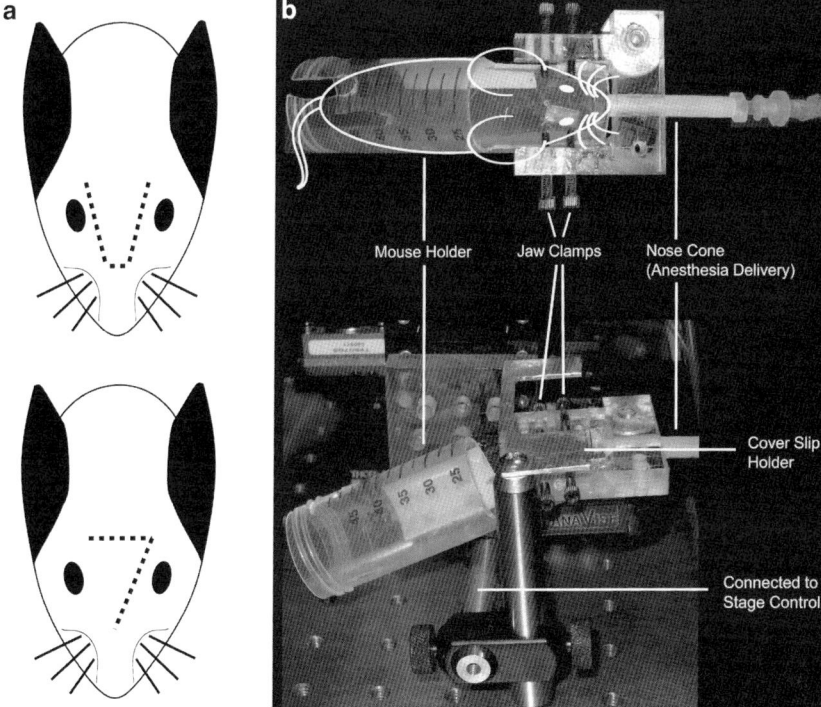

Fig. 1 (**a**) Scalp incision methods to expose calvarium for in vivo imaging. (**b**) Mounting the mouse for imaging on our in vivo bone marrow microscope. *Top view:* the body of the mouse fits inside a holder machined from a centrifuge tube, which is wrapped by heating tape (not shown) during imaging. Mouse position shown in *white outline*. Nose cone delivers anesthesia to the animal and two sets of jaw clamps hold its head in place. *Bottom view:* the animal holder is secured at an angle such that the calvarium lies flat for imaging. A No. 1 coverslip is attached to a holder and placed on the skull

3. In order to prepare the skull for imaging an incision needs to be made in the scalp. First, remove the hair over the area to be imaged (*see* **Note 11**). This can be done using a depilatory (*see* **Note 12**), razor blade, or small surgical scissors. When using a flexible razor blade, start scraping from the top of the head and move toward the nose, being careful not to cut the skin. The cut hair can be removed by wiping the head with a wet paper towel or alcohol wipe, making sure to avoid the eyes.

4. Once the hair has been removed a small flap of skin will need to be cut to expose the bone. There are a few ways in which the skin can be cut, but generally this is done so that upon suturing the scalp, minimal scarring occurs on the central region of the skull. Figure 1a demonstrates representative incisions. The easiest way to begin cutting the scalp is to pick up the skin between the ears with forceps and cut a small hole using surgical scissors. One side of the scissors can then be inserted under the skin to cut in the desired direction. The periosteum, or membrane over

the bone, is often attached to the skin so that the flap cannot be fully pulled away to expose the underside of the skin. The attachment should be cut with small scissors to allow the flap to be pulled away. The scalp should be washed with saline and wiped with foam tip applicators (*see* **Note 13**). Continue wiping the skull with fresh applicators as necessary until all hair and as much blood, as possible if present, are removed.

3.2.3 Mounting the Mouse

The animal holder for our in vivo microscopes is easily made. We use a disposable 50 ml centrifuge tube with machined, beveled openings for head and tail placement. The holder is then clamped at an angle such that the mouse head, resting on the holder's ledge, is near level and the calvarium is flat for imaging. Heating tape (connected to heating system) is wrapped around the tube to hold its temperature at 34 °C for the duration of imaging (*see* **Note 14**). After the mouse is placed in the heated tube with an exposed calvarium, the skin flap is pulled away from the skull and a drop of Methocel 2 % solution is placed on the exposed bone. Methocel, an ophthalmic lubricant with refractive index matched to water, keeps the exposed tissue hydrated and also helps keep the skin flap from flipping to its original position. Air bubbles are removed with an applicator. A coverslip holder rests a No. 1 coverslip on the mouse skull and further stabilizes the calvarium position with additional clamps, in metal or plastic, that hold snug at the jaw (Fig. 1b). A drop of water is then placed on the coverslip for use with water immersion objectives.

3.2.4 Imaging

Calvarial bone marrow cavities mostly reside in the frontal bones. The posterior cavities (closer to the parietal bones) are larger, flatter and therefore preferred for imaging; they are distributed parasagittally along the sagittal suture, along which the central vein and artery are also located. These cavities have a size range of 10^2 μm; some are interconnected at lower depths. Osteoblasts—large, flat (~10 μm thick), and irregularly shaped bone forming cells—line the bone's inner cavity surface (endosteum) in a discrete fashion. The cavities are highly vascularized. Blood vessels vary with size and architecture, from the mesh of capillary-like microvessels that line the endosteum to the larger sinusoids [16]. Hematopoietic and stromal cells pack the spaces unoccupied by the vasculature. Yellow marrow (fat cells) is absent as the calvarium is a flat bone.

Use of Bone Signal and Landmarks to Find the BM

A prominent landmark of the frontal bone marrow space, which also serves as a helpful orientation guide while imaging, is the intersection between the coronal and sagittal sutures (Fig. 2a). The coronal suture has a characteristic curvature and striations in the bone's endogenous second harmonic generation signal. This landmark

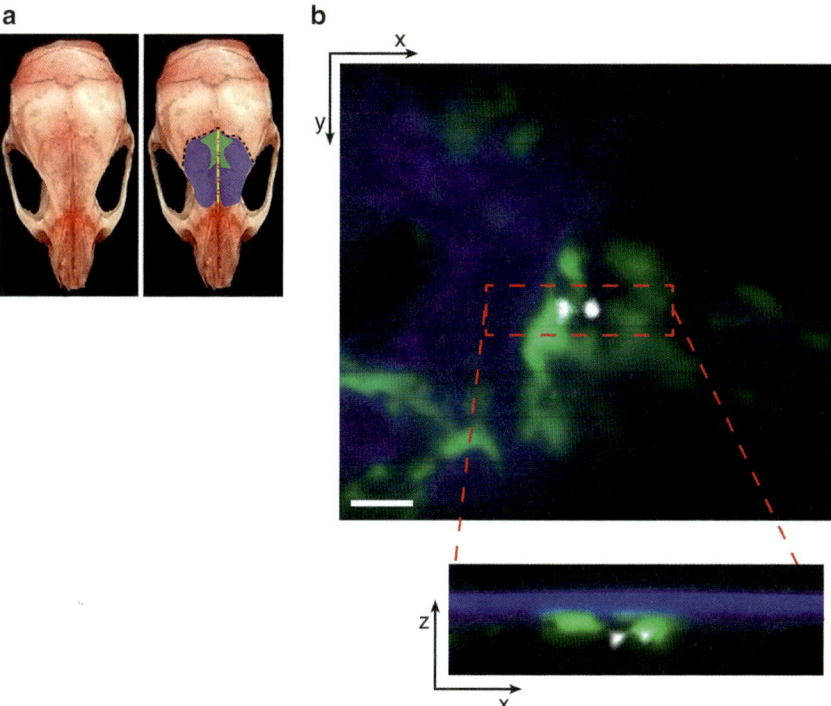

Fig. 2 (**a**) Bone marrow cavities and their landmark. *Black dotted line*: coronal suture. *Yellow dotted line*: sagittal suture. *Purple area*: parietal bones. *Green area*: location of imaging-friendly cavities. (**b**) In vivo image of an HSC, 24 h after systemic infusion into an irradiated mouse. *Blue*: second harmonic generation signal from bone collagen. *Green*: Col2.3-eGFP osteoblasts. *White*: DiD-labeled HSCs. The $X–Y$ plane is the imaging plane; the $X–Z$ (sagittal) plane shows the calvarial cortical bone covering the bone marrow cavity. Scale bar = 100 μm

may also be observed in the reflectance channel, if available on the microscope.

The search for the landmark and the bone marrow cavities can be made more efficient by adjusting the stage or mouse such that the approximate location of the landmark, i.e., the suture intersection, which is often discernable by careful inspection with naked eye, is placed close to the incident light path.

1. After the mouse is mounted and the coverslip secured on the skull, carefully move the stage such that the objective almost touches the coverslip (*see* **Note 15**).

2. Turn on the excitation source for the bone. Slowly move the stage away from the coverslip (*Z*-axis) until the bone surface signal comes to focus in the imaging software. Then scan the objective across the surface (*X–Y* axes plane).

3. Note presence of the sagittal suture that runs along the length of the calvarium. Once the sagittal suture is located, tracing its path will inevitably arrive at the coronal-sagittal suture intersection.

4. If available from the stage control, record the positional coordinates of the landmark. It allows easy return to the spot later if "lost" while navigating the bone marrow space.

5. Scanning from the landmark, bone marrow cavities are easily located along the sagittal suture, towards the nose.

Locating and Imaging the HSCs

HSCs are present in low numbers in vivo. Once the bone marrow cavities are located, the next task is to search for and image these cells and other bone marrow structures of interest (Fig. 2b). Exact protocols for imaging depend on intended outcome; however, the usual approach is as follows:

1. Switch on all excitation sources and detection channels. Start with a quick scan of the whole bone marrow area, count the number of HSCs and observe the general environment. Note the quality of fluorescent labeling and the maximum depth each fluorescence detection channel can reach. The minimum of these depth measurements estimates the percent of the bone marrow volume that is covered by imaging (~150 μm imaging depth into the cavities translates to a volumetric coverage of 40–60 %). Adjust power of the excitation source and gain of the detection channels such that it is optimal for all features. This step also allows time for the mouse to settle into final position (*see* **Note 16**).

2. Imaging proceeds in descending priority of what needs to be imaged. While with motorized, precision stage control it is theoretically possible to set up automated mapping with imaging "Z-stacks" that capture the whole bone marrow space, the "get all, sieve later" strategy tends to generate large amounts of unnecessary data and is taxing for both the animal and the imaging system.

3. The autofluorescence channel is used to distinguish fluorescently labeled HSCs and strongly autofluorescent cells prior to recording images of each HSC. No background correction is performed.

4. Images are assumed to *not* co-localize perfectly even if recorded in succession. If more than one image is needed for capture the HSCs and other relevant bone marrow structures, at least one channel's information is repeated between images so that any shift in position or tilt in the animal (due to breathing and heartbeats) can later be identified.

5. Z-Spacings of ≤5 μm are recommended when acquiring image "Z-stacks" of the calvarial bone marrow. Larger spacings (3–5 μm) are usually sufficient for capturing the locations of HSCs. Spacings between 1 and 3 μm, acquired in two photon imaging mode, can capture the fine details of the bone marrow microarchitecture, such as the microvessels at the endosteum–marrow interface.

Limits on Imaging

The imaging approach described above is due to the time constraint of in vivo imaging sessions, consequent of animal welfare concerns and gradual loss of fluorescence signals. While animals can remain continuously anesthetized, they will, over time, suffer from dehydration. For this reason, bone marrow imaging sessions should be limited to no more than 4 h from the start of anesthesia induction. Gradual loss of fluorescence signal is due to the inevitable clearance of the fluorescent probes, which are foreign materials, from the animals' circulation over time. The effect can be direct (vascular probes) or indirect, the latter by exhausting the supply of probes for labeling (fluorescently tagged antibodies). Another route to fluorescence signal loss is by photobleaching of the fluorophore. The length of retention and photostability of the probe are both specific to the probes' chemistry. Two-photon imaging reduces out-of-focus photobleaching (*see* **Note 17**).

3.2.5 Multiple Imaging Sessions (Treatment of the Animal)

The mouse can be imaged a few times within a week before significant scarring occurs. In order to image multiple times the wound will have to be sutured after each imaging session. Before suturing, wash the skull and skin flap with saline (*see* **Note 13**). Wipe the saline away with a foam-tipped applicator. Apply a "dollop" of triple antibiotic to the underside of the skin flap, fold the skin back to its proper position, and begin suturing. A number of videos describing suturing techniques are available on the internet [36, 37]. Acetaminophen (4.48 mg/ml in drinking water or 7 ml children's Tylenol in 43 ml water) can be placed in the cage water as an analgesic between imaging sessions. Upon imaging again, cut the sutures and reopen the skin, following Subheadings 3.2.2 through 3.2.4.

3.3 Image Processing

3.3.1 Looking at HSCs and the Bone Marrow in Different Anatomical Planes

Image "Z-stacks" are serial X–Y planes captured as a function of depth (Z) from the calvarial surface. In anatomical terms, X–Y plane images approximate traverse plane anatomical slices, off-axis by the tilt of the calvarium during imaging. Views of the orthogonal coronal (Y–Z) and sagittal (X–Z) planes may convey more information regarding the positional relationship of HSCs and relevant bone marrow structures.

Due to the time constraints of in vivo imaging, we generate these views after the imaging session, using the open source image processing software ImageJ or its derivative, FIJI, which is bundled with plugins [38]. Image "Z-stacks" can be opened as a single .tif file or imported as an image sequence, after which ImageJ should present the data visually as a "Stack". Commands described in the following steps are based on FIJI (ImageJA v.1.45b).

1. If ImageJ loads the data as separate images, run the command *Image>Stacks>Images to Stack* to create a new stack with opened images.

2. If dimensional information is not included in the imaging files, input the X, Y pixel size of the image "Z-stack" under

Image>Properties…, as "Pixel Width" and "Pixel Height." Input the Z spacing between consecutive *X–Y* planes as "Voxel Depth".

3. Run the command *Image>Stacks>Reslice[/]…*. to convert the image "Z-stack" to coronal or sagittal plane view. For coronal view, reslicing should start at *Top*, sagittal view, at *Left*. If the Z-spacing is different from the *X–Y* pixel dimensions, *Avoid Interpolation* should be unchecked to generate the view in accurate aspect ratios.

4. The new view, also an image stack, can be saved as a single .tif image or an image sequence. It (as well as the original image "Z stack") can also be exported as an .avi movie at a user-specified frame rate; the resolution, however, may not match the original image data's.

3.3.2 Looking at HSCs and the Bone Marrow in 3D: A Cautionary Note

The 3D Viewer plugin in ImageJ and FIJI allows conversion of an image "Z-stack" into 3D volumetric representation (*Plugins>3D Viewer, Display as Volume*). Automatic intensity thresholding involved in this conversion often results in the loss of smaller and/or dimmer features in the resulting 3D representation. We recommend the user perform the binarization step manually, channel by channel, to ensure fidelity of image features and input the binarized image "Z-stack" for 3D conversion.

4 Notes

1. Use of genetically incompatible donors and recipients will result in graft versus host disease when the immune system is reconstituted. Likewise, use of female bone marrow donors in male recipients, even in the syngeneic situation, may result in graft versus host disease, owing to recognition of Y chromosome proteins as foreign "other" proteins by the donor female immune cells. This situation can confuse experimental results as the mice die due to conditions other than those being experimentally tested.

2. Special care needs to be given to the choice of antibody conjugates for FACS when using transgenic mice carrying fluorescent proteins (FP) as the FP channel(s) will not be available for antibody detection. As always, a single color control for the FP (unlabeled FP BM cells) should be run on the FACS for compensation purposes. A good antibody color scheme that works well with either DsRed or eGFP is c-kit-APC, Sca-1-PE-Cy7, Lin positive-Pacific Orange, viability dye-DAPI.

3. We do not recommend the PKH family of lipophilic dyes, which require a special iso-osmotic medium Diluent C for cell labeling. While it has been reported that transfer of lipophilic carbocyanine membrane dyes (including Vybrant DiO, DiI,

DiD, and DiR) from the labeled cell to neighboring cells occurs in vivo [39], we have not found this phenomenon problematic in our intravital imaging experience. Membrane dye transfer has been correlated with the viability of the labeled cell [40] and points to the importance of taking care to eliminate unbound dye before injection and to handle cells carefully in order to maintain viability. We also caution against the use of amine-reactive CFSE for intravital cell tracking experiments of HSCs and other cell types. The coupling of CFSE to cell surface proteins may affect the cells' motility and ability to home.

4. In order for lipophilic carbocyanine dyes to insert well into cell membranes, there should be few other lipids around. Since HSC isolation procedures are generally done in 2 % FBS, after FACS collection there is not a requirement for extensive washing as there will be little FBS remaining. It is important to limit the number of centrifugations as well to limit cell loss.

5. This step is necessary to bind and sequester the unbound dye molecules, and to prevent the cells from clumping upon centrifugation. If cells clump after lipophilic carbocyanine dye labeling, it is virtually impossible to resuspend them.

6. Total combined volume injected into a mouse (from HSC suspension and vascular probe) should not exceed 10 % of the total blood volume, 1.5–2.0 ml, depending upon the size of the animal. If a probe to be injected is solubilized in water, add, at 10 % probe volume, 10× concentrated saline or PBS immediately before injection. Beware that when using a disposable syringe and needle system, a non-injectable "dead volume" (~150 μl) collects at the syringe/needle attachment site. Insulin syringes do not have this "dead volume"; however, their larger needle size renders injections more challenging to perform.

7. Post irradiation, tail vein injections are often more difficult than usual because the veins may collapse. In pigmented mice, visualizing the tail vein can be made easier by using a Seymour Light (Syris Scientific, Gray, ME), which is fitted with polarizing lens. Alternatively, the HSCs can be delivered through retro-orbital injection.

8. Variable autofluorescence among bone marrow cells and wavelength dependence of tissue optics makes fallible an attempt to resolve overlapping fluorescence via ratiometric correction (compensation). The full fluorescence excitation and emission spectra (not just the peak wavelengths) should be considered when electing optical channels and filters for imaging. One-photon spectra of many fluorescent probes are available for viewing online [41, 42]. Chudakov et al. [43] provide a good reference list for the optical properties of commonly used fluorescent proteins. Two-photon fluorescence spectra are usually equivalent to one photon's, but the excitation spectra can deviate

Table 3
Guide to dosages for fluorescent probes (*see* Note 18)

Target	Fluorescent probe	Dosage
Vasculature	Rhodamine–Dextran, 70k-2E6 MW	2.5–5.0 mg/kg mouse weight (0.05–0.1 mg/mouse) i.v. Immediately before imaging
	Qtracker Vascular Labels	3.5 nmol/kg mouse weight (70 pmol/mouse) i.v. Immediately before imaging
	Angiosense	0.1 μmol/kg mouse weight (2 nmol/mouse) i.v. Immediately before imaging
Cells	Hoechst 33342	10–30 mg/kg mouse weight (0.2–0.6 mg/mouse) i.v. Immediately before imaging
Bone	OsteoSense	0.2 μmol/kg mouse weight (4 nmol/mouse) i.v. 24 h before imaging
Cell markers, adhesion molecules and chemokines	Antibodies conjugated with: • Cy3, Cy5, Cy5.5, Cy7 • Alexa Fluor 488, Alexa Fluor 647	0.5–1.5 mg/kg mouse weight (10–30 μg/mouse) i.v. 24 h before imaging

significantly from the twofold approximation of wavelength values [44–46].

9. High molecular weight is needed for dextran-based vascular dyes as much of the bone marrow vasculature is sinusoidal and fenestrated and more leaky than other vessels throughout the body [47]. Angiosense and Qtracker vascular probes are specifically designed to remain in the circulation for hours in vivo.

10. The Hoechst exclusion "side population" assay was, in fact, an early method of HSC enrichment [48].

11. In addition to harboring bacteria and therefore capable of causing infection at the wound site, hair is highly autofluorescent under the microscope. For these reasons, hair should be eliminated as much as possible.

12. If using a depilatory, the procedure should be done 24 h before imaging as the cream will cause irritation to the skin.

13. Before washing the skull and scalp, make sure that the animal is on a piece of absorbent material that will not allow the saline to bead on the surface. This precaution will prevent the possibility of drowning the animal, which cannot breathe in any liquid. Foam-tipped applicators are preferable to cotton-tipped applicators as the cotton leaves fibers on the skull that will be detected upon imaging. Foam is also a good material for removal of stray cut hairs.

14. Due to its small size and compromised ability to self-regulate body temperature while anesthetized, the animal must be kept warm during the imaging session and while it recovers from anesthesia. Otherwise, it can expire due to *hypo*thermia. However, to avoid *hyper*thermia, the temperature control on the heater should be set at the temperature of the surface of the skin, about 34 °C.

15. The coverslip should rest snugly against the skull. When imaging smaller mice, a rolled Kimwipe can help steady the mouse in its holder; in our setup, for example, we push it into the animal holder below the mouse tail.

16. It is common for the mouse to sag slightly in the first 5–10 min of imaging as gravity pulls the body to the lowest position permitted by the animal holder. However, sudden, renewed restlessness in the animal and/or inability to remain in sufficient anesthesia depth may be due to uncomfortable mouse mounting, such as poor positioning of the neck and head. Nose wetting by Methocel, which disrupts proper breathing, may also be the cause. Prompt attention should be given to the animal. Depth of anesthesia should be checked frequently.

17. Photobleaching can be quick for some fluorescent probes, but the loss of signal remains a gradual process. A *sudden* loss of fluorescent signal is often due to drying of the water or Methocel above and below the coverslip.

18. We have used the dosages described in Table 3 for in vivo bone marrow imaging. They serve as starting guides; optimal dosages will depend on the imaging setup.

Acknowledgments

The authors wish to thank Raphaël Turcotte for the mouse skull shown in Fig. 2a and Drs. Luke Mortensen and Amir Schajnovitz for the image displayed in Fig. 2b. This work was funded by NIH HL097748, HL97794, and HL100402.

References

1. Sutherland HJ, Lansdorp PM, Henkelman DH, Eaves AC, Eaves CJ (1990) Functional characterization of individual human hematopoietic stem cells cultured at limiting dilution on supportive marrow stromal layers. Proc Natl Acad Sci U S A 87:3584–3588

2. Till JE, McCulloch EA (1961) A direct measurement of the radiation sensitivity of normal mouse bone marrow cells. Radiat Res 14:213–222

3. Nilsson SK, Johnston HM, Covedale JA (2001) Spatial localization of transplanted hematopoietic stem cells: inferences for the localizarion of stem cell niches. Blood 97:2293–2299

4. Kiel MJ, Yilmez OH, Iwahita T, Yilmaz OH, Terhorst C, Morrison SJ (2005) SLAM family receptors distinguish hematopoietic stem and progenitor cells and reveal endothelial niches for stem cells. Cell 121:1109–1121

5. Heissig B, Hattori K, Dias S, Friedrich M, Ferris B, Hackett NR, Crystal RG, Besmer P, Lyden D, Moore MA, Werb Z, Rafii S (2002) Recruitment of stem and progenitor cells from the bone marrow niche requires MMP-9 mediated release of kit-ligand. Cell 109:625–637

6. Avecilla ST, Hattori K, Heissig B, Tejada R, Liao F, Shido K, Jin DK, Dias S, Zhang F, Hartman TE, Hackett NR, Crystal RG, Witte L, Hicklin DJ, Bohlen P, Eaton D, Lyden D, de Sauvage F, Rafii S (2004) Chemokine-mediated interaction of hematopoietic progenitors with the bone marrow vascular niche is required for thrombopoiesis. Nat Med 10:64–71

7. Ding L, Saunders TL, Enikolopov G, Morrison SJ (2012) Endothelial and perivascular cells maintain haematopoietic stem cells. Nature 481:457–462

8. Sugiyama T, Kohara H, Noda M, Nagasawa T (2006) Maintenance of the hematopoietic stem cell pool by CXCL12-CXCR4 chemokine signaling in bone marrow stromal cell niches. Immunity 25:977–988

9. Sacchetti B, Funari A, Michienzi S, Di Cesare S, Piersanti S, Saggio I, Tagliafico E, Ferrari S, Robey PG, Riminucci M, Bianco P (2007) Self-renewing osteoprogenitors in bone marrow sinusoids can organize a hematopoietic microenvironment. Cell 131:324–336

10. Mendez-Ferrer S, Michurina TV, Ferraro F, Mazloom AR, Macarthur BD, Lira SA, Scadden DT, Ma'ayan A, Enikolopov GN, Frenette PS (2010) Mesenchymal and haematopoietic stem cells form a unique bone marrow niche. Nature 466:829–834

11. Calvi LM, Adams GB, Weibrecht KW, Weber JM, Olson DP, Knight MC, Martin RP, Schipani E, Divieti P, Bringhurst FR, Milner LA, Kronenberg HM, Scadden DT (2003) Osteoblastic cells regulate the haematopoietic stem cell niche. Nature 425:841–846

12. Kollet O, Dar A, Shivtiel S, Kalinkovich A, Lapid K, Sztainberg Y, Tesio M, Samstein RM, Goichberg P, Spiegel A, Elson A, Lapidot T (2006) Osteoclasts degrade endosteal components and promote mobilization of hematopoietic progenitor cells. Nat Med 12:657 661

13. Yoshihara H, Arai F, Hosokawa K, Hagiwara T, Takubo K, Nakamura Y, Gomei Y, Iwasaki H, Matsuoka S, Miyamoto K, Miyazaki H, Takahashi T, Suda T (2007) Thrombopoietin/MPL signaling regulates hematopoietic stem cell quiescence and interaction with the osteoblastic niche. Cell Stem Cell 1:685–697

14. Zhang J, Niu C, Ye L, Huang H, He X, Tong WG, Ross J, Haug J, Johnson T, Feng JQ, Harris S, Wiedemann LM, Mishina Y, Li L (2003) Identification of the haematopoietic stem cell niche and control of the niche size. Nature 425:836–841

15. Fujisaki J, Wu J, Carlson AL, Silberstein L, Putheti P, Larocca R, Gao W, Saito TI, Lo Celso C, Tsuyuzaki H, Sato T, Cote D, Sykes M, Strom TB, Scadden DT, Lin CP (2011) In vivo imaging of Treg cells providing immune privilege to the haematopoietic stem-cell niche. Nature 474:216–219

16. Lo Celso C, Fleming HE, Wu JW, Zhao CX, Miake-Lye S, Fujisaki J, Cote D, Rowe DW, Lin CP, Scadden DT (2009) Live-animal tracking of individual haematopoietic stem/progenitor cells in their niche. Nature 457:92–96

17. Carlson AL, Fujisaki J, Wu J, Runnels JM, Turcotte R, Celso CL, Scadden DT, Strom TB, Lin CP (2013) Tracking single cells in live animals using a photoconvertible near-infrared cell membrane label. PLoS One 8:e69257

18. Dutta P, Courties G, Wei Y et al (2012) Myocardial infarction accelerates atherosclerosis. Nature 487:325–329

19. Park D, Spencer JA, Koh BI, Kobayashi T, Fujisaki J, Clemens TL, Lin CP, Kronenberg HM, Scadden DT (2012) Endogenous bone marrow MSCs are dynamic, fate-restricted participants in bone maintenance and regeneration. Cell Stem Cell 10:259–272

20. Cao YA, Wagers AJ, Beilhack A, Dusich J, Bachmann MH, Negrin RS, Weissman IL, Contag CH (2004) Shifting foci of hematopoiesis during reconstitution from single stem cells. Proc Natl Acad Sci U S A 101:221–226

21. Wang X, Rosol M, Ge S, Peterson D, McNamara G, Pollack H, Kohn DB, Nelson MD, Crooks GM (2003) Dynamic tracking of human hematopoietic stem cell engraftment using *in vivo* bioluminescence imaging. Blood 102:3478–3482

22. Plett PA, Frankovitz SM, Orschell CM (2003) Distribution of marrow repopulating cells between bone marrow and spleen early after transplantation. Blood 102:2285–2291

23. Mazo IB, Gutierrez-Ramos JC, Frenette PS, Hynes RO, Wagner DD, von Andrian UH (1998) Hematopoietic progenitor cell rolling in bone marrow microvessels: parallel contributions by endothelial selectins and vascular cell adhesion molecule 1. J Exp Med 188:465–474

24. Storer JB (1966) Acute responses to ionizing radiation. In: Green EL (ed) Biology of the laboratory mouse, 2nd edn. Dover Publications, Inc., New York

25. Lo Celso C, Lin CP, Scadden DT (2011) In vivo imaging of transplanted hematopoietic stem and progenitor cells in mouse calvarium bone marrow. Nat Protoc 6:1–14

26. Ema H, Morita Y, Yamazaki S, Matsubara A, Seita J, Tadokoro Y, Kondo H, Takano H, Nakauchi H (2006) Adult mouse hematopoietic stem cells: purification and single-cell assays. Nat Protoc 1:2979–2987

27. Colmone A, Amorim M, Pontier AL, Wang S, Jablonski E, Sipkins DA (2008) Leukemic cells create bone marrow niches that disrupt the behavior of normal hematopoietic progenitor cells. Science 322:1861–1865

28. Lo Celso C, Scadden D (2007) Isolation and transplantation of hematopoietic stem cells (HSCs). J Vis Exp 2:157

29. Lassailly F, Foster K, Lopez-Onieva L, Currie E, Bonnet D (2013) Multimodal imaging reveals structural and functional heterogeneity in different bone marrow compartments: functional implications on hematopoietic stem cells. Blood 122:1730–1740

30. Sipkins DA, Wei X, Wu JW, Runnels JM, Cote D, Means TK, Luster AD, Scadden DT, Lin CP (2005) In vivo imaging of specialized bone marrow endothelial microdomains for tumour engraftment. Nature 435:969–973

31. Runnels JM, Carlson AL, Pitsillides C, Thompson B, Wu J, Spencer JA, Kohler JMJ, Azab A, Moreau A-S, Rodig SJ, Kung AL, Anderson KC, Ghobrial IM, Lin CP (2011) Optical techniques for tracking multiple myeloma engraftment, growth, and response to therapy. J Biomed Opt 16:011006–011013

32. Barrett O, Sottocornola R, Lo Celso C (2012) In vivo imaging of hematopoietic stem cells in the bone marrow niche. Methods Mol Biol 916:231–242

33. Ishii M, Egen JG, Klauschen F, Meier-Schellersheim M, Saeki Y, Vacher J, Proia RL, Germain RN (2009) Sphingosine-1-phosphate mobilizes osteoclast precursors and regulates bone homeostasis. Nature 458:524–528

34. Malide D, Metais JY, Dunbar CE (2012) Dynamic clonal analysis of murine hematopoietic stem and progenitor cells marked by 5 fluorescent proteins using confocal and multiphoton microscopy. Blood 120:e105–e116

35. Visnjic D, Kalajzic I, Gronowicz G, Aguila HL, Clark SH, Lichtler AC, Rowe DW (2001) Conditional ablation of the osteoblast lineage in Col2.3Δtk transgenic mice. Bone Miner Res 16:2222–2231

36. 8 Common Suture Techniques for Skin Closure (2012) http://www.youtube.com/watch?v=-ZWUgKiBxfk. Accessed 30 Sept 2013

37. Suture-Basic Technique 1 (2009) http://www.youtube.com/watch?v=6P0rYS6LeZw. Accessed 30 Sept 2013

38. Fiji Is Just ImageJ (2013) http://www.fiji.sc/Fiji. Accessed 30 Sept 2013

39. Lassailly F, Griessinger E, Bonnet D (2010) "Microenvironmental contaminations" induced by fluorescent lipophilic dyes used for noninvasive in vitro and in vivo cell tracking. Blood 115:5347–5354

40. Li P, Zhang R, Sun H, Chen L, Liu F, Yao C, Du M, Jiang X (2013) PKH26 can transfer to host cells in vitro and vivo. Stem Cells Dev 22:340–344

41. Invitrogen Fluorescence SpectraViewer (2013) http://www.lifetechnologies.com/us/en/home/life-science/cell-analysis/labeling-chemistry/fluorescence-spectraviewer.html. Accessed 26 Sept 2013

42. BD Fluorescence Spectrum Viewer (2013) http://www.bdbiosciences.com/research/multicolor/spectrum_viewer/index.jsp. Accessed 26 Sept 2013

43. Chudakov DM, Matz MV, Lukyanov S, Lukyanov KA (2010) Fluorescent proteins and their applications in imaging living cells and tissues. Physiol Rev 90:1103–1163

44. Bestvater F, Spiess E, Stobrawa G, Hacker M, Feurer T, Porwol T, Berchner-Pfannschmidt U, Wotzlaw C, Acker H (2002) Two-photon fluorescence absorption and emission spectra of dyes relevant for cell imaging. J Microsc 208:108–115

45. Drobizhev M, Makarov NS, Tillo SE, Hughes TE, Rebane A (2011) Two-photon absorption properties of fluorescent proteins. Nat Methods 8:393–399

46. Spiess E, Bestvater F, Heckel-Pompey A, Toth K, Hacker M, Stobrawa G, Feurer T, Wotzlaw C, Berchner-Pfannschmidt U, Porwol T, Acker H (2005) Two-photon excitation and emission spectra of the green fluorescent protein variants ECFP, EGFP and EYFP. J Microsc 217:200–204

47. Inoue S, Osmond DG (2001) Basement membrane of mouse bone marrow sinusoids shows distinctive structure and proteoglycan composition: a high resolution ultrastructural study. Anat Rec 264:294–304

48. Challen GA, Little MH (2006) A side order of stem cells: the SP phenotype. Stem Cells 24:3–12

Chapter 18

Immunodeficient Mouse Model for Human Hematopoietic Stem Cell Engraftment and Immune System Development

Ken-Edwin Aryee, Leonard D. Shultz, and Michael A. Brehm

Abstract

Immunodeficient mice engrafted with human immune systems provide an exciting model to study human immunobiology in an in vivo setting without placing patients at risk. The essential parameter for creation of these "humanized models" is engraftment of human hematopoietic stem cells (HSC) that will allow for optimal development of human immune systems. However, there are a number of strategies to generate humanized mice and specific protocols can vary significantly among different laboratories. Here we describe a protocol for the co-implantation of human HSC with autologous fetal liver and thymic tissues into immunodeficient mice to create a humanized model with optimal human T cell development. This model, often referred to as the Thy/Liv or BLT (bone marrow, liver, thymus) mouse, develops a functional human immune system, including HLA-restricted human T cells, B cells, and innate immune cells.

Key words Hematopoietic stem cells, SCID, Thymus, HSC, Humanized mice, BLT

1 Introduction

The study of human immunobiology is constrained by ethical and logistical concerns of working with patients [1, 2]. Given these complications, our basic understanding of many fundamental biological processes in humans has been shaped by experimental studies in animal models, particularly in rodents. However, many characteristics and properties of mammalian biological systems, particularly mouse and human immune systems, are species-specific, and these species differences often limit the translation of experimental findings from rodent studies to the clinic [3–5]. Moreover, the use of nonhuman primates for basic research in Europe and United States is severely restricted [6]. Immunodeficient mice that are engrafted with human hematopoietic stem cells (HSC) and that develop human immune systems or "humanized mice" are an attractive alternative to study immunobiology [7–9].

The ideal mouse strains for engraftment of human HSC are immunodeficient stocks of *scid*, *Rag1^{null}*, or *Rag2^{null}* mice [9] that

Kevin D. Bunting and Cheng-Kui Qu (eds.), *Hematopoietic Stem Cell Protocols*, Methods in Molecular Biology, vol. 1185, DOI 10.1007/978-1-4939-1133-2_18, © Springer Science+Business Media New York 2014

express targeted mutations within the IL-2 receptor common gamma chain (*IL2rγ*) gene. The *IL2rγ* chain is a critical component of the IL-2, IL-4, IL-7, IL-9, IL-15, and IL-21 receptors [10, 11]. The absence of this *IL2rγ* chain leads to severe impairments in adaptive immune system development and function and completely prevents NK cell development [12, 13]. Immunodeficient mice bearing a mutated *IL2rγ* gene support greatly enhanced engraftment of human HSC [13] and fetal thymic tissues [14] as compared with previous models.

Humanized mice can be generated using human HSC derived from a number of sources, including umbilical cord blood, G-CSF mobilized peripheral blood, bone marrow and fetal liver [15]. NOD-*Prkdc^{scid}IL2rg^{Tm1Wjl}* (NSG) mice engraft efficiently with human HSC and give rise to functional human immune systems [16, 17]. However human T cell development in immunodeficient mice injected with HSC is hindered by the absence of human thymic epithelium that is required for T cell education [18]. This problem can be overcome by co-implanting human fetal thymic tissues with autologous fetal liver derived HSC [19]. This model, originally referred to as the SCID-hu mouse, has been modified over recent years and allows for optimal generation of functional human T cells that are HLA-restricted [20–23]. This chapter describes a current technique to generate these Thy/Liv or "BLT" (bone marrow, liver thymus) mice as practiced by our lab.

2 Materials

2.1 Fetal Tissue Preparation

1. Wash Buffer: RPMI 1640 (Gibco, Life technologies, Grand Island, NY USA) supplemented with 1 % Penicillin–Streptomycin (Gibco, Life Technologies, Grand Island, NY USA), 2.5 µg/mL Fungizone (Gibco, Life Technologies, Grand Island, NY USA) and 10 µg/mL Gentamicin (APP Pharmaceuticals, Lake Zurich, IL, USA). Keep washing buffer sterile and chilled on ice. The buffer can be stored at 4 °C.

2. Quenching Buffer: Wash Buffer supplemented with 3 % fetal bovine serum (FBS, Atlanta Biologicals, Lawrenceville, GA, USA).

3. Human fetal liver and fetal thymus tissue (Advanced Bioscience Resources, Alameda, CA) (*see* **Note 1**).

4. 100 mm plastic petri dishes (BD Falcon, Franklin Lakes, NJ, USA), sterile.

5. No. 21 disposable scalpel (Feather Safety Razor Co., LTD, Kita-Ku, Osaka, Japan).

6. 50 mL centrifuge tubes (BD Falcon, Franklin Lakes, NJ, USA).

7. Liver digest buffer (Gibco, Life technologies, Grand Island, NY, USA).

8. Water bath.

9. Parafilm.

10. Nutating mixer (VWR international, Randor, PA, USA).

11. Laboratory tape or alternatively rubber bands.

12. 85 mm tissue grinder homogenizer cup with size 50 metal sieve (Sigma-Aldrich, St Louis, MO, USA).

13. 10 mm Syringe.

14. Histopaque-1077 (Sigma-Aldrich, St Louis, MO, USA).

2.2 CD3 T Cell Depletion

1. Human CD3 microbead kit (Miltenyi Biotec, Auburn, CA, USA) (*see* **Note 2**).

2. MidiMACS Separator (Miltenyi Biotec, Auburn, CA, USA).

3. MACS multistand (Miltenyi Biotec, Auburn, CA, USA).

4. LD columns (Miltenyi Biotec, Auburn, CA, USA).

5. MACS Buffer: PBS supplemented with 2 % FBS, 1 mM EDTA. Sterilize buffer through a vacuum flask with a 0.2 μm filter.

2.3 Flow Cytometry Analysis

1. FACS buffer: phosphate buffered saline (PBS) supplemented with 1 % FBS, and 0.1 % Sodium azide (*see* **Note 3**).

2. Human CD29 FITC, Clone K20 (Beckman Coulter, Marseille, France).

3. HLA-A2 FITC, Clone BB7.2; Human CD29 PE, Clone MAR4; Human CD34 PE, Clone 581; Human CD3 PE, Clone UCHT1; Human CD45 APC, Clone HI30; Human CD45 APC-H7, Clone 2D1; Human CD3 FITC, Clone UCHT1; Human CD20 APC, Clone 2H7; Human CD4 Pacific Blue, Clone RPA-T4; Human CD8 PE, PTA-T8; Mouse Ly5 PerCP, Clone 30-F11 (BD Biosciences, San Jose, CA, USA).

2.4 Reagents and Materials for Tissue Implant

1. Immunodeficient mice: Our laboratory primarily uses NOD-*Prkdc*^{scid} *IL2rg*^{Tm1Wjl} (NSG) mice between 8 and 12 weeks of age (*see* **Note 4**).

2. Ketamine–Xylazine solution (store at 4 °C), Working concentration is 15 mg/mL of Ketamine and 2 mg/mL of xylazine, (Hospira, Lake Forest, IL, USA) diluted in PBS. Dosage per mouse is 150 and 10 mg/kg of body weight, respectively.

3. Buprinex SR (slow release), 1 mg/mL (Zoopharm, Laramie, WY, USA). Dosage per mouse is 2.4 mg/kg. Store at 4 °C.

4. Cefazolin–Gentamicin (Sandoz, Princeton, NJ, USA/APP Pharmaceuticals, Lake Zurich, IL, USA): Working concentration is 8.3 mg/mL of cefazolin and 2 mg/mL gentamicin. Dosage per mouse is 0.83 and 0.2 mg, respectively. Store working stock in 1 mL aliquots at –20 °C.

5. Sterile surgical draping.

6. Betadine surgical scrub: 10 % Povidone–iodine (Purdue Pharma L.P., Stamford, CT, USA).

7. 70 % Ethanol: Mix 70 mL of absolute ethanol with 30 mL of deionized water.

8. Sterile physiological (0.85 %) saline: 10 mL bottle.

9. 100-mm plastic petri dishes (BD Falcon, Franklin Lakes, NJ, USA), sterile.

10. 30-mm plastic petri dishes (BD Falcon, Franklin Lakes, NJ, USA), sterile.

11. Ohaus portable balance (Fisher Scientific, Pittsburgh, PA, USA).

12. Sterile Curity gauze sponges 4″×4″ (Covidien, Mansfield, MA, USA).

13. No. 21 disposable scalpel.

14. Electric clipper with size 40 blade.

15. 10 mL syringe with 20-G needle.

16. 1 mL syringe with 25-G needle.

2.5 Surgical Instruments

Sterilized by placing in an instant sealing sanitization pouch (Fisher Scientific, Pittsburgh, PA, USA) and autoclaving before use

1. Microdissecting forceps curved (small).

2. Dressing forceps (large).

3. Operating scissors: straight, sharp (Large).

4. Microdissecting scissors: straight, sharp (small).

5. 5.16-G Trocar.

6. Olsen-Hegar needle holder.

7. Surgical suture.

8. Autoclip wound clip applicator and 9-mm autoclips.

3 Methods

This protocol involves the extensive manipulation of human tissues and cells. Extreme care should be taken when handling the specimens and all work should be done in a standard laminar flow hood. All unused tissues and waste should be disposed of using protocols approved by an Institutional Biosafety Committee. All mouse surgeries should be done using protocols approved by an Institutional Animal Care and Use Committee.

3.1 Fetal Tissue Preparation

1. Place fetal liver and thymus tissues into 100 mm plastic petri dishes with 10 mL of Wash Buffer. Reserve medium that tissues were shipped in to combine with liver preparation (*see* below).

Using a scalpel cut one piece of liver and one piece of thymus of sufficient size to prepare the quantity of 1 mm³ pieces needed to implant all recipient mice. Place pieces in separate 50 mL conical tubes in 30 mL of Wash Buffer (*see* Subheading 3.3, **step 3** below) and keep on ice until transplant (*see* **Note 5**). In addition, dissect 1 mm³ pieces of liver and thymus for optional flow cytometric analysis to evaluate HLA-A2 status (*see* Subheading 3.4, **step 1**). Dissect the remaining fetal liver tissue into 4–5 mm³ pieces that are of sufficient size to pass through the bore of a 25 mL pipette. Combine liver pieces and Wash Buffer with reserved liver shipping medium into two 50 mL tubes and centrifuge (400×*g* for 10 min at 4 °C).

2. Thaw Liver Digest Buffer at room temperature. Resuspend liver tissue pellets from both tubes in a total of 25 mL of Liver Digest Buffer and combine into one 50 mL conical tube. Seal tube lid tightly with Parafilm. Place the 50 mL conical tube with digesting liver tissues on a Nutator in a 37 °C incubator for 20–25 min. Secure tube to platform with laboratory tape or rubber bands.

3. Decant contents into a size 50 mesh in the 85 mm tissue grinder homogenizer cup that is sitting in a fresh 100 mm petri dish. Using a 10 mL syringe plunger, homogenize tissue through the sieve. Rinse the sieve well with 25 mL of Quenching Buffer to stop digestion reaction and split volume equally in two 50 mL conical tubes (approximately 60 mL total volume) (*see* **Note 6**).

4. Allow cell suspension to settle for 3 min to preferentially pellet the fetal liver hepatocytes Each tube should have a large pellet that is brown in color and that contains primarily nucleated cells and some RBCs. Carefully remove supernatant (leukocyte rich) without disturbing the hepatocyte pellet and spit volume equally into four fresh 50 mL conical tubes. Bring total volume to 30 mL per tube using Quenching Buffer.

5. Underlay with 12 mL of Histopaque in each tube by slowly pipetting the Histopaque into bottom of tube. Ensure a clear interface is maintained between the Histopaque and the cell-containing medium. Centrifuge the layered tubes at 400×*g* for 30 min at 18–20 °C.

6. Carefully harvest cells along the interface, collecting all visible cells. Transfer interface cells to fresh 50 mL conical tubes and add at least 4 volumes of Quenching Buffer. Centrifuge the tubes at 400×*g* for 10 min at 18–20 °C. Resuspend the pellets in a total volume of 10 mL Quenching Buffer and determine viable cell number. Keep recovered cells on ice. Reserve 200 µL of cells for flow cytometric evaluation (*see* Subheading 3.4).

3.2 Depletion of CD3⁺ T Cells from Fetal Liver Hematopoietic Cells

For the depletion step we use reagents from Miltenyi Biotec and follow the manufacturer's recommendations (*See* **Note 3**)

1. Resuspend recovered cells in 1 mL RPMI and perform viable cell counts.

2. Reserve 100,000 cells for flow cytometric analysis to validate depletion of CD3 T cells and to quantify the CD34⁺ cells (*see* Subheading 3.4, **steps 2, 3**).

3. Based on results of CD34⁺ cell analysis, resuspend recovered liver cells at 4.0×10^5 CD34⁺ cells per mL in Wash Buffer. Keep cells on ice until injection.

3.3 Preparation of Mice for Surgery and Tissue Implant

1. Irradiate recipient mice by whole body gamma irradiation (a ^{137}Cs radioactive source is most commonly used). For the NSG mouse strain, our laboratory use a conditioning dose of 200 cGy, which is normally well tolerated by adult NSG mice. Tissues implant can be performed immediately after irradiation. However HSC injection must be performed at 4–24 h after irradiation (*see* **Note 7**).

2. Using fetal thymic and liver pieces from Subheading 3.1, **step 1** above, place the tissues and medium into a 100 mm petri dish. Dissect fetal tissues into 1 mm³ pieces using a scalpel.

3. Determine mouse weight using balance and use ear punch to give each mouse a unique identifier.

4. Inject each mouse with 0.1 mL of cefazolin–gentamicin antibiotic mixture subcutaneously (*see* **Note 8**).

5. Anesthetize the mice by intraperitoneal injection of ketamine–xylazine solution at a dose of 150 and 10 mg/kg, respectively. When the mice are fully anesthetized, proceed to the next step. This generally takes between 10 and 15 min.

6. Place anesthetized mouse on sterile draping with left side facing up. Shave the left side of the mouse from the level of the shoulder joint to that of the hip joint with an electric clipper. Scrub the left side of the mouse 6 times alternating between betadine and 70 % alcohol (Fig. 1a).

7. Use a pair of operating scissors and dressing forceps to make a left-side skin incision (1.5 cm long) ventral to the spine and between the last rib and the hip joint. Loosen connective tissue under the skin using the blunt side of the operating scissors. Make a 0.5 cm incision in a longitudinal direction in the abdominal wall (through muscle wall) (Fig. 1b).

8. Using a pair of straight microdissecting forceps, grasp the fatty tissue at the base of the kidney and gently expose the kidney through the incision. Do not apply direct pressure to the kidney to prevent damaging the organ. Secure the kidney by grasping the renal vessels and ureter just below kidney between

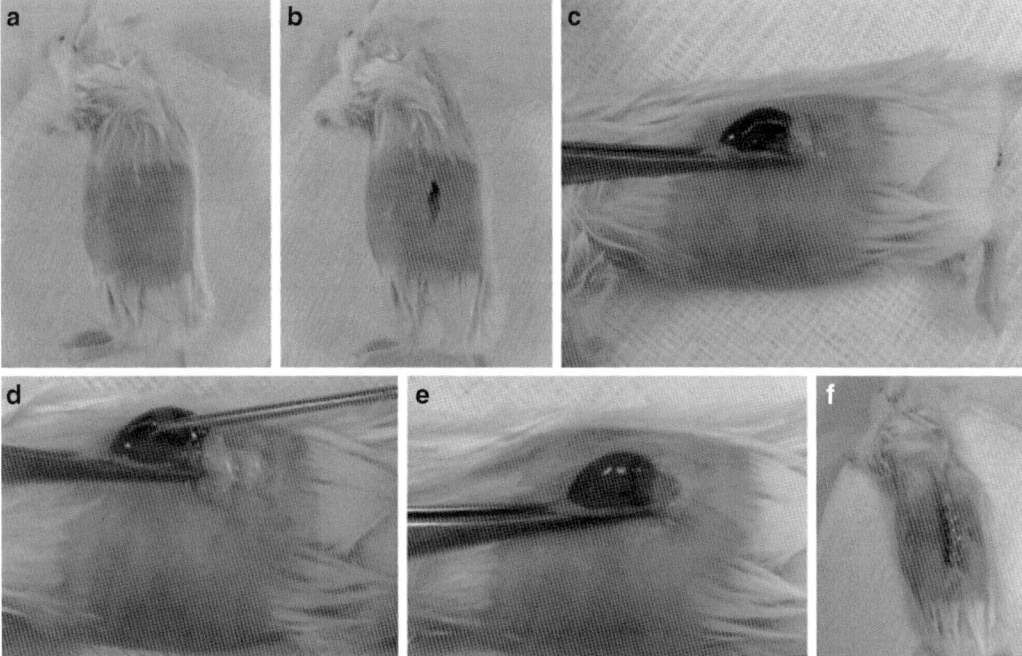

Fig. 1 Surgical procedure for implantation of human fetal thymus and liver. (**a**) Fur is removed from anesthetized NSG mice and (**b**) a 0.5 cm incision in a longitudinal direction in the abdominal wall (through muscle wall).
(**c**) Using a pair of straight microdissecting forceps, expose kidney through the incision. (**d**) Secure the kidney
by grasping the renal vessels and ureter just below kidney between the flaps of skin with a pair second pair of
straight microdissecting forceps. Make a small (1–2 mm) incision in kidney capsule at the posterior lateral side
with a scalpel. Insert the trocar loaded with human tissue as far as possible through the incision in the kidney
capsule and staying as superficial as possible. Push the plunger piece of the trocar to expel the loaded tissue.
(**e**) Remove trocar from kidney. Gently pull apart the two cut edges of the muscular wall using two pairs of
microdissecting forceps. The kidney will retract into the peritoneal cavity. (**f**) Close the abdominal wall with two
separate sutures, using the Olsen-Hegar needle holder and surgical suture. Cut the ends of the suture and
close the knots. Close the skin incision with three autoclips, using an autoclip wound-clip applicator. The mice
are then placed on a warming tray and observed until they awake from the anesthesia

the flaps of skin with a second pair of straight microdissecting
forceps. Throughout the surgical procedure, keep the exposed
kidney moist, by frequently applying sterile PBS to tissue surface
with a 10 mL syringe with an 18-G needle (Fig. 1c).

9. Place the trocar plunger inside the trocar barrel. Using a pair
of curved dissecting forceps, place a 1 mm^3 piece of fetal thymus on the tip of the trocar and aspirate it inside by drawing
back the plunger piece of the trocar. Place a 1 mm^3 piece of
fetal liver and aspirate it inside using the same technique.

10. Make a small (1–2 mm) incision in kidney capsule at the
posterior lateral side with a scalpel. Insert the trocar loaded
with human tissue as far as possible through the incision in the

kidney capsule and staying as superficial as possible. The kidney capsule is delicate and caution should be taken to not rupture. Push the plunger piece of the trocar to expel the loaded tissues. Remove trocar from kidney. To prevent tissues from sliding out with the trocar, light pressure can be applied to the kidney and against the tissues using a pair of straight microdissecting forceps as the trocar is removed (Fig. 1d, e).

11. Gently pull apart the two cut edges of the muscular wall using two pairs of microdissecting forceps. The kidney will retract into the peritoneal cavity. Close the abdominal wall with two separate sutures, using the Olsen-Hegar needle holder and surgical suture. Cut the ends of the suture and close the knots. Close the skin incision with three autoclips, using an autoclip wound-clip applicator (Fig. 1f).

12. Once the mice recover from the anesthesia and are active, dose each mouse with 2.4 mg/kg buprenex SR (1 mg/mL) by intraperitoneal injection. Check the mice daily for surgical would healing and remove autoclips, 5–7 days after surgery.

13. Inject 2×10^5 CD34$^+$ fetal liver hematopoietic cells isolated earlier (*see* Subheading 3.2, **step 3**) through the lateral tail vein at least 4 h after irradiation (*see* **Notes 9, 10**).

14. At 12 weeks after tissue transplants run a FACS analysis on the peripheral blood of the mice to validate human immune cell engraftment.

3.4 Flow Cytometry Analysis

1. For the optional HLA-A2 FACS, mechanically dissociate the saved liver and thymus tissue pieces to make a suspension, keeping tissues separate. Stain with anti-CD29 FITC (1:50) as a FITC control, and stain test sample with anti-HLA-A2 FITC (1:50).

2. To validate the CD3$^+$ T cell depletion, stain the saved cells from Subheading 3.2, **step 3** with anti-CD45 PE (1:50) as PE control, anti-CD45 APC (1:50) as an APC control and stain the test sample with anti-CD3 PE (1:50) and anti-CD45 APC (1:50).

3. To determine the CD34$^+$ cell levels, stain the saved cells with anti-CD45 PE (1:50) as PE control, anti-CD45 APC (1:50) as APC control and stain the test sample with anti-CD34 PE (1:50) and anti-CD45 APC (1:50).

4. To determine the engraftment of human lymphocytes, stain 120 µL of blood with the following antibody mixture: CD45 APC-H7 (1:100), CD3 FITC (1:50), CD20 APC (1:25), CD8 PE (1:50), CD4 Pacific Blue (1:50), and mLy5 PerCP (1:100) (Fig. 2a–d).

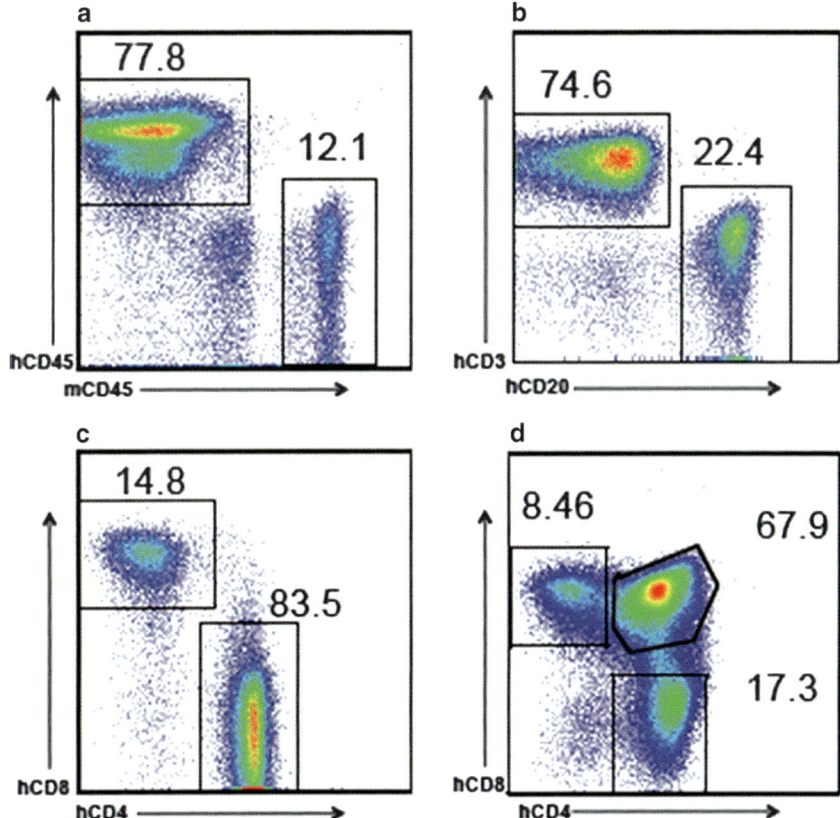

Fig. 2 Evaluation of human cell chimerism. At 12 weeks post-implant human immune cell chimerism levels can be evaluated in the peripheral blood. Our standard panel includes antibodies specific for mouse CD45, human CD45, human CD3, human CD20, human CD4, and human CD8. (**a**) The gating strategy for this panel includes first evaluating percentages of human CD45⁺ and mouse CD45⁺ cells. (**b**) Human CD45⁺ cells are evaluated for expression of human CD3 to enumerate T cells and human CD20 to enumerate B cells and (**c**) then the T cells populations are further defined by CD4 and CD8 expression. (**d**) In addition the human thymocytes recovered from the thymic organoid can be evaluated by expression of human CD4 and CD8

4 Notes

1. The gestational ages of the human tissues are between 16 and 20 weeks.

2. Injection of mature human T cells into immunodeficient mice such as the NSG mice results in the development of a severe xenogeneic graft-versus-host disease (GVHD). For this reason we use a magnetic-bead based approach to deplete T cells from the liver cell preparation. The frequency of human T cells within the recovered cells generally falls between 0.3 and 0.5 %

of total cells. Alternatively CD34+ HSC can be isolated by a positive selection approach, which will exclude T cells and other cell types from the final cell preparation [21]. We routinely use the CD3 depletion approach as accessory cells (CD34 negative and CD3 negative) present within the liver cell preparation have been shown to facilitate engraftment of human HSC in immunodeficient mice [19].

3. We substituted FBS for BSA.

4. NSG mice are available from the Jackson Laboratory. The mice are maintained on medicated (sulfamethoxazole/trimethoptrim) water to prevent infections. An alternative strain for engraftment is NOD-*scid* mice.

5. In our experience we generally receive tissues that are sufficient to implant 30–40 mice but the size of the tissues can be variable. It is critical to have sufficient liver tissue for HSC recovery. The recovery of HSC can vary with the gestational age of the tissues.

6. For optimal cell recovery from the liver tissues, ensure that the liver has been completely disaggregated and that the sieve has been extensively rinsed.

7. The specific dose of irradiation used will be dependent on the specific strain being used as a recipient. Mice homozygous for the *scid* mutation are radiosensitive, while $Rag1^{null}$ or $Rag2^{null}$ mice are more radioresistant and require higher irradiation doses for optimal conditioning. Each laboratory will need to determine the optimal dose based on the mouse strain and the specific irradiation source.

8. The use of antibiotics is at the discretion of the laboratory. However, procurement of the fetal tissues is not done under sterile conditions, and in our experience, pretreatment with antibiotics reduces the risk of infection and premature death of engrafted mice.

9. Cryopreserving/thawing the fetal thymus before transplantation into the mice has been reported to support human thymopoiesis and T-cell reconstitution while clearing preexisting thymocytes from the thymic graft [24].

10. Another modification for generating BLT mice involves transplantation of the autologous fetal thymic and liver tissue under the renal capsules of the mice. After 3 weeks, the engrafted mice are sublethally irradiated and injected through their tail veil with the autologous cryopreserved/thawed CD34+ cells isolated from the fetal liver [23].

Acknowledgements

We thank Jamie Kady, Meghan Dolan, Pamela St Louis, Linda Paquin, Michael Bates, Bruce Gott, and Allison Ingalls for excellent technical assistance. This work was supported by National Institutes of Health research grants AI046629 and grants from the Juvenile Diabetes Research Foundation International and the Helmsley Charitable Trust. The contents of this publication are solely the responsibility of the authors and do not necessarily represent the official views of the National Institutes of Health.

References

1. Davis MM (2008) A prescription for human immunology. Immunity 29:835–838
2. Davis MM (2012) Immunology taught by humans. Sci Transl Med 4:117fs112
3. Mestas J, Hughes CC (2004) Of mice and not men: differences between mouse and human immunology. J Immunol 172:2731–2738
4. Seok J, Warren HS, Cuenca AG et al (2013) Genomic responses in mouse models poorly mimic human inflammatory diseases. Proc Natl Acad Sci U S A 110:3507–3512
5. Driver JP, Serreze DV, Chen YG (2011) Mouse models for the study of autoimmune type 1 diabetes: a NOD to similarities and differences to human disease. Semin Immunopathol 33: 67–87
6. Wadman M (2013) Time called on chimp work. Nature 495:289–290
7. Brehm MA, Shultz LD, Greiner DL (2010) Humanized mouse models to study human diseases. Curr Opin Endocrinol Diabetes Obes 17:120–125
8. Brehm MA, Powers AC, Shultz LD et al (2012) Advancing animal models of human type 1 diabetes by engraftment of functional human tissues in immunodeficient mice. Cold Spring Harb Perspect Med 2:a007757
9. Shultz LD, Ishikawa F, Greiner DL (2007) Humanized mice in translational biomedical research. Nat Rev Immunol 7:118–130
10. Sugamura K, Asao H, Kondo M et al (1996) The interleukin-2 receptor gamma chain: its role in the multiple cytokine receptor complexes and T cell development in XSCID. Annu Rev Immunol 14:179–205
11. Rochman Y, Spolski R, Leonard WJ (2009) New insights into the regulation of T cells by gamma(c) family cytokines. Nat Rev Immunol 9:480–490
12. Cao X, Shores EW, Hu-Li J et al (1995) Defective lymphoid development in mice lacking expression of the common cytokine receptor gamma chain. Immunity 2:223–238
13. Shultz LD, Lyons BL, Burzenski LM et al (2005) Human lymphoid and myeloid cell development in NOD/LtSz-scid IL2R gamma null mice engrafted with mobilized human hemopoietic stem cells. J Immunol 174: 6477–6489
14. Stoddart CA, Maidji E, Galkina SA et al (2011) Superior human leukocyte reconstitution and susceptibility to vaginal HIV transmission in humanized NOD-scid IL-2Rgamma(−/−) (NSG) BLT mice. Virology 417:154–160
15. Shultz LD, Brehm MA, Garcia-Martinez JV et al (2012) Humanized mice for immune system investigation: progress, promise and challenges. Nat Rev Immunol 12:786–798
16. Brehm MA, Cuthbert A, Yang C et al (2010) Parameters for establishing humanized mouse models to study human immunity: analysis of human hematopoietic stem cell engraftment in three immunodeficient strains of mice bearing the IL2rγnull mutation. Clin Immunol 135: 84–98
17. Brehm MA, Bortell R, Diiorio P et al (2010) Human immune system development and rejection of human islet allografts in spontaneously diabetic NOD-Rag1null IL2rgammanull Ins2Akita mice. Diabetes 59:2265–2270
18. Shultz LD, Saito Y, Najima Y et al (2010) Generation of functional human T-cell subsets with HLA-restricted immune responses in HLA class I expressing NOD/SCID/IL2r gamma(null) humanized mice. Proc Natl Acad Sci U S A 107:13022–13027
19. Covassin L, Jangalwe S, Jouvet N et al (2013) Human immune system development and survival of NOD-scid IL2rgamma (NSG) mice

engrafted with human thymus and autologous hematopoietic stem cells. Clin Exp Immunol 174:372–388

20. Brainard DM, Seung E, Frahm N et al (2009) Induction of robust cellular and humoral virus-specific adaptive immune responses in human immunodeficiency virus-infected humanized BLT mice. J Virol 83:7305–7321

21. Lan P, Tonomura N, Shimizu A et al (2006) Reconstitution of a functional human immune system in immunodeficient mice through combined human fetal thymus/liver and CD34+ cell transplantation. Blood 108:487–492

22. McCune JM, Namikawa R, Kaneshima H et al (1988) The SCID-hu mouse: murine model for the analysis of human hematolymphoid differentiation and function. Science 241: 1632–1639

23. Melkus MW, Estes JD, Padgett-Thomas A et al (2006) Humanized mice mount specific adaptive and innate immune responses to EBV and TSST-1. Nat Med 12:1316–1322

24. Kalscheuer H, Danzl N, Onoe T et al (2012) A model for personalized in vivo analysis of human immune responsiveness. Sci Transl Mede 4:125ra130

Chapter 19

Homing and Migration Assays of Hematopoietic Stem/Progenitor Cells

Xi C. He, Zhenrui Li, Rio Sugimura, Jason Ross, Meng Zhao, and Linheng Li

Abstract

Hematopoietic stem and progenitor cells (HSPCs) reside mainly in bone marrow; however, under homeostatic and stressed conditions, HSPCs dynamically change their location—either egressing from bone marrow and getting into circulation, a process of mobilization; or coming back to the bone marrow, the homing process. How to analyze these two processes will be critical for understanding the behavior of HSPCs. Here we provide an experimental protocol to monitor and analyze homing and migration of HSPCs.

Key words Hematopoietic stem and progenitor cells, Homing, Lodging, Migration

1 Introduction

In this chapter we introduce three interrelated protocols that can facilitate assaying homing and migration of HSPCs. Part I focuses on the homing and lodging assay. Fluorescently labeled bone marrow or LSK (Lin$^-$Sca-1$^+$c-Kit$^+$) cells were transplanted into the mouse model and assessed for the ability to home to or lodge within the bone marrow (primarily a property of hematopoietic stem cells—HSCs). Total bone marrow was labeled with 5- (and 6) carboxyfluorescein diacetate succinimidyl ester per manufacturer's instructions and transplanted into irradiated recipient mice for homing assay and into non-irradiated mice for the lodging assay. At different time points, recipient-derived BM and spleen cells were evaluated by flow cytometry for donor cells. Part II focuses on dynamically monitoring HSPC under the homeostatic state or during the migrating process using the ex vivo imaging stem cells (EVISC) system. Part III focuses on measuring migration behavior of HSPC in vitro using Transwell insert system.

Kevin D. Bunting and Cheng-Kui Qu (eds.), *Hematopoietic Stem Cell Protocols*, Methods in Molecular Biology, vol. 1185, DOI 10.1007/978-1-4939-1133-2_19, © Springer Science+Business Media New York 2014

2   Materials

2.1   Homing and Lodging Assay by Combining CFDA-SE Labeling and Quantification Using Flow Cytometry

1. 1× PBS/2 % FBS.

2. Hemolytic buffer: NH4Cl 8.3 g/L, NaHCO₃ 1.0 g/L, EDTA 37 mg/L.

3. Carboxyfluorescein diacetate succinimidyl ester (CFDA-SE).

4. Water Incubator.

5. Refrigerated Centrifuge.

6. 5 mL Polypropylene Round Bottom Falcon Tubes.

2.2   Ex Vivo Imaging Stem Cells (EVISC)

1. LE Agarose (BioExpress, Cat. No. E-3120-500); DMEM (1×).

2. 35 mm glass bottom culture dishes (uncoated).

3. LSM-710 two photon microscope (Zeiss)
Scl-tTA:H2B-GFP mice are Dox-chased for 90 days [1, 2]. The best way is to transplant bone marrow derived from Scl-tTA:H2B-GFP mice into non-GFP transgenic mice. After stabilization (1–2 months), start chasing for 2–3 months, thus reducing nonspecific background as Tre-H2BGFP mainly leaks to osteocyte without activation by Scl-driven tTA.

Under stressed conditions:

4. 5FU (Sigma-Aldrich, Cat. No. F6627).

5. Scl-tTA:H2B-GFP mice are Dox-chased for 90 days [1, 2]. Four days before EVISC, inject 5FU into mice once via tail vein at 150–300 µg/g body weight [3]. The other materials as described above.

2.3   In Vitro Migration Assay on Sorted LSK/Flk2-/CD34-Cells Using Transwell

1. Transwell plate + inserts.

2. DMEM (no phenol red) + 5 % FBS.

3. Recombinant SDF-1 (dissolved in PBS).

4. Trypan blue.

5. Hemocytometer.

6. 24-well plate.

3   Methods

3.1   Homing and Lodging Assay by Combining CFDA-SE Labeling and Quantification Using Flow Cytometry

1. Harvest bone marrow from the femur and tibia of the desired mice according to standard techniques.

2. Lyse the red blood cells (RBCs) from the sample using hemolytic buffer (NH₄Cl 8.3 g/L, NaHCO₃ 1.0 g/L, EDTA 37 mg/L).

3. Count the number of cells post lysis of RBCs.

4. Concentrate the cells by centrifuging at $500 \times g$ for 5 min and aspirate the supernatant.

5. Resuspend the cells in 5 μM CFDA staining solution (prepared according to manufacturer's specifications).

6. Incubate the cells for 5 min at room temperature.

7. Concentrate the cells by centrifuging at $500 \times g$ for 5 min and aspirate the supernatant.

8. Resuspend the cells in 3 mL PBS.

9. Repeat **steps 7–8** two more times.

10. IV inject 1×10^6 cells into irradiated recipient mice for homing per recipient mice; IV inject 1×10^7 cells into nonirradiated mice for lodging assays per recipient mice.

11. After 6 h, harvest the bone marrow from one femur and tibia of the recipient mice. The other femur and tibia will be fixed for image analysis.

12. Harvest the spleens from these same mice; a portion of spleens will be used for preparation of single cell suspension using standard techniques; a portion of spleens will be fixed for image analysis.

13. Lyse the red blood cells from single cell suspension.

14. Process the samples on the flow cytometer, according to manufacturer's instructions.

15. Repeat **steps 11–14** after 18 h.

16. Analyze the data using FlowJo (Tree Star) or a similar flow cytometry data analysis package.

A two dimensional dot plot can be used to compare the number of cells that lodged in each tissue type, while a single dimensional histogram can demonstrate cell division times.

3.2 Ex Vivo Imaging Stem Cells (EVISC)

1. Sacrifice Scl-tTA:H2B-GFP mice that have been Dox-chased for 90 days with or without 5-FU treatment. Isolate the femurs from the mice and soak in PBS/2 % FBS until ready to use.

2. For transplanted purified HSCs, irradiate recipient mice with 10 Gray. 10–12 h later, inject into the tail vein 50,000 HSCs (Flk2⁻LSK cells) purified from BM cells from actin-eGFP transgenic mice. 4–6 h after transplantation, sacrifice the mice. Isolate the femurs from the mice and soak in PBS/2 % FBS until ready to use.

3. Prepare 0.5 % agarose gel dissolved in DMEM media (phenol-red free) and let it cool down to appropriate temperature (*see* **Note 1**).

4. While the 0.5 % agarose gel is cooling down, section the femur in the vertical orientation with the coronal plane (2–3 mm height), embed bone (especially trabecular bone) vertically in

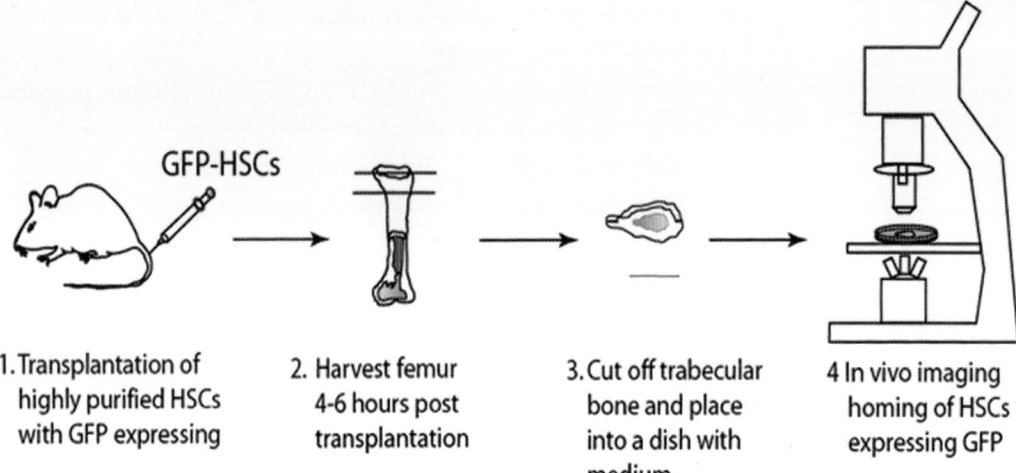

GFP-HSCs

1. Transplantation of highly purified HSCs with GFP expressing

2. Harvest femur 4-6 hours post transplantation

3. Cut off trabecular bone and place into a dish with medium

4 In vivo imaging homing of HSCs expressing GFP

Fig. 1 Diagram showing the process of ex vivo imaging of stem cells (EVISC)

0.5 % agarose in DMEM medium (phenol-red free) 35 mm glass bottom culture dish, with the sectioned plane facing down to the coverslip glass bottom, allowing the laser beam to penetrate into the bone marrow from the open end of bone/bone marrow. Diagram showing the process of EVISE (Fig. 1) (*see* **Note 2**).

5. Apply 2–4 mL cooled 0.5 % agarose gel to the femur section and make sure that the femur section is soaked in the gel. Let the agarose gel polymerize on the bench for 10 min (*see* **Note 3**).

6. Detect live GFP^{hi} LRCs images under a confocal laser scanning *LSM-710* two photon microscope (Carl Zeiss).

3.3 In Vitro Migration Assay on Sorted LSK/ Flk2-/CD34-Cells Using Transwell

1. Harvest BM following standard protocol and process for nucleated, single-cell suspension.

2. Count cells on the Quanta to determine number of cells and concentration.

3. Stain with antibody panel for LSK/CD34/Flk2 (as follows) and wash with PBS + 2 % FBS. Stain with Lin-PECy5, Sca1-PECy7, cKit-APC, CD34-Fitc, and Flk2-PE.

4. Prepare Transwell plate by placing a Transwell insert into an empty well, inverting the plate, using the insert as a guide to draw a grid on the bottom of the plate, similar to the example to the right (Fig. 2) (*see* **Note 4**).

5. Aliquot 600 μL of media + SDF-1 (100 ng/mL) per well for each assay and media alone for negative controls. Pre-warm in 37 °C incubator.

6. Single-pass sort 1,000 LSK/CD34-/Flk2-cells directly into Transwell inserts containing 100 μL of media alone.

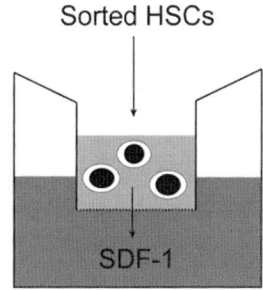

Fig. 2 Illustration of Transwell migration assay to test the response of HSPCs to SDF1

7. Place inserts containing cells into wells and incubate the plate at 37 °C, 5 % CO_2 for 3 h (*see* **Note 5**).

8. Remove the inserts, add 20 μL of trypan blue to the lower chamber, and allow the plate to sit at room temperature for 5 min. Using an inverted microscope, proceed to count viable cells that have migrated.

9. As a control, remove all media from each insert and transfer to a 1.5 mL microcentrifuge tube. Add 10 μL of trypan blue and allow to sit at room temperature for 5 min. Load a hemocytometer and count viable cells (to determine viability of unmigrated fraction).

4   Notes

1. Cooling down the 0.5 % agarose gel is very important to avoid destruction of bone marrow as well as to prevent HSC activity reduction.

2. The orientation for the bone section in the EVISC system can also be in sagittal way, thus obtaining different view of bone and bone marrow.

3. Apply the 0.5 % agarose gel before it starts to polymerize.

4. This is simply a guide to aid in counting migrated cells. All migrated cells will fall within this area.

5. Note—for inhibitor treatment, pre-incubate the plate with inserts at 37 °C, 5 % CO_2 for 45 min–1 h, then proceed with migration assay.

Acknowledgements

This work was supported by Stowers Institute for Medical Research.

References

1. Sugimura R, He XC, Venkatraman A et al (2012) Noncanonical Wnt signaling maintains hematopoietic stem cells in the niche. Cell 150: 351–365

2. Wilson A, Laurenti E, Oser G et al (2008) Hematopoietic stem cells reversibly switch from dormancy to self-renewal during homeostasis and repair. Cell 135:1118–1129

3. Zhao M, Ross JT, Itkin T et al (2012) FGF signaling facilitates postinjury recovery of mouse hematopoietic system. Blood 120: 1831–1842

Part VI

Genetic Modification

Retroviral Transduction of Murine and Human Hematopoietic Progenitors and Stem Cells

Marioara F. Ciuculescu, Christian Brendel, Chad E. Harris, and David A. Williams

Abstract

Genetic modification of cells using retroviral vectors is the method of choice when the cell population is difficult to transfect and/or requires persistent transgene expression in progeny cells. There are innumerable potential applications for these procedures in laboratory research and clinical therapeutic interventions. One paradigmatic example is the genetic modification of hematopoietic stem and progenitor cells (HSPCs). These are rare nucleated cells which reside in a specialized microenvironment within the bone marrow, and have the potential to self-renew and/or differentiate into all hematopoietic lineages. Due to their enormous regenerative capacity in steady state or under stress conditions these cells are routinely used in allogeneic bone marrow transplantation to reconstitute the hematopoietic system in patients with metabolic, inflammatory, malignant, and other hematologic disorders. For patients lacking a matched bone marrow donor, gene therapy of autologous hematopoietic stem cells has proven to be an alternative as highlighted recently by several successful gene therapy trials.

Genetic modification of HSPCs using retrovirus vectors requires ex vivo manipulation to efficiently introduce the new genetic material into cells (transduction). Optimal culture conditions are essential to facilitate this process while preserving the stemness of the cells. The most frequently used retroviral vector systems for the genetic modifications of HSPCs are derived either from Moloney murine leukemia-virus (Mo-MLV) or the human immunodeficiency virus-1 (HIV-1) and are generally termed according to their genus gamma-retroviral (γ-RV) or lentiviral vectors (LV), respectively. This chapter describes in a step-by-step fashion some techniques used to produce research grade vector supernatants and to obtain purified murine or human hematopoietic stem cells for transduction, as well as follow-up methods for analysis of transduced cell populations.

Key words HSPC isolation, HSPC transduction, γ-Retroviral vectors, Lentiviral vectors, Gene therapy

1 Introduction

HSPCs give rise to all mature blood lineages over the lifetime of the animal. Intense research has been carried out in the field of hematopoiesis during the past decades due to the high clinical

Marioara F. Ciuculescu and Christian Brendel have contributed equally to this chapter.

Kevin D. Bunting and Cheng-Kui Qu (eds.), *Hematopoietic Stem Cell Protocols*, Methods in Molecular Biology, vol. 1185, DOI 10.1007/978-1-4939-1133-2_20, © Springer Science+Business Media New York 2014

relevance in transplantation-based therapies in leukemia and other congenital blood disorders. These studies led to the identification of a discrete stem and progenitor cell (HSPCs) hierarchy which allows for the prospective isolation of these cells based on cell surface immunophenotype. The most commonly isolated HSPC populations used for in vitro manipulation are $CD34^+$ or $CD34^+CD38^-$ cells in the human system, and lineage-negative (lin$^-$) or lineage-negative/Sca-1$^+$/c-Kit$^+$ (LSK) cells in the murine setting. Using multiple surface markers leads to higher enrichment of HSPCs, but at the expense of total cell yield, higher costs, and time required for the procedure. Purification of HSPCs using the aforementioned markers yields functionally heterogeneous populations which can be further subdivided according to subtle differences in surface marker profiles, functional properties, or gene expression patterns. Nonetheless, these purification methods effectively enrich for HSPCs, which are a good starting material for further manipulation using retroviral vectors in vitro. It is worth mentioning that many alternatives to isolate HSPCs do exist and that new ones are added frequently which are based on the expression of other surface markers (e.g., $CD133^+$ in humans [1]) or $CD150^+$/$CD48^-$ in mice [2] and specific properties such HOECHST-dye exclusion (the so-called side population [3]) or low reactive oxygen species levels (ROS-low, isolated as Rhodamine123-low cells [4]) to name just a few.

One of the most essential components for successful gene therapy or various types of laboratory research is efficient, stable and reproducible expression of the therapeutic gene or gene of interest. To this end researchers make use of the natural properties of retroviruses to stably integrate genetic material into a host cell genome, thus mediating permanent vector-directed transgene expression and transmission of the vector genome into the transduced cell progeny. The most frequently used retroviral vector systems are genetically engineered γ-retroviruses (γ-RV) or lentiviruses (LV).

Retroviral particles consist of an enveloped protein shell (retroviral capsid) that contains the viral genome [5]. The retroviral genome is linear, nonsegmented, single stranded RNA-(ssRNA) [6]. RVs are classified as simple oncogenic retroviruses (e.g., Moloney murine leukemia virus) or complex-type retroviruses (e.g., lentiviruses). Simple retroviruses contain *gag*, *pol* and *env* genes which encode the corresponding proteins. Each protein has a specific function during the retroviral lifecycle as summarized in Table 1. In contrast, lentiviruses encode two essential regulatory factors *rev* and *tat* as well as several nonessential accessory proteins in addition to *gag*, *pol*, *env*.

The retroviral life cycle begins with the binding of viral envelope molecules to specific host receptors. After receptor uptake, the viral core is released into host cell cytoplasm through the fusion of viral and cellular lipid membranes or via an endosomal escape

Table 1
Function of retroviral genes

Gene	Full name	Function
gag	Group associated antigen	Codes for viral core structural proteins
pol		Codes for reverse transcriptase, integrase, and viral protease
env	Envelope	Codes for viral envelope proteins
rev	Regulator of expression of virion proteins	Binds to rev responsive element (RRE) in lentiviral genome, essential for nuclear export of unspliced viral genomic RNA
tat	HIV trans-activator	Binds to TAR element in virus long terminal repeat, acts as a transcriptional activator

mechanism [7, 8]. Reverse transcription of the viral ssRNA into dsDNA [9–11] takes place simultaneously and is completed in the cytoplasm of the host cell. The dsDNA integrates into the host cell genome after breakdown of the nuclear membrane during cell cycle (γ-RV) or after actively being shuttled into the nucleus in non-dividing cells (LV). The integrated genome becomes a permanent part of the host cell genome adopting a proviral form. In this context the provirus is transcribed by RNA polymerase II which gives rise to several viral protein coding mRNAs as well as the full length viral genomic RNA. After translation and assembly of the individual viral components on the cell membrane viral particles containing viral genomic RNA can bud and escape from the cell and infect other cells [12].

Retroviruses are able to transduce a great variety of different cell types through receptor mediated uptake. The efficiency of transduction largely depends on the presence of the cognate receptor on the target cell surface. To obtain optimal transduction efficiencies diverse viral envelope proteins are available that can be used to pseudotype vector particles. The most frequently used pseudotypes for γ-retroviral vectors are ecotropic-MLV envelope (Eco, targets mouse and rat cells), amphotropic MLV-envelope (Ampho, targets multiple species, including mouse and human) or gibbon ape leukemia virus envelope (GALV, targets primate cells, including human) (reviewed in [13]). The almost universal pseudotype used for lentiviruses is the vesicular stomatitis virus glycoprotein envelope (VSV-G) owing to its broad spectrum of target cells and species and the physical stability allowing for the concentration of viral supernatants by ultracentrifugation. Alternative envelope proteins for lentiviral vectors are derived from simian endogenous retrovirus (RD114), lymphocytic choriomeningitis virus (LCMV) or measles virus (MV), each one exhibiting a specific tropism for certain cell types (reviewed in [14]).

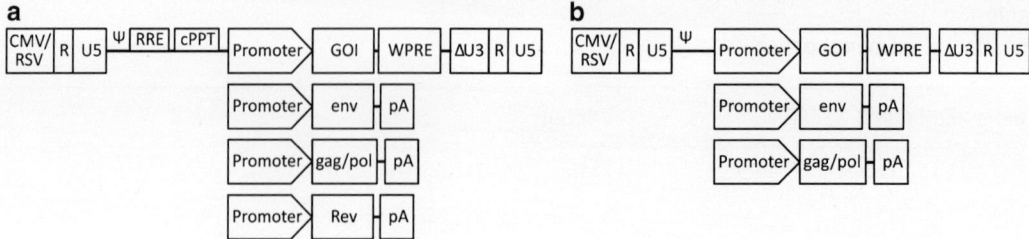

Fig. 1 Plasmid systems for generating retrovirus vectors. (**a**) Four plasmid system used for transfection of HEK-293T cells during production of lentiviral vector particles. (**b**) γ-retroviral three plasmid system. On top the transfer plasmid harboring the GOI, below plasmids providing the accessory proteins in trans: *gag, pol, env* (and *rev*). Abbreviations: *CMV* cytomegalovirus promoter, *RSV* Rous sarcoma virus promoter, Ψ packaging signal, *RRE* rev responsive element, *cPPT* central polypurine tract, *GOI* gene of interest, *WPRE* woodchuck hepatitis virus posttranscriptional regulatory element, *env* envelope protein gene, *gag/pol* viral capsid proteins and enzymes, *pA* polyadenylation signal

In order to use retroviruses as transfer vectors they have been genetically engineered to render them replication defective and to allow for the inclusion of a gene(s) of interest. These modifications include the development of the split packaging design, where genes encoding all essential viral components are provided on separate plasmids for vector production. As a result the transfer vector is devoid of all protein coding reading frames giving space to include a recombinant expression cassette. For γ-retroviral vectors a three plasmid system is used (Fig. 1). One plasmid contains the transfer vector, which harbors the gene(s) of interest flanked by the viral LTRs, a second plasmid contains the gag and pol reading frames, while the envelope is encoded on a third plasmid [15].

A very similar three plasmid system is available for lentiviral vectors, but here the packaging plasmid containing the *gag-pol* reading frames additionally encodes for the *rev* and *tat*-proteins (second generation lentiviral systems), [16, 17]. A newer generation of lentiviral vectors has been introduced which uses a four plasmid system (Fig. 1) [18]. Here the *rev*-coding region is provided on a separate plasmid and tat sequences are completely deleted. *Tat* is dispensable in this generation of lentiviral vectors due replacement of the HIV-U3 region within the 5'LTR with a *tat*-independent heterologous promoter such as from Cytomegalovirus (CMV) or Rous sarcoma virus (RSV).

As a second layer of additional safety modifications the self-inactivating vector (SIN) design has been introduced for both the γ-retroviral and lentiviral vectors. A major deletion of the viral promoter sequences within the U3 region of the LTR abolishes transcriptional activities of integrated vectors, thus preventing generation of full length viral genomic RNA, rendering the vector replication incompetent. The abrogation of viral LTR-driven promoter activity necessitates the inclusion of an internal heterologous promoter to express the vector encoded transgene. Different promoters can be chosen to obtain the desired expression levels or provide restricted expression in specific cell types, the most prominent ones being derived from CMV, Spleen Focus Forming Virus (SFFV), Elongation Factor 1 α (EF1α), or Phosphoglycerate Kinase (PGK) to name just a few.

A convenient method to monitor vector mediated transgene expression is to coexpress a selectable marker protein along with the gene of interest. This can be accomplished by use of an internal ribosomal entry site (IRES) between the gene of interest and the reporter coding region. These virus derived sequences allow for translation reinitiation of a second reading frame from a polycistronic mRNA (reviewed in [19]). Viruses use these elements to bypass shutdown of cap-dependent translation as a consequence of cellular antiviral responses. As an alternative, 2A peptide sequences can be used (reviewed in [20]). This type of element is available from multiple different viruses and sequential coupling of several of them or the combination with an IRES element enables the expression of polycistronic mRNAs. The stoichiometry of transgene expression in bicistronic or polycistronic vectors is difficult to predict and needs to be empirically determined. Generally the second transgene is less efficiently translated than the first one, and in most cases the reporter is placed in second position. A long list of useful reporters exists, among them fluorescent proteins, antibiotic resistance genes or cell surface markers. Available fluorescence proteins cover the full range of wavelengths detectable in FACS or fluorescence microscopy and the palette of colors is being continuously expanded. Resistance markers have the advantage that a pure population of transduced cells can be obtained upon antibiotic selection, while expression of a cell surface marker such as signaling defective truncated versions of CD34 or low affinity nerve growth factor receptor (LNGFR) allow for quick and easy magnetic assisted cell selection (MACS) [21, 22]. However, both vector systems have limitations in their packaging capacity. Using γ-retroviral vectors, expression cassettes of about up to 8 kb can be efficiently packaged, while lentiviral vectors can efficiently package roughly an additional 2 kb (e.g., up to ≈10 kb). Although no strict size limitation exists, there is a negative correlation between insert size and the amount of infectious viral particles produced. As a rule of thumb the larger the insert the lower the vector titer is.

Transgene expression and vector titers can be increased by incorporation of the woodchuck hepatitis virus posttranscriptional regulatory element (WPRE) downstream of the transgene. The WPRE stabilizes the mRNA through secondary structures resulting in an up to fivefold increase in gene expression [23, 24].

Vector production occurs by cotransfection of the transfer vector and the packaging plasmids into a producer cell line, such as HEK-293T cells. Alternatively, stable packaging cell lines are available for γ-retroviral vectors which constitutively express the viral accessory proteins (gag, pol, env) necessitating transfection of the transfer plasmid alone [25, 26]. The supernatant containing the vector particles is collected on subsequent days and cleared by ultrafiltration. This supernatant can be directly used to transduce target cells, stored at −80 °C for later use or concentrated, a step routinely performed when working with VSV-G-pseudotyped vector particles.

Considering the sometimes low titers of retroviral vector preparations, much work has been done to enhance the transduction efficiency particularly of difficult to transduce cell types. Improved transduction of murine HSPC was first reported utilizing "coculture" of target cells with vector producer cell lines [27]. Subsequent clinically relevant improvements were reported [28], where co-localization of ecotropic retroviral particles with the target cells mediated by fibronectin leads to enhanced transduction of HSPCs. Recombinant fibronectin fragments, CH-296 (RetroNectin™) can interact with different adhesion domains, such as integrins (very late antigens 4 and 5 (VLA4 and VLA5)) and proteoglycans, and moreover can mediate co-localization of ecotropic and amphotropic retroviral particles to the cells via a heparin-binding domain [29]. Furthermore, it has been reported that nonhuman primate CD34+ cells showed less apoptosis and increased proliferation in the presence of CH-296 [30]. Coating of tissue vessels with CH-296 has become a standard method adopted in many clinical transduction protocols as well as in basic research laboratories.

Progress in the development of retroviral vectors and their application has led to the initiation of several gene therapy trials during the past two decades. Examples are X-linked and ADA-severe combined immunodeficiencies [31–34], Wiskott-Aldrich-Syndrome [35, 36], X-linked Chronic Granulomatous Disease [37], Adrenoleukodystrophy [38], and more recently Metachromatic Leukodystrophy [39]. Despite the occurrence of several cases of severe adverse events these trials represent a major breakthrough in the treatment of serious, life-threatening diseases utilizing HSPCs as targets and serve as basis for further developments in this field.

Current research is focusing on reducing the genotoxicity and improving the expression properties of vectors by shielding the vector genome from its chromosomal environment using "insulators" in the virus long terminal repeats. Substantial improvements from early attempts have been made after the identification of minimal essential sequences conferring enhancer blocking activity as well as chromatin boundary function [40–42]. This approach may also limit the impact of the vector on genes adjacent to proviral integration sites, thus reducing the risk of a potentially harmful dysregulation of cellular genes. Another approach to improve gene expression include the use of the so-called chromatin opening elements (UCOEs) to prevent epigenetic silencing and reduce position effect variegation (PEV), a problem which is particularly important when working with embryonic stem cells, but is also observed in other cell types. PEV contributes to loss of expression or heterogeneous expression between different cell clones [43–45]. Following is a list of specific reagents and protocols utilized in viral production and cell transduction analysis.

2 Materials

2.1 Reagents

1. Phosphate Buffer Saline (PBS).
2. Hank's Balanced Salt Solution (HBSS).
3. Dulbecco's modified Eagle's medium (DMEM).
4. Iscove's modified Dulbecco's medium (IMDM).
5. Fetal Calf Serum (FCS, Biowest, Nuaillé, France).
6. Penicillin–Streptomycin stock solution-100× (pen–strep).
7. Wash/Staining Buffer: PBS, 2 % (v/v) FCS, 1 % (v/v) pen–strep.
8. Red blood cells (RBC) lysing buffer (BD Biosciences, San Jose, CA).
9. Trypsin–EDTA stock solution-0.05 %.
10. Cell dissociation buffer (Life Technologies).
11. Calcium Phosphate transfection kit containing $CaCl_2$, H_2O, HBS (Life Technologies, Grand Island, NY, USA).
12. Histopaque-1083.
13. Dynabeads (Life Technologies, Grand Island, NY, USA).
14. Gelatin: 1 % gelatin in dH_2O.
15. RetroNectin™: 60–80 µg of FN in 2 mL PBS (Takara, Otsu Shiga, Japan).
16. Hexadimethrine bromide (Polybrene 1 µL/mL, stock solution: 8 mg/mL, Sigma-Aldrich Corp. St. Louis, MO, USA).
17. 5 Fluorouracil (5-FU, stock solution: 68–80 mg/mL): dissolved in HBSS, to a final concentration 10 mg/mL.
18. Chloroquine (25 mM, Sigma-Aldrich Corp. St. Louis, MO, USA).
19. 1 M Hepes buffer solution.
20. Calf Serum (CS, Thermo Scientific).
21. L-Glutamine-100× (Life Technologies).
22. StemSpan SFEM (STEMCELL Technologies Inc., Canada).
23. Polyethyleneimine (linear from Polysciences MW ~25.000; branched from Sigma, MW ~25,000).

24. Neomycin sulfate.

25. 1 M NaOH.

26. Recombinant murine stem cell factor (rmSCF, Peprotech).

27. Recombinant human granulocyte colony stimulating factor (rhGCSF, Peprotech).

28. Recombinant human Thrombopoietin (rhTPO, Peprotech).

29. Bovine serum albumin 95 % (BSA).

30. Dynal MPC-L magnet (Invitrogen, Carlsbad, CA, USA).

2.2 Retroviral Production and Titration Media

1. Retroviral Growth Media: DMEM, 10 % (v/v) CS, 1 % (v/v) pen/strep.

2. Retroviral Transfection Media: retroviral growth media supplemented with 25 µM chloroquine solution.

3. Retroviral Harvest Media: DMEM, 10 % (v/v) FCS, 1 % (v/v) pen/strep, 0.67 mL 1 M NaOH, 1 % (v/v) Hepes, 1 % (v/v) L-Glutamine.

4. Retroviral Titering Media: DMEM, 10 % (v/v) CS, 1 % (v/v) pen/strep, 8 mg/mL hexadimethrine bromide (Polybrene).

2.3 Cell Culture Media

1. Cell Growth Media: DMEM, 10 % (v/v) FCS, 1 % (v/v) pen–strep.

2. Bone marrow cell (BMC) culture media: IMDM, 10 % (v/v) FCS, 1 % (v/v) pen/strep supplemented with 100 ng/mL rmSCF, 100 ng/mL rhGCSF, and 100 ng/mL rhTPO.

3. Alternatively serum-free culture medium can be used for the cultivation of mouse and human HSPCs: StemSpan SFEM supplemented with 100 ng/mL SCF (species matched), 100 ng/mL hTPO (interspecies cross-reactive), 100 ng/mL hFLT3-L (interspecies cross-reactive), and 20 ng/mL IL-3 (species matched).

2.4 Antibodies

1. Monoclonal antibodies: B220 (Pacific Blue, clone RA-6B2), Gr-1 (APC-Cy7, clone RB6-8C5), CD45.1 (PerCP-Cy), CD45.2 (APC) all from BD Biosciences, San Jose CA, CD3 (PE-Cy7, clone 145-2C11), Sca-1 (FITC, clone D7), c-Kit (APC, clone 2B8), Streptavidin (PE), CD34 (APC, clone 4H11) eBioscience, San Diego, mCD45 (Brilliant-Violet 421, clone 30-F11), Biolegend, San Diego, hCD45 (PerCP-Cy5.5, clone 2D1) eBioscience, San Diego, hCD (PE-Cy7, clone HIB19) Biolegend, San Diego, CD33 (PE, P67.7) eBioscience, San Diego, hCD3 (APC, OKT3), eBioscience, San Diego, human and mouse Fc-Block, Miltenyi Biotech, Auburn, fixable viability dye eFluor780, eBioscience, San Diego, CA, USA.

2. Biotin-labeled antibodies for lineage depletion (lin⁻): Gr-1 (clone RB6-8C5), Mac-1 (clone M1/70), Ter119 (clone TER-119),

B220 (clone RA-6B2), CD5 (clone 53-7.3), and CD8a (clone 53-6.7), all from BD Biosciences, San Jose, CA, USA.

3. Mouse lineage cell depletion kit, Miltenyi Biotech, Auburn, CA, USA.

4. CD34 MicroBead Kit, Mitenyi Biotech, Auburn, CA, USA.

2.5 Cell Lines

1. HEK-293T cells (ATCC, Manassas, VA, USA).

2. NIH-3T3 cells (ATCC, Manassas, VA, USA).

3. HT 1080 cells (ATCC, Manassas, VA, USA).

2.6 Mice

1. C57BL/6J (No. 000664, Jackson Laboratory, Bar Harbor, ME, USA).

2. B6.SJL-Ptprc^a-Pep3^b/BoyJ (No. 002014, Jackson Laboratory, Bar Harbor, ME, USA).

3. NOD.Cg-Prkdc^{scid} Il2rg^{tm1Wjl}/SzJ (No. 005557, Jackson Laboratory, Bar Harbor, ME, USA).

3 Methods

3.1 Retroviral Supernatant Production

Production of retroviral vector particles is performed by co-transfection of retroviral plasmids into human embryonic kidney (HEK)-293T cells. Expression of the SV40-large T-antigen in this cell line mediates episomal replication of plasmids harboring the SV40-ori, which is present on plasmids used for vector production. A schematic representation of retroviral supernatant production is shown in Fig. 2. Efficient transfection of 293T cells is the key to obtain high titer vector supernatants and is often the first step in a series of optimizations that may be required. Here, we describe two methods for the transfection of HEK-293T for vector production, which usually give comparable results. Due to the many factors influencing the outcome of the process, such as cell passage and transfectability, plasmid quality, and even minor factors like temperature during calcium phosphate precipitate formation, it is associated with a significant degree of variability and is difficult to control.

3.1.1 HEK- 293 Cells Culture

1. Grow HEK-293T cells to 70–80 % confluence in growth media on 10-cm tissue-culture-treated dishes (TCT). Optional: TCT can be gelatinized using 0.1 % gelatin for 5 min at 37 °C if detachment of cells during virus production procedure is observed.

2. Remove media from the HEK-293T cells and rinse with 5 mL PBS. Discard the PBS and add 1 mL of 0.05 % trypsin.

3. Incubate until the cells are detached. Inactivate the enzymatic reaction by adding a minimum of 2 mL growth media. Pipet until a single cell suspension is obtained.

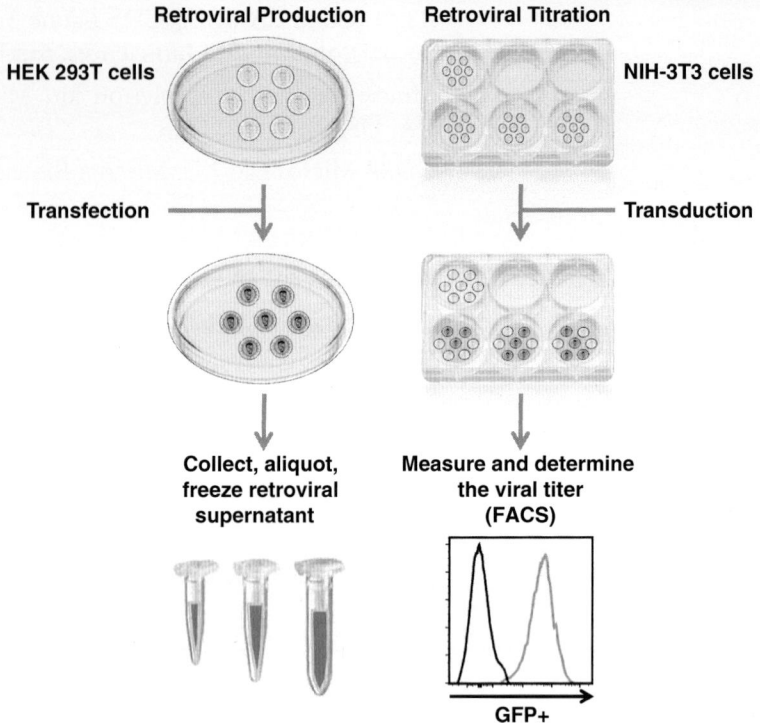

Fig. 2 Experimental timeline of retrovirus vector production and titration. *HEK 293T* Human Embryonic Kidney Cells, *NIH 3T3* 3T3 fibroblast cell line

4. Spin the cells down at $400 \times g/5$ min and decant supernatant. Resuspend cell pellet in transfection media, add to transfection cocktail (*see* Subheading 3.1.2) and plate to a density of 50–70 %.

3.1.2 Transfection of HEK-293T for Vector Production using Calcium Phosphate (CaPO₄)

1. Thaw the transfection reagents to room temperature (RT).

2. Using a sterile reaction tube add water, plasmids and mix with $CaCl_2$ (*see* Table 2).

3. Add HBS and bubble air through the mixture using a portable Pipet-Aid (BD, Bioscience, Bedford, MA) for 45 s to 1 min.

4. To obtain an efficient formation of DNA–calcium phosphate precipitate, incubate the transfection cocktail for 45 min at RT.

5. (*Day 0*): Mix cell suspension (*see* Subheading 3.1.1, **step 4**) with transfection cocktail and divide among the plates to achieve a 50–70 % cell density. Incubate at 37 °C, 5 % CO_2 overnight.

6. (*Day 1-morning*): Replace the transfection media with 5 mL fresh virus-harvest media and incubate.

7. (*Day 1-evening*): After 8–10 h collect the viral supernatant containing the vector particles from the plate into 50 mL

Table 2
Transfection mixture for retrovirus or lentivirus production

γ-Retroviral plasmids	
Construct	**µg/1 plate**
Retroviral plasmid	8–10 µg[a]
GAG/POL plasmid	10 µg
Envelope (Eco/Ampho/GALV)	3 µg
CaCl$_2$	36 µL
H$_2$O	Fill to 300 µL
HBS	300 µL

Lentiviral plasmids	
Construct	**µg/1 plate**
Lentiviral vector	8–10 µg[a]
pCDNA3.1 gag/pol	5 µg
RSV-Rev	2.5 µg
VSV-G	2.5 µg
CaCl$_2$	36 µL
H$_2$O	Fill to 300 µL
HBS	300 µL

The amount of plasmid listed indicated microgram (µg) of DNA needs/10 cm plate
[a]The amount of plasmid should be determined depending on the size of construct

Falcon tube. Refresh media by adding 5–6 mL of harvest media. Keep the collected supernatant at 4 °C, overnight.

8. (*Day 2-morning*): Perform collection of viral particles up to 3 days post transfection, in 12–24 h intervals, and keep cool or at 4 °C. Continue collecting viral supernatant as described at **step 5** until three collections are obtained depending on cell viability.

9. (*Day 4*): Combine the viral supernatant and filter the supernatant through 0.45 µm filter unit. If desired proceed to viral particle concentration (*see* **step 9**).

10. Viral supernatants can be used immediately or stored at −80 °C for later purposes. Optional: viral supernatant can be concentrated using an ultracentrifuge rotor at $70,000 \times g$ for 2 h, 4 °C or $13,233 \times g$ overnight at 4 °C in a Beckman XL-90 using an appropriate swinging bucket rotor (e.g., SW28 or SW32Ti).

11. After spin, resuspend the viral pellet in a small volume of IMDM, 1 % BSA (e.g., to obtain a 100–300-fold concentration). Combine the first and second rinse together, mix, then aliquot in smaller volumes (suitable working volumes: 10–100 µL), freeze, and store at –80 °C (*see* **Note 1**).

12. Proceed to titration step (*see* Subheading 3.2) using a small aliquot of viral supernatant, e.g., 40 µL for concentrated virus, 200 µL for unconcentrated virus.

3.1.3 Transfection of HEK-293T for Vector Production Using Polyethyleneimine (PEI)

In this procedure plasmid DNA entry into the cell is augmented by complexation with PEI. Both linear and branched PEI have been used with similar success. The method is simple, relatively robust, and cost-effective.

1. To prepare PEI solution for transfection create a 20 mg/mL stock solution in water (heating to ~80 °C is required for linear PEI). Adjust to pH 7, filter-sterilize, and make a working solution of 1 mg/mL by diluting with sterile water. Store at 4 °C.

2. For transfection mix plasmids as shown in Table 2, in 1 mL of DMEM (no additives) per 10-cm plate.

3. Add 3 µL of PEI solution per µg of plasmid DNA, mix immediately, and incubate for 20–30 min, RT. The ratio of PEI to DNA is a critical factor and different ratios may be tested for optimization of transfection.

4. Apply the transfection mix gently to the cells and incubate overnight.

5. Twelve hours to 24 h after transfection replace medium with 5–6 mL collection medium.

6. Continue with **step 6** of Subheading 3.1.2.

3.2 Titration of Viral Supernatants

The concentration of infectious vector particles in the supernatant can be determined by titration on a permissive "standard" cell line. This is done by applying serial dilution of the retroviral supernatant to the culture media of an appropriate cell line, species matched to the retroviral envelope (e.g., NIH-3T3 for ecotropic envelope). The following steps are describing the titration of retroviral supernatants on NIH-3T3 cells.

1. Plate ≈100,000 NIH-3T3 cells well into 6-well TCT dishes in 2 mL growth medium (*see* **Note 2**).

2. 24 h later perform cell count of one or two unused wells for titer calculation.

3. (Day 0): Aspirate media from seeded cells and add complete growth media supplemented with Polybrene at a final concentration of 8 µg/mL to each well (1.8 mL per well). Perform 10-fold serial dilutions of vector supernatant; starting with 200 µL for unconcentrated and 50 µL for concentrated retrovi-

ral supernatant (as shown Fig. 2). When finished, incubate at 37 °C/5 % CO_2.

4. (Day 1): Remove the supernatant from the wells and refresh media by adding 3 mL of growth media/well.

5. (Day 3): 48 h post-viral transduction, perform FACS analysis and examine transduction efficiency (e.g., via GFP or tCD34).

6. Wash with PBS and then detach cells using 1 mL of 0.05 % trypsin and resuspend cell in 2 mL PBS.

7. Transfer 500 μL cell suspension into a flow tube and perform flow cytometry analysis.

The concentration of infectious viral particles/mL is determined using the following formula (*see* **Note 3**):

$$infectious\ particles\ /\ mL = (\#\ of\ cells\ at\ time\ of\ transduction)$$
$$\times (\%\ of\ fluorescent\ cells) \times (factor\ of\ dilution) / (total\ volume(mL))$$

3.3 Hematopoietic Stem Cell Isolation

3.3.1 Isolation of Human CD34 Cells

CD34 positive HSPCs can be isolated from bone marrow aspirates, cord blood or mobilized peripheral blood. The frequency of $CD34^+$ cells in these samples normally ranges from 0.1 % to 4 % of all nucleated cells. Depending on the source and upstream handling of the sample the technical procedures for enrichment of $CD34^+$ cells differ. Generally, washing the cells precedes the enrichment of mononuclear cells by Ficoll. Finally, after immunolabeling the cells magnetic assisted cell sorting (MACS) is performed to isolate the $CD34^+$ fraction. As very reliable commercially available kits do exist, we recommend using these kits and following the manufacturer's instructions. Then proceed to transduction (*see* Subheading 3.4).

3.3.2 HSPC Enrichment After 5-FU Induction of Donor Mice

Proliferating cells are eliminated by application of the nucleotide analog 5-Fluorouracil (5-FU). This step mainly eliminates the bulk fraction of late progenitor cells and as a consequence quiescent HSPCs, which are not affected by 5-FU treatment, become enriched in the mononuclear fraction.

1. Inject 150 mg/kg of 5-FU intravenously.

2. 4 days post injection, mice are euthanized. Harvest bones (femur, tibia and iliac crest) from donor mice and place in PBS, at RT.

3. Under sterile conditions crush bones with mortar and pestle, and rinse the cells with PBS. Filter the cell suspension through 40 nm cell strainer and spin down at $400 \times g$, 5 min, RT.

4. Resuspend cell pellet in PBS (e.g., 5 mL/5 mice).

5. Perform a Ficoll by layering the cell suspension over an equal volume of Histopaque-1083, centrifuge at $450 \times g$, for 20 min, RT without brake. Cells are separated according to density.

Table 3
Biotin-labeled antibody cocktail

Monoclonal antibody
CD3 (PE-Cy7)
B220 (Pac Blue)
Gr-1 (APC-Cy7)
CD45.1 (PerCP-Cy)
CD45.2 (APC)
Sca-1 (FITC)
c-Kit (APC)
Streptavidin (PE)

The volume listed indicate microliter (μL) of biotin-labeled antibody per 1×10^8 cells

6. Collect interphase layer of cells into a 50 mL conical tube containing 30 mL ice cold washing buffer. The interphase layer contains the low-density bone marrow cells (LDBM) including the HSPCs.

7. Centrifuge the cells at $400 \times g$, 5 min, RT, discard the supernatant and resuspend the cells into desired volume of bone marrow culture media.

8. Plate the cells (≈ 0.5–1×10^6/mL) into a non-tissue-culture-treated (NTC) dishes, at 37 °C, 5 % CO_2, for 24 h. If desired perform retroviral transduction (*see* **Note 4**).

3.3.3 HSPC Isolation from Bone Marrow Cells

Low-density bone marrow cells are isolated using Histopaque-1083 as previously described (*see* Subheading 3.3.2, **steps 2–6**). Lineage positive cells are depleted using lineage specific biotin-conjugated antibodies and a Dynal MPC-L magnet (Invitrogen, Carlsbad, CA).

1. Perform bone marrow harvest and Ficoll (*see* Subheading 3.3.2, **steps 2–6**).

2. Resuspend the pellet in PBS (1 mL/mouse), save an aliquot of 100 μL cell suspension for future controls, and determine cell number and viability (≈ 0.5–2.5×10^7 cells/mouse); all manipulations from here on should be done at 4 °C. According to the LDBM cell number, prepare a mix of biotin-labeled antibodies (*see* Table 3). Add 112.4 μL antibody mix/1×10^8 cells, place the labeled cells on rotator, in cold room, at low speed for 45 min.

3. After incubation wash cells and antibody with ice-cold wash buffer. Aliquot a sample of this suspension for flow cytometry (undepleted control). Centrifuge the cell suspension at $400 \times g$,

4 °C, 5 min, and resuspend pellet in 1.5 mL of PBS 2 % FCS and 1 % pen/strep.

4. Add cells to a tube of prewashed Dynabeads ($1 \text{ mL}/1 \times 10^8$ cells, equals 4 beads/cell) in a 15 mL conical tube and place back on the rotator at low speed for 45 min.

5. Fill to 7 mL with cold staining buffer and place tube on Dynal MPC-L magnet. After 2–3 min collect lineage negative cell suspension (in solution) with the tube still mounted on the magnet. Do not disturb pellet of labeled cells sticking to the tube wall. Repeat the procedure up to four times to increase purity of lineage negative cells.

6. Count the depleted cells using a cell counter, then centrifuge the cells at $400 \times g$, 4 °C for 5 min, decant, and resuspend the cell pellet in 1–2 mL of staining buffer (≈ 0.5–1×10^6 cells/mouse).

7. To the resuspended cells and control (*see* **step 2**) add Streptavidin (PE), c-Kit (APC), and Sca-1 (FITC) at a concentration of $6 \text{ μL}/1 \times 10^7$ cells, mix, and keep it on ice for 30 min, in dark.

8. Single color/isotope controls should be set up for FACS compensation from pre-depletion sample. As well as viability controls.

9. Add 10–15 mL of cold PBS, 2 % FCS, 1 % pen/strep to remove unbound antibody, and spin down at $400 \times g$, 4 °C, and 5 min. Decant, filter the cell suspension through 40 μm cell strainer into FACS tube, and proceed to flow sorting.

10. After sorting, count the viable cells (hemocytometer), centrifuge at $400 \times g$, 4 °C, 5 min, decant, then resuspend the pellet in cytokine-supplemented BMC culture media (rhGCSF, rhTPO, rmSCF), (≈ 2–7×10^4 cells/mouse).

11. Day 0: Plate sorted LSK cells into NTC (e.g., at a density of 0.5–2×10^6 cells/mL).

3.3.4 Alternative Protocol: Lineage Depletion Using Commercially Available Kits

A simple alternative to the above mentioned method is to use the Miltenyi mouse lineage cell depletion kit according to the manufacturer's instructions. Similar kits are also available from other suppliers, and all of them follow the same principle. After depletion the user can either directly proceed to viral transduction (*see* Subheading 3.4) or continue to **step 5** of Subheading 3.3.2 for isolation of LSK cells.

3.4 Cultivation and Retroviral Transduction of Hematopoietic Stem Cells

After purification of the desired population (human CD34$^+$, mouse lineage negative or LSK cells) the cells are cultivated in cytokine supplemented medium. The type and concentration of added cytokines has a major impact on the resulting cell phenotype, proliferation, biology, and success of viral transduction. A large number of different cultivation conditions have been published, unfortunately often using different read out assays either in vitro and/or in vivo.

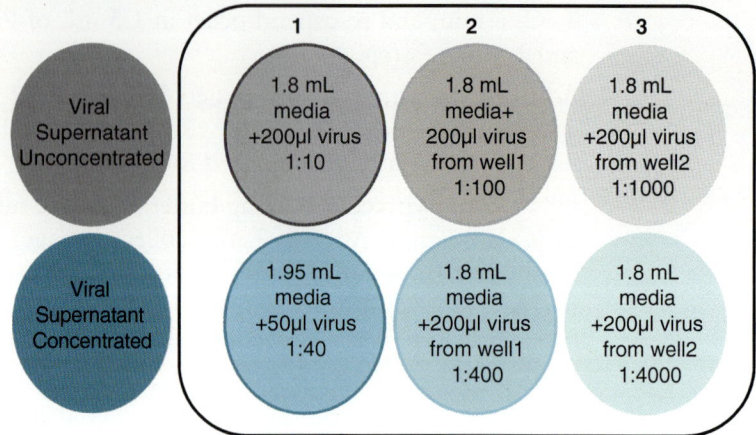

Fig. 3 Schematic representation of viral serial dilution used to determine supernatant titer

No comprehensive study exists to date comparing them all, particularly not in the context of viral transduction. Consequently, there is no state of the art cultivation method for laboratory purposes and the cultivation conditions described here may be replaced by the experimentator's conditions of choice most suitable for their downstream application. Nonetheless, newer protocols mostly use serum-free culture conditions to reduce inter-experimental variability. Furthermore, for clinical gene therapy high-dose cytokine supplementation using SCF (100 or 300 ng/mL), Flt3-L (100 or 300 ng/mL), TPO (100 ng/mL), and IL-3 (60 ng/mL) has been established.

Depending on the desired target cell retroviral copy number and vector titer, HSPCs can be transduced once or more. It is worth mentioning that the efficiency of transduction significantly differs between cell lines used for titration and primary HSPCs. Furthermore, the cytokine mix as well as the species and tissue origin of HSPCs has a great impact on transduction. For example freshly isolated human mobilized peripheral blood CD34+ cells are known to be difficult to transduce, so multiplicity of infection (MOI) of 50–100 are used to achieve ~40–90 % of transduction. Bone marrow derived CD34+ cells or murine HSPCs are easier to transduce, and similar results can be obtained by applying an MOI of 20–50. The protocol below describes a double transduction procedure, followed by a 24 h recovery (as outlined in Fig. 3). The procedure is suitable to transduce HSPCs isolated by any of the previously described methods using either γ-retroviral or lentiviral vectors. For lentiviral vectors, which are able to transduce non-dividing cells the duration of the total procedure can be reduced by prestimulation for only 24 h and by performing a single round of transduction. However, if the cells need to be FACS-sorted, at least 48 h time should be given after transduction to allow for marker transgene expression.

1. (*Day 2*): 48 h post-isolation count the plated cells from *Day 0* (*see* Subheading 3.3.3, **step 11**) and determine viability. If debris has accumulated wash in PBS at low speed (~150×g) for 5 min and discard supernatant. Resuspend cells in cytokine supplemented growth media.

2. Pretreat NTC well plates with RetroNectin™ 4–20 µg/square cm diluted in PBS, and incubate 2 h at 37 °C. After incubation, remove RetroNectin™ and save (*see* **Note 4**). If working with ecotropic γ-retroviral vectors continue with pre-loading of vector particles to RetroNectin™ (**step 3**), in all other applications proceed to **step 4**.

3. Add virus to the well and allow the plate to sit in the incubator for 30 min (*see* **Note 5**).

4. *1st Transduction*: Add cells in an appropriate volume of growth media for the selected cell culture plate. If working with ecotropic γ-retroviral vectors continue with **step 5**, in all other cases add suitable virus volume to the cell suspension and mix. Incubate overnight (6–12 h).

5. *2nd Transduction*: (*Day 3*) Aspirate the supernatant without removing the cells. Refill to original volume by adding fresh viral supernatant and media supplemented with cytokines. Return to the incubator overnight.

6. (*Day 4*): Aspirate the viral supernatant and replace with fresh BMC growth media.

7. (*Day 5*): Detach the cells using cell dissociation buffer; count after spinning down at 400×g, 5 min. Pass the cell suspension through a 40 µm cell strainer and proceed to FACS analysis and sort for cell surface marker or fluorescence expression. Alternatively sorting of transduced cells can be omitted and downstream experiments can be performed using the bulk population.

8. Fluorescent sorted cells can be used for in vitro or in vivo experiments as described in Fig. 4.

3.5 In Vitro Experiments

In many cases the relevant scientific questions can be answered by in vitro experiments, which saves the time and costs associated with animal work. Given the diversity of possible in vitro assays, a comprehensive description is beyond the scope of this chapter. Thus, only an overview will be given. Frequently used assays aiming to identify HSPC content include colony-forming cell assays (CFC) in semisolid medium (measures progenitor cell frequency/activity), or the long-term culture initiating cell (LTC-IC) and cobblestone area forming cell assay (CAFC). The latter two are used to monitor the presence and frequency of more immature HSPCs and require cultivation for 5–10 weeks on a stromal cell layer.

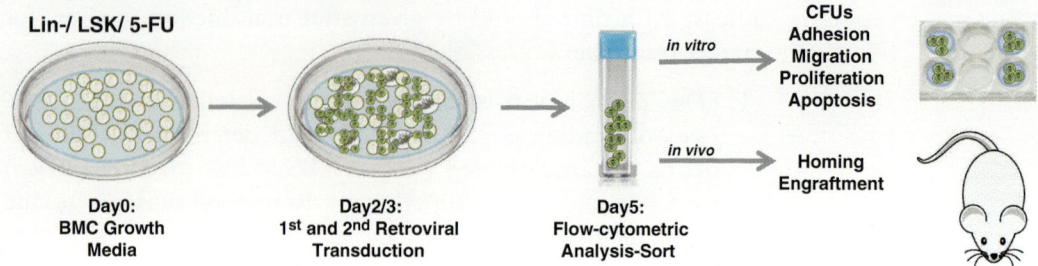

Fig. 4 Schematic representation of retrovirus vector transduction on RetroNectin™ and in vitro *and* in vivo analysis. *Lin⁻* lineage negative fraction, *LSK* Lin⁻Sca-1⁺c-Kit⁺, *5-FU* cells isolated post 5-Fluorouracil treatment, *CFU* colony-forming units

3.6 Recipient Preparation and Transplantation

Generally, recipient mice have to be conditioned by irradiation or chemotherapy to allow for efficient engraftment of donor cells. The sensitivity of mice to these kinds of treatments may differ between strains, age, and gender of the animals. Also, there is variability between the irradiation devices used, so that a dose response curve may need to be established. Engraftment of donor cells in the mouse/ mouse setting can be monitored by exploitation of a naturally occurring polymorphism in the CD45 surface antigen. Leucocytes in the mouse strain C57BL/6 express the CD45.2 isoform, which can be discriminated via specific antibodies from the CD45.1 isoform expressed on the leukocytes of the congenic mouse strain BoyJ. The number of the cells required for the transplantation varies significantly depending on the cells used and the pre-treatment *in vitro*.

3.6.1 Transplantation of Murine Cells from C57BL/6 into BoyJ Recipients

1. Recipient mice (BoyJ) are lethally irradiated up to 24hrs in advance (*see* **Note 6**).

2. Count HSPCs from (see subheading 3.4, **step 8**) spin down for 5 min, $400 \times g$. Resuspend cells in PBS, (*see* **Note 7**).

3. Inject $5 \times 10^5 - 2 \times 10^6$ transduced cells/ mouse- when assessing homing experiments. For engraftment purposes transplant $5 \times 10^4 - 1 \times 10^5$ cells/ mouse together with helper cells ($5 \times 10^5 - 1 \times 10^6$ cells/ mouse). Prepare the cells suspension in up to 300 μL and administer intravenously via tail vein injection.

3.6.2 Transplantation of Human Cells into NSG Recipients

NOD.Cg-Prkdc^scid Il2rg^{tm1Wjl}/SzJ (NSG-mice) carry both the severe combined immune deficiency (scid) mutation and a knock-out of the interleukin-2 receptor gamma chain in the NOD/ShiLtJ background. As a result, NSG mice lack mature T, B, and NK cells, and are deficient in cytokine signaling. Human hematopoietic stem cells can be efficiently engrafted in this mouse strain after mild conditioning, although the lineage output is heavily biased towards B-lymphocytes [46]. In case low numbers of cells are transplanted, female mice should be chosen as recipients as they have been shown to better engraft human cells.

1. Recipient mice are sublethally irradiated with 2.5Gy within 24 h of transplantation. Due to the severe immunodeficiency and conditioning Neomycin or Baytril-supplemented water should be provided thereafter.

2. Harvest the cells from the culture or FACS-sort. Pellet at $400 \times g$ for 5 min. Resuspend the pellet in up to 300 µl per mouse in plain IMDM or PBS for transplantation. As recipient mice are sublethally irradiated, remaining endogenous hematopoietic activity competes with transplanted cells. As a consequence the total applied cell dose will be reflected by the degree of human cell engraftment. Application of more than 2×10^5 cells is recommended, and up to 10^7 cells per mouse can be administered.

3. Proceed to tail vein injection.

3.6.3 Transplantation of Murine Cells from C57BL/6 into BoyJ Recipients

1. Recipient mice (BoyJ) are lethally irradiated up to 24 h in advance (*see* **Note 7**).

2. Count HSPCs from (*see* Subheading 3.4, **step 8**) spin down for 5 min, $400 \times g$. Resuspend cells in PBS, (*see* **Note 8**).

3. Inject 5×10^5–2×10^6 transduced cells/mouse- when assessing homing experiments. For engraftment purposes transplant 5×10^4–1×10^5 cells/mouse together with helper cells (5×10^5–1×10^6 cells/mouse). Prepare the cells suspension in up to 300 µL and administer intravenously via tail vein injection.

3.7 Donor/Recipient Engraftment Analysis

Transplanted HSPC cells engraft into the bone marrow of conditioned animals and will continuously produce all mature blood lineages. Engraftment occurs sequentially, it can first be observed in the short lived granulocytic lineage, while it is delayed in the long lived T-cell fraction. Blood counts normalize around 6 ± 2 weeks post transplantation; accordingly it may be helpful to perform analysis of donor-derived leukocytes around this time point.

3.7.1 Analysis of NSG Mice Transplanted with Human Cells

1. Take a blood sample according to the procedure established at your institute. Collect the blood (50–200 µL/mouse) using micro-capillary tubes into an EDTA coated micro-containers. Optional: Perform total blood counts.

2. Lyse the erythrocytes with hypotonic 1× RBC-lysis buffer (1 mL of 1× lysis buffer for 100 µL of blood). Incubate 10 min, at room temperature. Pellet the cells at $400 \times g$, 5 min. Resuspend the cells in wash buffer, spin down $400 \times g$, 5 min.

3. Add 1 µL of antibody per sample: human FcR-block, mouse-FcR-block, hCD45 (PercP-Cy5.5), hCD33 (PE), mCD45

(Brilliant-Violet 421), hCD19 (PE-Cy7), hCD3 (APC), and fixable viability dye eFluor780. Incubate for 30 min, at 4 °C in the dark.

4. Wash the cells using 1 mL of PBS via centrifugation at $400 \times g$, 5 min. Resuspend the cell pellet in 350 µL of PBS. Proceed to flow cytometry measurement. Analyze fluorescence and surface marker expression.

3.7.2 Analysis of BoyJ Mice Transplanted with Murine Cells: C57BL/6

1. For blood samples perform as described in **steps 1** and **2** of Subheading 3.7.1.

2. Add 1 µL of each antibody (for up to 1×106 cells/tube): B220 (PacificBlue), Gr-1 (APC-Cy7), CD45.1 (PerCP-Cy5.5), CD45.2 (APC), all from BD Biosciences, San Jose CA, CD3 (PE-Cy7), eBioscience, San Diego, CA, and incubate, 25–30 min, 4 °C, dark (*see* **Note 8**). Optional: Block the unspecific binding by adding blocking antibody (CD16/CD32—Fc III/II R, clone 2.4G2).

3. For FACS analysis *see* Subheading 3.7.1, **step 4**.

4 Notes

1. It is recommended to aliquot the retroviral supernatant in small volumes. When thawing, retrovirus vectors can lose up to 50 % of viral activity with each thaw.

2. When determining the viral titer, we usually plate three wells for each virus concentration and an extra three wells for controls (two wells are for cell counting).

3. The calculation of viral titers should be done using dilutions resulting in transduction efficiencies between 5 and 20 %. At high transduction rates the titer will be underestimated due to multiple integration events in each cell.

4. RetroNectin™ incubation can be 2 h, 37 °C, 5 % CO_2 prior the transduction or overnight at 4 °C. We reuse the RetroNectin™ 3 times.

5. Typically we transduce 500,000 cells/well in a six well plate.

6. We normally use a split dose of 7Gy and 4Gy, 3–4 h apart.

7. When homing of cells to the bone marrow is assessed, it is recommended to label the cells with specific dyes such as Vibrant DiD or CellTrace CFSE. This will help to achieve a better readout.

8. Since applications of antibodies vary, it is recommended to titer the reagents according to the manufacturer's instructions for optimal results.

Acknowledgments

This manuscript is dedicated to past and present members of the Williams's laboratory, who have helped to develop many methods described herein. Thanks to Dr. Mathilde Gavillet, Jenna Wood for assistance and proofing the manuscript. This work was supported by NIH grants (R01 CA113969, R24 DK099808, and R01 DK062757) and the German Academic Exchange Service grant (DAAD D/12/03783).

References

1. Yin AH, Miraglia S, Zanjani ED et al (1997) AC133, a novel marker for human hematopoietic stem and progenitor cells. Blood 90:5002–5012

2. Kiel MJ, Yilmaz OH, Iwashita T et al (2005) SLAM family receptors distinguish hematopoietic stem and progenitor cells and reveal endothelial niches for stem cells. Cell 121:1109–1121

3. Goodell MA, Brose K, Paradis G et al (1996) Isolation and functional properties of murine hematopoietic stem cells that are replicating in vivo. J Exp Med 183:1797–1806

4. Spangrude GJ, Johnson GR (1990) Resting and activated subsets of mouse multipotent hematopoietic stem cells. Proc Natl Acad Sci U S A 87:7433–7437

5. Osten P, Grinevich V, Cetin A (2007) Viral vectors: a wide range of choices and high levels of service. Handb Exp Pharmacol 178:177–202

6. Coffin JM, Hughes SH, Varmus HE (1997) The interactions of retroviruses and their hosts. In: Coffin JM, Hughes SH, Varmus HE (eds) Retroviruses. Cold Spring Harbor, New York

7. Sommerfelt MA (1999) Retrovirus receptors. J Gen Virol 80:3049–3064

8. Overbaugh J, Miller AD, Eiden MV (2001) Receptors and entry cofactors for retroviruses include single and multiple transmembrane-spanning proteins as well as newly described glycophosphatidylinositol-anchored and secreted proteins. Microbiol Mol Biol Rev 65:371–389, table of contents

9. Temin HM, Mizutani S (1970) RNA-dependent DNA polymerase in virions of Rous sarcoma virus. Nature 226:1211–1213

10. Risco C, Menendez-Arias L, Copeland TD et al (1995) Intracellular transport of the murine leukemia virus during acute infection of NIH 3T3 cells: nuclear import of nucleocapsid protein and integrase. J Cell Sci 108:3039–3050

11. Fassati A, Goff SP (1999) Characterization of intracellular reverse transcription complexes of Moloney murine leukemia virus. J Virol 73:8919–8925

12. Buchschacher GL et al (2004) Safety considerations associated with development and clinical application of lentiviral vector systems for gene transfer. In: Buchschacher G (ed) Current genomics, vol 5. Bentham Science Publishers, Sharjah, UAE, pp 19–35

13. Maetzig T, Galla M, Baum C et al (2011) Gammaretroviral vectors: biology, technology and application. Viruses 3:677–713

14. Cronin J, Zhang XY, Reiser J (2005) Altering the tropism of lentiviral vectors through pseudotyping. Curr Gene Ther 5:387–398

15. Naldini L, Blomer U, Gallay P et al (1996) In vivo gene delivery and stable transduction of nondividing cells by a lentiviral vector. Science 272:263–267

16. Dull T, Zufferey R, Kelly M et al (1998) A third-generation lentivirus vector with a conditional packaging system. J Virol 72:8463–8471

17. Mautino MR, Morgan RA (2002) Gene therapy of HIV-1 infection using lentiviral vectors expressing anti-HIV-1 genes. AIDS Patient Care STDS 16:11–26

18. Schambach A, Galla M, Modlich U et al (2006) Lentiviral vectors pseudotyped with murine ecotropic envelope: increased biosafety and convenience in preclinical research. Exp Hematol 34:588–592

19. Martinez-Salas E (1999) Internal ribosome entry site biology and its use in expression vectors. Curr Opin Biotechnol 10:458–464

20. Szymczak AL, Vignali DA (2005) Development of 2A peptide-based strategies in the design of multicistronic vectors. Expert Opin Biol Ther 5:627–638

21. Fehse B, Uhde A, Fehse N et al (1997) Selective immunoaffinity-based enrichment of CD34+ cells transduced with retroviral vectors containing an intracytoplasmatically truncated version of the human low-affinity nerve growth factor receptor (deltaLNGFR) gene. Hum Gene Ther 8:1815–1824

22. Fehse B, Richters A, Putimtseva-Scharf K et al (2000) CD34 splice variant: an attractive marker for selection of gene-modified cells. Mol Ther 1:448–456

23. Schambach A, Bohne J, Baum C et al (2006) Woodchuck hepatitis virus post-transcriptional regulatory element deleted from X protein and promoter sequences enhances retroviral vector titer and expression. Gene Ther 13:641–645

24. Zufferey R, Donello JE, Trono D et al (1999) Woodchuck hepatitis virus posttranscriptional regulatory element enhances expression of transgenes delivered by retroviral vectors. J Virol 73:2886–2892

25. Buchschacher GL Jr (2001) Introduction to retroviruses and retroviral vectors. Somat Cell Mol Genet 26:1–11

26. Hu WS, Pathak VK (2000) Design of retroviral vectors and helper cells for gene therapy. Pharmacol Rev 52:493–511

27. Williams DA, Lemischka IR, Nathan DG et al (1984) Introduction of new genetic material into pluripotent haematopoietic stem cells of the mouse. Nature 310:476–480

28. Moritz T, Patel VP, Williams DA (1994) Bone marrow extracellular matrix molecules improve gene transfer into human hematopoietic cells via retroviral vectors. J Clin Invest 93:1451–1457

29. Hanenberg H, Xiao XL, Dilloo D et al (1996) Colocalization of retrovirus and target cells on specific fibronectin fragments increases genetic transduction of mammalian cells. Nat Med 2:876–882

30. Donahue RE, Sorrentino BP, Hawley RG et al (2001) Fibronectin fragment CH-296 inhibits apoptosis and enhances ex vivo gene transfer by murine retrovirus and human lentivirus vectors independent of viral tropism in nonhuman primate CD34+ cells. Mol Ther 3:359–367

31. Aiuti A, Cattaneo F, Galimberti S et al (2009) Gene therapy for immunodeficiency due to adenosine deaminase deficiency. N Engl J Med 360:447–458

32. Hacein-Bey-Abina S, Hauer J, Lim A et al (2010) Efficacy of gene therapy for X-linked severe combined immunodeficiency. N Engl J Med 363:355–364

33. Gaspar HB, Cooray S, Gilmour KC et al (2011) Long-term persistence of a polyclonal T cell repertoire after gene therapy for X-linked severe combined immunodeficiency. Sci Transl Med 3:97ra79

34. Gaspar HB, Cooray S, Gilmour KC et al (2011) Hematopoietic stem cell gene therapy for adenosine deaminase-deficient severe combined immunodeficiency leads to long-term immunological recovery and metabolic correction. Sci Transl Med 3:97ra80

35. Boztug K, Schmidt M, Schwarzer A et al (2010) Stem-cell gene therapy for the Wiskott-Aldrich syndrome. N Engl J Med 363:1918–1927

36. Aiuti A, Biasco L, Scaramuzza S et al (2013) Lentiviral hematopoietic stem cell gene therapy in patients with Wiskott-Aldrich syndrome. Science 341:1233151

37. Ott MG, Schmidt M, Schwarzwaelder K et al (2006) Correction of X-linked chronic granulomatous disease by gene therapy, augmented by insertional activation of MDS1-EVI1, PRDM16 or SETBP1. Nat Med 12:401–409

38. Cartier N, Hacein-Bey-Abina S, Bartholomae CC et al (2009) Hematopoietic stem cell gene therapy with a lentiviral vector in X-linked adrenoleukodystrophy. Science 326:818–823

39. Biffi A, Montini E, Lorioli L et al (2013) Lentiviral hematopoietic stem cell gene therapy benefits metachromatic leukodystrophy. Science 341:1233158

40. Ramezani A, Hawley TS, Hawley RG (2008) Combinatorial incorporation of enhancer-blocking components of the chicken beta-globin 5'HS4 and human T-cell receptor alpha/delta BEAD-1 insulators in self-inactivating retroviral vectors reduces their genotoxic potential. Stem Cells 26:3257–3266

41. Gaussin A, Modlich U, Bauche C et al (2012) CTF/NF1 transcription factors act as potent genetic insulators for integrating gene transfer vectors. Gene Ther 19:15–24

42. Koldej RM, Carney G, Wielgosz MM et al (2013) Comparison of insulators and promoters for expression of the Wiskott-Aldrich syndrome protein using lentiviral vectors. Hum Gene Ther Clin Dev 24:77–85

43. Pfaff N, Lachmann N, Ackermann M et al (2013) A ubiquitous chromatin opening element prevents transgene silencing in pluripotent stem cells and their differentiated progeny. Stem Cells 31:488–499

44. Zhang F, Frost AR, Blundell MP et al (2010) A ubiquitous chromatin opening element (UCOE) confers resistance to DNA

methylation-mediated silencing of lentiviral vectors. Mol Ther 18:1640–1649

45. Zhang F, Thornhill SI, Howe SJ et al (2007) Lentiviral vectors containing an enhancer-less ubiquitously acting chromatin opening element (UCOE) provide highly reproducible and stable

transgene expression in hematopoietic cells. Blood 110:1448–1457

46. McDermott SP, Eppert K, Lechman ER et al (2010) Comparison of human cord blood engraftment between immunocompromised mouse strains. Blood 116:193–200

Chapter 21

Lentiviral Gene Transduction of Mouse and Human Hematopoietic Stem Cells

Niek P. van Til and Gerard Wagemaker

Abstract

Lentiviral vectors can be used to genetically modify a broad range of cells. Hematopoietic stem cells (HSCs) are particularly suitable for lentiviral gene augmentation, because these cells can be enriched with relative ease from mouse bone marrow and human hematopoietic sources, and in principle require relatively limited cell numbers to completely reconstitute the hematopoietic system in vivo. Furthermore, lentiviral vectors are very efficient if pseudotyped with broad tropism envelope proteins. This chapter focuses on gene modification by the use of self-inactivating third-generation human immunodeficiency virus-derived lentiviral vectors for ex vivo HSC modification for both mouse and human application.

Key words Hematopoietic stem cells, Lentiviral vectors, Ex vivo

---

1 Introduction

Lentiviral vectors based on the human immunodeficiency virus (HIV) have been used for more than a decade to genetically modify almost any type of cell. These lentiviral vectors are efficient, are able to transduce both dividing and nondividing cells [1, 2], do not require pretreatment with cytokines for transduction of hematopoietic stem cells (HSCs) [3], and have a relatively low genotoxic risk compared to gammaretroviral vectors [4].

In this chapter we focus on the in vitro manipulation of HSCs by third-generation self-inactivating lentiviral vectors (LV) [5–8] for potential transplantation in vivo. Third-generation lentiviral vectors are a four-plasmid-based system, in which the transfer vector containing the expression cassette has been separated from the required structural proteins, i.e., gag/pol (containing reverse transcriptase and integrase), another plasmid with *Rev*, and one containing the envelope protein. These structural proteins are divided over three packaging plasmids to reduce the risk of replication-competent virus generation by homologous recombination. The transfer vector transcript in this system is expressed from a chimeric

Kevin D. Bunting and Cheng-Kui Qu (eds.), *Hematopoietic Stem Cell Protocols*, Methods in Molecular Biology, vol. 1185, DOI 10.1007/978-1-4939-1133-2_21, © Springer Science+Business Media New York 2014

enhancer/promoter region of the 5′ long-terminal repeat (LTR) and in combination with a deleted U3 region (ΔU3) of the 3′ LTR that is self-inactivating. This ΔU3 region is copied to the 5′LTR during reverse transcription, resulting in a 5′LTR with very little to no residual promoter activity [9]. Hence, the internal promoter cassette can be used to provide constitutive active or cell-specific expression from a eukaryotic promoter. Lentiviral vectors are commonly produced in cell lines with high transfectability, e.g., human embryonic kidney (HEK) 293T cells. To transduce HSCs efficiently, LV virus particles are vesicular stomatitis virus G-protein (VSV-g) pseudotyped, which provides a broad tropism to transduce many different cell types. In the case of HSCs, both mouse and human HSCs can be effectively transduced up to 100 % by a single overnight incubation without pretreatment with cytokines.

We have successfully used this protocol for therapeutic LV transduction of hematopoietic stem cells in several mouse models for enzyme deficiencies and a lysosomal storage disease [3, 10, 11]. Furthermore, LV vectors based on the same self-inactivating third-generation lentiviral vector backbone have most recently been used in two clinical trials for metachromatic leukodystrophy (MLD) [12] and Wiskott–Aldrich syndrome [13] and resulted in efficient transduction and reconstitution of gene-modified HSCs in patients.

2   Materials

2.1   Production of Lentiviral Vectors

1. HEK 293T cells (ATCC CRL-11268, or other cells with high transfectability).

2. For calcium phosphate transfection prepare 2.5 M calcium chloride ($CaCl_2$) solution in ultrapure water. Sterile filter with 0.22 μm pore.

3. 4-(2-Hydroxyethyl) piperazine-1-ethanesulfonic acid (HEPES) solution containing 100 mM HEPES (Sigma, product no. H4034), 281 NaCl, 1.5 mM Na_2HPO_4, pH 7.09–7.12 in ultrapure water. Sterile filter with 0.22 μm pore (*see* **Note 1**).

4. DMEM culture medium consists of Dulbecco's modified Eagle's medium (DMEM, 500 mL) with 10 % fetal calf serum (FCS), 5 mL (100×) 10,000 units of penicillin and 10,000 μg of streptomycin per mL (Gibco), and 5 mL L-glutamine (200 mM, Gibco).

5. Trypsin–EDTA (Life Technologies).

6. Plasmids pMDLg/pRRE (Addgene: 12251), pMD-VSV-G (Addgene: 12259), pRSV-REV (Addgene: 12253), and a transfer vector, e.g., pRRLSIN.cPPT.PGK-GFP.WPRE (Addgene: 12252) dissolved in ultrapure water. This transfer vector contains a chimeric promoter driving the lentiviral vector

transcript expression, in this case Rous sarcoma virus (RSV) promoter (RRL backbone) [5, 7]. This plasmid also contains the central polypurine tract (cPPT), the promoter cassette containing the human phosphoglycerate kinase (PGK) promoter, green fluorescent protein (GFP), and the Woodchuck post-translational regulatory element (WPRE). Inclusion of the cPPT increases the transport of the proviral genome to the nucleus [14]. The WPRE stabilizes the RNA transfer vector transcript, thereby increasing titer and transgene transcript-driven expression from the internal promoter cassette after genomic integration [15, 16]. Alternative promoters providing similar constitutive active expression as the PGK promoter are elongation factor 1α (intron containing) and a short version (enhancer-less) [17]. If higher constitutive active expression is required spleen focus-forming virus-based elements are optional [17, 18].

7. T175 culture flasks for tissue culture of HEK293T cells.

8. Polystyrene tissue culture dish (14.5 cm diameter, CELLSTAR: cat. no. 639160).

9. Corning 50 mL centrifuge tubes (Sigma-Aldrich).

10. Nalgene SFCA cellulose acetate membrane (0.45 μm pore size) filter units with 150 mL funnel (Thermo Scientific, cat. no. 155-0045). Funnels for collection of larger volumes of vector batches are also available.

11. Ultracentrifuge (Optima LE-80 K, Beckman Coulter, or similar), SW32Ti rotor, or alike with buckets and autoclaved polyallomer centrifuge tubes (25 × 89 mm, Beckman Coulter, cat. no. 326823). Clean the centrifuge buckets with 70 % ethanol, and leave to dry in a laminar flow cabinet. Autoclaved (or sterile) polypropylene Eppendorf tubes (0.5 mL).

12. HeLa cells for titration (ATCC® CCL-2). Other cell types can also be used and may be less or more susceptible to transduction by lentiviral vectors.

2.2 Quantitative PCR to Determine the Vector Copy Number

1. Genomic DNA extraction kit.

2. HIV-U3 forward primer, 5′-CTGGAAGGGCTAATTCAC TC-3′, and HIV reverse primer, 5′-GGTTTCCCTTTCGCT TTCAG-3′. The forward primer anneals to the ΔU3 region in the 3′LTR and the reverse primer in the primer-binding site (PBS) region of the lentiviral vector. After reverse transcription the ΔU3 region is copied to the 5′LTR, and a 274 bp fragment can be amplified by qPCR [19].

3. SYBR Green Master mix (Roche or any other supplier) and 96-well PCR plates for Applied Biosystems (Kisker Biotech).

4. ABI PRISM 7900 HT sequence detection system (Applied Biosystems) or any other real-time quantification device.

2.3 Culture and Transduce Hematopoietic Progenitor Cells

1. Hanks' + HEPES buffer (H + H): 1 pack Hanks' Balanced Salts (Sigma, product no. H1387) for 10 L, 1 g Streptomycin sulfate (Fisiopharma, Italy) and 1×10^6 units (sodium-) penicillin G (Astellas Pharma B.V., The Netherlands), 25 g HEPES. Fill up to 10 L, and add 2.25 mL 10 N NaOH. Check if osmolarity is ~300 mM (300–310 mM), and sterile-filter with a 0.2 μm filter (Nalgene 1,000 mL, rapid top filter with flow bottle, aPES membrane, Thermo Scientific, cat no. 597-4520). Optional: Use cell culture-grade phosphate-buffered saline instead of H + H.

2. Hematopoietic medium: Use serum-free medium for HSC culture, e.g., StemSpan medium (StemCELL Technologies) or StemMACS HSC Expansion Media XF (Miltenyi), or other HSC medium are optional.

3. Add for mouse HSC culture the cytokine murine stem cell factor (mSCF) 100 ng/mL, human FMS-like tyrosine kinase 3-ligand (hFlt3-L) 50 ng/mL, and murine thrombopoietin (mTPO) 10 ng/mL. For human HSCs use hSCF 100 ng/mL and hTPO 10 ng/mL instead of the mouse cytokines (*see* **Note 2**).

3 Methods

3.1 Lentiviral Vector Production and Concentration

1. In this protocol lentiviral vector production is performed by calcium phosphate transfection. Commercial calcium phosphate transfection kit may also be purchased (e.g., CAPHOS, Sigma), but it is relatively easy to prepare all the solutions (*see* **Note 3**). Alternatively, other effective transfection methods may be used, e.g., transfection with polyethylenimine (PEI) or lipid-based formulations.

2. For calcium phosphate transfection, at the first day, remove all medium in the afternoon from one fully confluent T175 flask, and trypsinize the HEK293T cells with 5 mL trypsin–EDTA solution (*see* **Notes 4** and **5**). Incubate at 37 °C for a few minutes until the cells detach. Add 3 mL DMEM culture medium to inactivate the trypsin, and resuspend the cells thoroughly. Divide the cell suspension over four Ø14.5 culture dishes with 15 mL DMEM culture medium. Make sure that the cells are dispersed evenly. Alternatively, count and seed 1.5×10^7 HEK293T cells per culture dish.

3. The second day, replace the medium around noon. About 2 h later, for one dish combine the plasmids pMDLg/pRRE (13 μg), pMD-VSV-G (7 μg), and pRSV-REV (5 μg) and the transfer vector pRRLSIN.cPPT.PGK-GFP.WPRE (20 μg), and add up to 900 μL ultrapure water. Subsequently, add 100 μL 2.5 M CaCl$_2$ solution, mix, and leave for 10 min. Then add the DNA mixture to 1 mL HEPES dropwise by vortexing. Add immediately to the HEK293T cells, and disperse the cells evenly.

4. The third day, replace the culture medium in the morning. Do not add fluid/medium directly onto the cells.

5. At day 4, harvest the viral vector supernatant and centrifuge in 50 mL tubes for 5 min at 1,600 rpm ($300 \times g$) to dispose of cellular debris. Days 5 and 6 can also be used to harvest vector supernatant, but vector quality and titers will drop substantially.

6. Filtrate the supernatant with 0.45 μm SFCA cellulose acetate membrane filter units.

7. Fill ultracentrifuge polyallomer tubes up to 35 mL with the vector supernatant and centrifuge for 2 h at 20,000 rpm ($49,000 \times g$) at 4 °C. Ultracentrifugation is necessary to remove the FCS-containing culture medium and concentrate the lentiviral vector batches to high titers.

8. Dissolve the lentiviral vector pellet in PBS or HSC medium (concentrate ±150×, and if low numbers of HSCs are transduced dissolving the lentiviral vector pellet in HSC medium is recommended), aliquot in 0.5 mL tubes, and store at –80 °C until further use.

9. Determine the titer of the LV vector batch. Dilute the concentrated vector batch 100×. Seed HeLa cells (other cells are optional) at 2×10^5 per 10 cm² well in 2 mL DMEM culture medium. Then add serially diluted vector batch, i.e., 500, 50, and 5 μL to the wells, and replace the medium the next day.

Leave the cells for at least 5 days for the provirus to integrate, then trypsinize, and use the cells for flow cytometry. Calculate the titer (transducing units per mL) by multiplying by the dilution factor (100×), the number of cells seeded (2×10^5), and (1 mL/volume vector supernatant administered). For lentiviral vectors without fluorescent markers use quantitative PCR to determine the titer.

3.2 Quantitative PCR to Determine Titer

1. Extract genomic DNA from the transduced cells according to the manufacturer's protocol. Check genomic DNA quality with a spectrophotometer (Nanodrop, Thermo Scientific).

2. Measure the product of the quantitative PCR in an ABI PRISM 7900 HT sequence detection system (Applied Biosystems) or a similar device for PCR product quantification.

3. Perform the qPCR reaction on 100 ng of genomic DNA with SYBR Green PCR Master Mix (Applied Biosystems) with primers HIV-U3-FW and HIV-RV. For every reaction use 12.5 (2×) SYBR Green Master + 20 pmol HIV-U3-FW + 20 pmol HIV-RV, and add up to 21 μL with ultrapure water. Prepare one mix for all your samples, and pipette into the 96-well dish. Finally, add 100 ng genomic DNA template (4 μL of 25 ng/μL genomic DNA solution) to 96-well dishes.

A standard line for LV integrations can be prepared by transducing HeLa cells at a low multiplicity of infection (MOI ~0.05) with an LV vector containing a fluorescent marker. By using a low MOI the average vector copy number (VCN) will approach ~1 per cell, because transduction of cells occurs at random. The transduced GFP-positive cells can be obtained by fluorescence-activated cell sorting (FACS) to obtain a population of cells with an average 1 VCN per cell. Expand these cells, viable freeze, and also extract genomic DNA for qPCR. To obtain a standard line for mouse HSCs transduced mouse 3T3 cells can be sorted, and for human HSCs sorted fibroblasts can be used.

4. The following qPCR program can be used for the ABI PRISM 7900 HT sequence detection system (Applied Biosystems): 2 min 50 °C, 10 min 95 °C, then repeat 35 cycles of 15 s 95 °C and 1 min 62 °C, and finally 4 °C. This can be adapted for different real-time PCR machines. A dissociation curve can be included at the end of the qPCR program: continue with 15 s at 95 °C, 15 s at 60 °C, and 15 s at 95 °C.

5. Analyze samples with SDS 2.2.2 software or other real-time PCR software.

3.3 Selection of Mouse Hematopoietic Progenitor Cells

1. Use concentrated vector batches with titers higher than 1×10^8 TU/mL. Lower titers starting from 1×10^7 TU/mL can be used but may affect the transduction efficiency.

2. For transduction of mouse HSCs, extract donor bone marrow from both femurs and tibias of mice and prepare a single-cell suspension. Total mouse bone marrow cell suspensions can be used for transduction, but the HSC population can also be further enriched to reduce the lentiviral vector dose required (*see* Subheading 3).

3. To enrich mouse HSC progenitors approximately tenfold purify by lineage depletion (Lin⁻) according to the manufacturer's protocol (BD Biosciences or other providers) and continue at Subheading 3.4.

4. Optional: Enrich hematopoietic progenitor cells further for Sca-1⁺ and c-Kit⁺ cells with FACS.

5. For transduction of human HSCs from human umbilical cord blood (UCB) cells, bone marrow aspirates, or mobilized peripheral blood isolate mononucleated cells by Ficoll density gradient centrifugation (1.077 g/cm²; Nycomed Pharma AS, Oslo, Norway).

6. Use CD34 MicroBead Kit UltraPure (Miltenyi) or MACs technology (Miltenyi) to enrich CD34⁺ cells according to the manufacturer's protocol. Optional: Obtain CD34⁺ cells with a FACS sorter.

7. Optional: Sort for CD34⁺CD38⁻ cells for highly purified HSC progenitor populations.

3.4 Transduction of Hematopoietic Stem Cell Progenitors

1. Suspend the Lin⁻ mouse cells or CD34⁺ cells in serum-free medium with cytokines (*see* Subheading 2) (*see* **Note 6**). Maintain a cell density of at least 1×10^6 cells/mL. Transduce at a multiplicity (MOI) of 2 aiming at an average of 1 vector copy per cell (VCN). If higher average VCNs are required increase the MOI accordingly. Use 24-well dishes for up to 3×10^6 cells per mL or 6-well dishes up to 12×10^6 in 4 mL.

2. Transduce the HSCs overnight. Of note: Addition of retronectin to immobilize the HSCs or charged compounds such as polybrene or DEAE-dextran are *not* required for efficient transduction.

3. Resuspend the cells in the HSC medium, and transfer all medium from the wells to a 12 mL tube.

4. Add another 1 mL H + H or PBS, and transfer to the same tube under Subheading 3.

5. Add up to 12 mL, and centrifuge at 1,600 rpm for 5 min.

6. Discard supernatant, add 12 mL H + H or PBS, resuspend the HSCs, and centrifuge for another 5 min at 1,600 rpm at room temperature.

7. Discard the supernatant, dissolve in H + H or PBS, and leave at room temperature for injection into preconditioned mice. If further culture is required dissolve the HSCs into HSC medium instead of buffer.

8. To determine the transduction efficiency keep the cultured HSCs in culture for at least 5 days before collecting the cells for genomic DNA extraction. If lentiviral vectors with fluorescent markers are used determine the transduction efficiency by flow cytometry. Additionally, confirm the transduction efficiency by quantitative PCR as described above under Subheading 3.2 to control for effective transduction.

4 Notes

1. HEPES buffer for transfection pH has to be strictly between pH 7.09 and 7.12.

2. Other cytokines than described above are obsolete for efficient transduction of HSCs, although further developments in inhibition of stem cell differentiation are not excluded.

3. When making new components for calcium phosphate precipitation always check the transfection efficiency by comparing the previous batch. Use a lentiviral vector with a fluorescent marker. Prepare one dish with both HEPES batches, check the fluorescence signal with a fluorescence microscope, and determine and compare the titers. In general, titer has to be $>1 \times 10^6$ HeLa transducing units/mL to be considered for further use.

4. Use HEK 293T cells of low passage number, in general below passage 20.

5. Maintain the quality of the HEK293T cells, replate the cells twice a week, and do not overgrow the cells.

6. Wash your HSCs thoroughly to remove any remaining sodium azide after HSC enrichment and before transduction.

Acknowledgements

This work was supported by the European Commission's 5th, 6th, and 7th Framework Programs, Contracts QLK3-CT-2001-00427-INHERINET, LSHB-CT-2004-005242-CONSERT, 222878-PERSIST, and 261387-CELL-PID, and by the Netherlands Organization for Health Research ZonMW, program grants 431-00-016 and 434-00-010.

References

1. Naldini L, Blomer U, Gallay P, Ory D, Mulligan R, Gage FH et al (1996) In vivo gene delivery and stable transduction of nondividing cells by a lentiviral vector. Science 272:263–267

2. Kafri T, Blomer U, Peterson DA, Gage FH, Verma IM (1997) Sustained expression of genes delivered directly into liver and muscle by lentiviral vectors. Nat Genet 17:314–317

3. van Til NP, Stok M, Aerts Kaya FS, de Waard MC, Farahbakhshian E, Visser TP et al (2010) Lentiviral gene therapy of murine hematopoietic stem cells ameliorates the Pompe disease phenotype. Blood 115:5329–5337

4. Montini E, Cesana D, Schmidt M, Sanvito F, Ponzoni M, Bartholomae C et al (2006) Hematopoietic stem cell gene transfer in a tumor-prone mouse model uncovers low genotoxicity of lentiviral vector integration. Nat Biotechnol 24:687–696

5. Dull T, Zufferey R, Kelly M, Mandel RJ, Nguyen M, Trono D et al (1998) A third-generation lentivirus vector with a conditional packaging system. J Virol 72:8463–8471

6. Guenechea G, Gan OI, Inamitsu T, Dorrell C, Pereira DS, Kelly M et al (2000) Transduction of human CD34+ CD38- bone marrow and cord blood-derived SCID-repopulating cells with third-generation lentiviral vectors. Mol Ther 1:566–573

7. Zufferey R, Dull T, Mandel RJ, Bukovsky A, Quiroz D, Naldini L et al (1998) Self-inactivating lentivirus vector for safe and efficient in vivo gene delivery. J Virol 72:9873–9880

8. Miyoshi H, Smith KA, Mosier DE, Verma IM, Torbett BE (1999) Transduction of human CD34+ cells that mediate long-term engraftment of NOD/SCID mice by HIV vectors. Science 283:682–686

9. Logan AC, Haas DL, Kafri T, Kohn DB (2004) Integrated self-inactivating lentiviral vectors produce full-length genomic transcripts competent for encapsidation and integration. J Virol 78:8421–8436

10. van Til NP, de Boer H, Mashamba N, Wabik A, Huston M, Visser TP et al (2012) Correction of murine Rag2 severe combined immunodeficiency by lentiviral gene therapy using a codon-optimized RAG2 therapeutic transgene. Mol Ther 20:1968–1980

11. Huston MW, van Til NP, Visser TP, Arshad S, Brugman MH, Cattoglio C et al (2011) Correction of murine SCID-X1 by lentiviral gene therapy using a codon-optimized IL2RG gene and minimal pretransplant conditioning. Mol Ther 19:1867–1877, Research Support, Non-U.S. Gov't

12. Biffi A, Montini E, Lorioli L, Cesani M, Fumagalli F, Plati T et al (2013) Lentiviral hematopoietic stem cell gene therapy benefits metachromatic leukodystrophy. Science 341:1233158

13. Aiuti A, Biasco L, Scaramuzza S, Ferrua F, Cicalese MP, Baricordi C et al (2013) Lentiviral hematopoietic stem cell gene therapy in

patients with Wiskott-Aldrich syndrome. Science 341:233151

14. Sirven A, Pflumio F, Zennou V, Titeux M, Vainchenker W, Coulombel L et al (2000) The human immunodeficiency virus type-1 central DNA flap is a crucial determinant for lentiviral vector nuclear import and gene transduction of human hematopoietic stem cells. Blood 96: 4103–4110

15. Zufferey R, Donello JE, Trono D, Hope TJ (1999) Woodchuck hepatitis virus posttranscriptional regulatory element enhances expression of transgenes delivered by retroviral vectors. J Virol 73:2886–2892

16. Schambach A, Bohne J, Baum C, Hermann FG, Egerer L, von Laer D et al (2006) Woodchuck hepatitis virus post-transcriptional regulatory element deleted from X protein and promoter sequences enhances retroviral vector titer and expression. Gene Ther 13:641–645

17. Schambach A, Bohne J, Chandra S, Will E, Margison GP, Williams DA et al (2006) Equal potency of gammaretroviral and lentiviral SIN vectors for expression of O6-methylguanine-DNA methyltransferase in hematopoietic cells. Mol Ther 13:391–400

18. Hildinger M, Eckert HG, Schilz AJ, John J, Ostertag W, Baum C (1998) FMEV vectors: both retroviral long terminal repeat and leader are important for high expression in transduced hematopoietic cells. Gene Ther 5:1575–1579

19. van Til NP, Markusic DM, van der Rijt R, Kunne C, Hiralall JK, Vreeling H et al (2005) Kupffer cells and not liver sinusoidal endothelial cells prevent lentiviral transduction of hepatocytes. Mol Ther 11:26–34

Chapter 22

High-Throughput Genomic Mapping of Vector Integration Sites in Gene Therapy Studies

Brian C. Beard, Jennifer E. Adair, Grant D. Trobridge, and Hans-Peter Kiem

Abstract

Gene therapy has enormous potential to treat a variety of infectious and genetic diseases. To date hundreds of patients worldwide have received hematopoietic cell products that have been gene-modified with retrovirus vectors carrying therapeutic transgenes, and many patients have been cured or demonstrated disease stabilization as a result (Adair et al., Sci Transl Med 4:133ra57, 2012; Biffi et al., Science 341:1233158, 2013; Aiuti et al., Science 341:1233151, 2013; Fischer et al., Gene 525:170–173, 2013). Unfortunately, for some patients the provirus integration dysregulated the expression of nearby genes leading to clonal outgrowth and, in some cases, cancer. Thus, the unwanted side effect of insertional mutagenesis has become a major concern for retrovirus gene therapy. The careful study of retrovirus integration sites (RIS) and the contribution of individual gene-modified clones to hematopoietic repopulating cells is of crucial importance for all gene therapy studies. Supporting this, the US Food and Drug Administration (FDA) has mandated the careful monitoring of RIS in all clinical trials of gene therapy. An invaluable method was developed: linear amplification mediated-polymerase chain reaction (LAM-PCR) capable of analyzing in vitro and complex in vivo samples, capturing valuable genomic information directly flanking the site of provirus integration. Linking this method and similar methods to high-throughput sequencing has now made possible an unprecedented understanding of the integration profile of various retrovirus vectors, and allows for sensitive monitoring of their safety. It also allows for a detailed comparison of improved safety-enhanced gene therapy vectors. An important readout of safety is the relative contribution of individual gene-modified repopulating clones. One limitation of LAM-PCR is that the ability to capture the relative contribution of individual clones is compromised because of the initial linear PCR common to all current methods. Here, we describe an improved protocol developed for efficient capture, sequencing, and analysis of RIS that preserves gene-modified clonal contribution information. We also discuss methods to assess dominant/overrepresented gene-modified clones in preclinical and clinical models.

Key words Retrovirus, Gene therapy, Integration, PCR, Computation, High-throughput sequencing

1 Introduction

A number of different types of retrovirus vectors including gammaretrovirus, lentivirus, foamy virus, and alpharetroviruses have been used for genetic modification of hematopoietic cells

Kevin D. Bunting and Cheng-Kui Qu (eds.), *Hematopoietic Stem Cell Protocols*, Methods in Molecular Biology, vol. 1185, DOI 10.1007/978-1-4939-1133-2_22, © Springer Science+Business Media New York 2014

in preclinical studies. Thus far, only gammaretroviruses and lentiviruses have been translated to human clinical trials for hematopoietic cell gene therapy. While these clinical studies have finally fulfilled the early promise of hematopoietic cell gene therapy, permanent modification of the genome by provirus integration is not a benign event and can lead to altered cellular behavior. Furthermore, the nucleotide sequences included in the vector, such as promoter and enhancer elements, can also dysregulate nearby genes and impact clonal behavior. These factors can cause clonal dominance (overrepresentation of a single gene-modified clone within a pool of multiple gene-modified clones) or monoclonality (identification of a single gene-modified clone contributing all detectable gene-modified cell hematopoiesis), both of which can be observed in the setting of normal or malignant hematopoiesis [1–4]. For this reason, identifying specific retrovirus integration sites (RIS) and identifying overrepresented clones with a presumed selective advantage in vivo is powerful method to evaluate the safety of retrovirus-mediated gene therapy.

To begin to address these issues, early studies using retrovirus vectors for gene transfer studies used fairly straightforward methods for provirus detection and even rudimentary copy number analysis by Southern blot and provirus-specific polymerase chain reaction (PCR) [5]. These methods were limited in sensitivity and specificity and gave limited information regarding the genomic DNA flanking the provirus. More specifically, retrovirus integration, broadly termed "insertional mutagenesis," can subsequently result in a variety of issues, including proto-oncogene activation, tumor suppressor inactivation/suppression, or microRNA dysregulation (reviewed in ref. 6). To address this, methods were developed that would allow for simultaneous detection of provirus and surrounding genomic information. Inverse PCR [7], and to a lesser extent splinkerette PCR [8], made possible for the first time reliable detection of genomic DNA flanking sites of provirus integration. Inverse PCR detected up to 40 RIS from bone marrow colony-forming unit (CFU) cells in a rhesus macaque (*Macaca mulatta*) [9–11], but this method did not allow for a detailed analysis of complex in vivo samples and suffered from relatively low sensitivity. A major advance to the field of retrovirus gene therapy was the development of linear amplification-mediated (LAM)-PCR [12, 13]. LAM-PCR allowed for the detection of numerous rare RIS from complex DNA samples, such as bulk peripheral blood white blood cells (WBCs). LAM-PCR proved to be a very effective method for RIS detection, but was a lengthy process that required laborious sequencing of individual clonal sequences. The rapid detection of RIS was not accomplished until directly coupling bulk LAM-PCR product cloning and sequencing, often referred to as shotgun sequencing [14]. This began the era of high-throughput RIS detection and was a precursor to the most advanced techniques used today. All techniques currently applied require identification of a known sequence some distance from the site of provirus integration such that PCR amplification of a sufficient length of intervening genomic DNA can be accomplished to identify the locus of integration. LAM-PCR initially relied on digestion of genomic DNA with restriction enzymes known to have high frequency recognition sites within the genome of interest followed by ligation of a known linker cassette; however, this introduced bias into the procedure by only retrieving clones with restriction digest sites within certain ranges of provirus integration. To reduce inherent bias, optimized LAM-PCR techniques have been developed eliminating all use of restriction sites for DNA fragmentation [15]. Another technique developed by Ronen et al. utilizes the Mu transposon to provide a known sequence from which to anchor a primer for amplification of intervening genomic DNA [16]. Each of these approaches have further improved RIS retrieval by multiple methods, including adaptation of a solid-phase-based enrichment of initial PCR products and coupling massively paralleled sequencing that allows for millions of sequence reads [16, 17]. More recently, we have adapted additional methods to whole genome sequencing techniques, termed modified genome sequencing (MGS)-PCR [1, 18], that allow for additional clonal population analysis not previously possible and that improve the reliability of dominant clone identification (Fig. 1). The resulting protocol, described herein, is short, sensitive, and quality-controlled for identification of truly overrepresented clones rather than those amplified by bias in PCR efficiency.

2   Materials

2.1 Genomic DNA Preparation

Gentra Puregene Blood Kit (QIAGEN; Hilden, Germany) or any other method for high-quality genomic DNA isolation.

2.2 DNA Fragmentation and Polishing

1. Covaris apparatus for focused acoustic DNA fragmentation (Covaris; Woburn, MA, USA).

2. Fragment end polishing kit such as End-It DNA End-Repair kit (Epicentre; Madison, WI, USA).

2.3 Preparation of Linker Cassette and Ligation

1. Linker cassette oligonucleotides:

   Linker Cassette T-end Upper (LCTU):

   5′-GAC CCG GGA GAT CTG AAT TCA GTG GCA CAG CAG TTA GG-3′

   Linker Cassette T-end Lower Short (LCTLS):

   5′-CCT AAC TGC TGT GCC ACT GAA TTC AGA TC-3′

Modified Genome Sequencing (MGS)-PCR

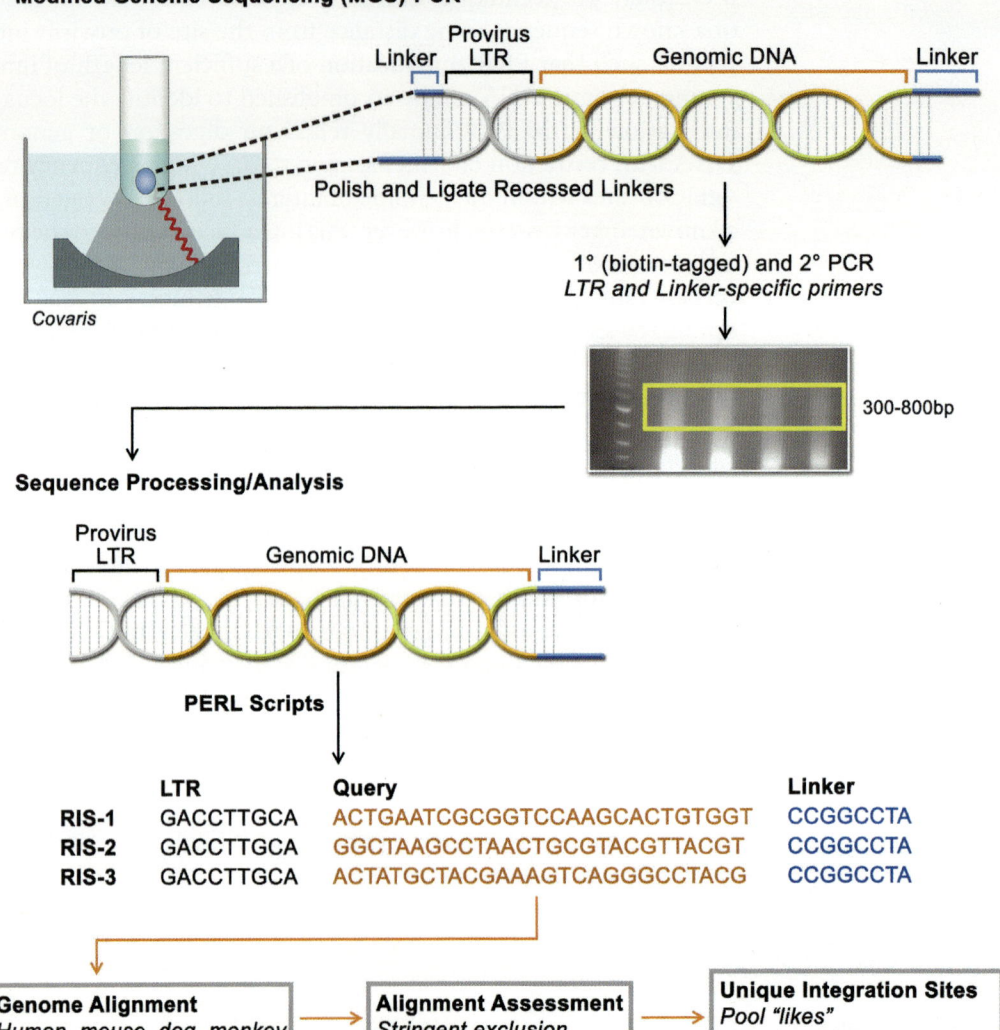

Fig. 1 Schematic Representation of MGS-PCR Amplification and Processing of Provirus Integration Sites. *Upper left*, focused acoustic shearing of DNA. *Upper right*, desired DNA fragment containing flanking linker cassettes (*blue*), retrovirus LTR (*gray*), and intervening DNA "query" (*yellow/orange*) following processing and prior to purification and sequencing. *Lower right*, representative agarose gel to purify the desired DNA fragment lengths for DNA sequencing. *Lower left*, sequencing and processing schematic of DNA sequence data of provirus integration sites

Linker Cassette Annealing Protocol (Linker Master Mix):

40 μL LCTU [1 mM].

40 μL LCTLS [1 mM].

110 μL 250 μM Tris pH 7.5.

20 μL 50 mM MgCl₂.

(a) Heat block or water bath set to 95 °C.

(b) Mix the reagents outlined above. Multiple tubes can be made in parallel.

(c) Place the tubes at 95 °C for 5 min.

(d) Remove samples from heat and allow to cool to room temperature.

(e) Aliquot the samples and freeze at –20 °C for storage.

2. Fast-Link™ DNA Ligation kit (Epicentre) or similar DNA ligation kit.

3. Heat block or water bath set to 70 °C.

4. MinElute PCR Purification kit (QIAGEN) or similar PCR purification kit.

5. Microcentrifuge.

2.4 Small DNA Fragment Removal by Solid-Phase Reversible Immobilization

1. AMPure XP beads (Beckman Coulter; Brea, CA, USA) or similar solid-phase reversible immobilization (SPRI) beads.

2. 70 % Ethanol.

3. QIAGEN Buffer EB (QIAGEN) or similar elution buffer (10 mM Tris–Cl, pH 8.5).

4. Magnetic Particle Concentrator (MPC).

5. Vortex device.

2.5 First, Standard Exponential PCR

1. Molecular biology grade H_2O.

2. Thermocycler.

3. Taq DNA Polymerase (Invitrogen/Life Technologies; Grand Island, NY, USA) or similar DNA polymerase.

4. 10× PCR Buffer (200 mM Tris–HCl, pH 8.4, 500 mM KCl).

5. 50 mM Magnesium Chloride.

6. 10 mM dNTP mix (PCR grade).

7. Primary PCR Primers (one linker cassette-specific and one LTR-specific, biotin-tagged):

MSCV-specific-1: 5′-[Biotin]-TCC TCC GAT AGA CTG CGT CG-3′.

MND-specific-1: 5′-[Biotin]-CAA CTG TTC TTG GCC CTG AG-3′.

Lentivirus-specific-1: 5′-[Biotin]-AGC TTG CCT TGA GTG CTT CA-3′.

Foamy virus-specific-1: 5′-[Biotin]-ACC GAC TTG ATT CGA GAA CC-3′.

Alpharetrovirus-specific-1: 5′-[Biotin]- GCA ATA CTC TTG TAC GTA GTG-3′.

Linker cassette-specific-1: 5'-GAC CCG GGA GAT CTG AAT TC-3'.

All oligonucleotide primers are purchased from commercially available sources and have been designed to be optimum for RIS capture (*see* **Note 1**).

2.6 Biotin-Streptavidin PCR Product Capture

1. Dynabeads M-280 Streptavidin (Invitrogen/Life Technologies) or similar beads for capture of biotinylated oligonucleotides.

2. 1 mg/mL bovine serum albumin (BSA).

3. Formamide solution (95 % formamide in 10 mM EDTA, pH 8.2).

4. 10× Binding buffer (0.1 % Triton X-100, 100 mM NaCl, 50 mM Tris, pH 7.5, 5 mM EDTA, pH 8.0).

5. MPC.

6. Thermomixer.

7. MinElute PCR Purification kit (QIAGEN) or similar PCR purification kit.

8. Microcentrifuge.

2.7 Nested Exponential PCR

1. Molecular biology grade H_2O.

2. Thermocycler.

3. Taq DNA Polymerase (Invitrogen/Life Technologies) or similar DNA polymerase.

4. 10× PCR Buffer (200 mM Tris–HCl, pH 8.4, 500 mM KCl).

5. 50 mM Magnesium Chloride.

6. 10 mM dNTP mix (PCR grade).

7. Second, nested PCR primers (one linker cassette-specific nested primer and one barcoded, LTR-specific nested primer):

MSCV-specific-2: 5'-XX-[Barcode]-GCC TCT TGC TGT TTG CAT CC-3'.

MND-specific-2: 5'-XX-[Barcode]-GGC AGG AAC TGC TTA CCA CA-3'.

Lentivirus-specific-2: 5'-XX-[Barcode]-AGT AGT GTG TGC CCG TCT GT-3'.

Foamy virus-specific-2: 5'-XX-[Barcode]-GCT AAG GGA GAC ATC TAG TG-3'.

Alpharetrovirus-specific-2: 5'-XX-[Barcode]-TTG GTG TGC ACC TGG GTT GA-3'.

Linker cassette-specific-2: 5'-XX- GAT CTG AAT TCA GTG GCA CAG-3'.

XX: refers to any sequence that needs to be included at the terminus of sequences for various high-throughput sequencing platforms.

Barcode: Any unique identifying sequence that will allow for sorting of resulting sequence reads. There are numerous publically available and common lists that have been tested for optimum resolution of resulting sequencing reads (e.g., Roche MID 1-100).

All oligonucleotide primers are purchased from commercially available sources and have been designed to be optimum for RIS capture (*see* **Note 1**).

2.8 Separation of PCR Products in an Agarose Gel

1. E-Gel pre-cast ethidium bromide 2 % agarose gels (Invitrogen/Life Technologies) or similar 2 % agarose gel with ethidium bromide for visualization of resolved DNA.

2. 6× Orange Loading Dye (New England Biolabs; Ipswich, MA, USA) or similar Orange G DNA Loading dye.

3. Molecular grade H_2O.

4. E-gel electrophoresis apparatus or similar agarose gel electrophoresis tank.

   (a) If using electrophoresis tank, Tris/Borate/EDTA (TBE) buffer (89 mM Tris, 89 mM boric acid, 2 mM EDTA, pH 8.3).

5. Ultraviolet light.

6. Sterile, DNase-free razor blades.

7. QIAquick gel extraction kit (QIAGEN) or similar agarose gel extraction kit.

2.9 Purification of Products and DNA Sequencing

1. AMPure XP beads or similar SPRI beads.

2. 70 % Ethanol.

3. QIAGEN Buffer EB (QIAGEN) or similar elution buffer (10 mM Tris–Cl, pH 8.5).

4. Magnetic Particle Concentrator (MPC).

5. Vortex device.

6. Access to IonTorrent® Next-Generation sequencing platform (Invitrogen/Life Technologies) or other similar next-generation sequencing platform.

2.10 Analysis of Sequences for Identification of RIS and Relative Clonal Contribution

Computer with internet access and 64-bit Microsoft Excel.

3 Methods

The methods outlined below describe (1) initial considerations for RIS analysis, (2) a minimally biased PCR-based method for amplification of RIS, (3) high-throughput sequencing and analysis of

sequencing data, and (4) assessment and quality control of dominant (overrepresented) gene-modified clones.

3.1 Initial Considerations for RIS Analysis

Factors to consider prior to initiating RIS analysis include the level
of gene transfer in the source sample to be used for RIS analysis,
the quantity of genomic DNA available for RIS analysis, and the
ability to manage resulting sequencing data. These considerations
are discussed in Subheadings 3.1.1–3.1.3.

3.1.1 Impact of the Level of Gene Transfer on RIS Capture Efficiency

As next-generation sequencing platform technologies advance, the
sensitivity of RIS amplification procedures also improves (*see* **Note 2**).
However, even with hundreds of millions of sequencing reads possible on single chip platforms, the level of gene marking in the
source sample to be analyzed will dramatically affect the resulting
RIS integration profiles. Generally speaking, gene marking levels
>5 % of all genomes in the sample will yield reasonable results with
no other considerations required given reasonable amounts of
DNA can be isolated (*see* Subheading 3.1.2). If the level of gene
marking in the source sample to be analyzed is <5 %, and especially
<1 %, of all genomes, sorting the gene-modified cell population by
any vector-specific marker or phenotype is ideal. It should be noted
that if cells are sorted based on provirus transgene expression (e.g.,
green fluorescent protein expression sorting by flow cytometry),
poorly expressing or silenced gene-modified clones may be missed.

3.1.2 Impact of the Quantity of Genomic DNA Available on Efficiency of RIS Analysis

In addition to the considerations concerning the levels of gene-
modified cells, the total number of cells from which gDNA can be
isolated is also of importance. The ideal quantity of starting
genomic DNA is 1.5–3 μg per sample; however, as little as 100 ng
can be used. The number of cells, bulk or sorted, required to
obtain this amount of gDNA following extraction varies based on
the target cell population to be analyzed. Included in Table 1 is a

Table 1
Cell numbers from blood and bone marrow cell sources for extraction of 3 μg gDNA

Cell source	Input cell number reliably yielding 3 μg gDNA[a] (10^6)
Bulk peripheral blood or bone marrow WBCs	0.5
Bulk peripheral blood granulocytes	5–7
Bulk peripheral blood mononuclear cells	0.3
Sorted peripheral blood or bone marrow mononuclear cells (CD34[+], CD3[+], CD20[+])	0.5

[a]Based on gDNA extraction using the GENTRA Puregene Blood kit or the QIAamp DNA blood kit (both
from QIAGEN)

reference range of cell numbers for several human blood and bone marrow cell sources which reliably yield 1.5–3 µg of genomic DNA based on our laboratory's experience. It should be noted that required cell numbers can also vary based on the species of origin and the method of gDNA extraction being used. If cell numbers of interest are limiting, whole genome amplification (i.e., REPLI-g; QIAGEN) of isolated gDNA can be used to render sufficient quantities of input DNA; however, bias within the amplification procedure may affect RIS capture efficiency, especially in source samples with low gene marking constituted by a number of low-frequency clones (polyclonality).

3.1.3 Managing Large Sequencing Data Files

In order to provide a comprehensive overview of the RIS profile the most advanced RIS amplification protocols must be coupled with massively paralleled sequencing. As sequencing technologies have advanced to provide hundreds of millions of reads in a single chip platform, managing large sequencing data sets is an absolute necessity. With this, cataloging and storage concerns also have to be considered. In later sections a general overview of computer scripting required to process data sets is addressed, with mention of publicly available tools for accomplishing data processing. Although it will not be explicitly detailed in Subheading 3, a relational database to allow for easy comparison and extraction of desired datasets is just as important as automated processing, and ideally the two can be linked.

3.2 Minimally Biased, Modified Genomic Sequencing (MGS)-PCR Method for Amplification of RIS

The amplification of provirus integration sites using the MGS-PCR method comprises the following major procedures: preparation of gDNA from source cell sample (*see* Subheading 3.2.1), fragmentation and polishing of gDNA (*see* Subheading 3.2.2), ligation of linker cassette and small fragment removal (*see* Subheading 3.2.3), standard exponential PCR to amplify target RIS (*see* Subheading 3.2.4), solid-phase capture of first PCR products (*see* Subheading 3.2.5), nested PCR (*see* Subheading 3.2.6), agarose gel resolution and purification of products for next-generation sequencing (*see* Subheading 3.2.7), and final purification of products for next-generation sequencing (*see* Subheading 3.2.8).

3.2.1 Sample Collection and DNA Isolation

Generally any common method for gDNA extraction is acceptable provided the resulting gDNA remains relatively intact, and contains minimal protein contamination for accurate DNA quantitation. In the case that a lower limit of DNA cannot be met (*see* below) for a particular gene-modified cell population, a crude extraction protocol can be utilized to maximize DNA yield as follows:

1. Suspend cells in 90 µL of molecular grade, DNase-free H_2O containing 19 Units/mL proteinase K.

2. In a thermocycler, heat the sample to 56 °C for 2 h, then inactivate proteinase K by increasing the temperature to 99 °C for 10 min.

3. Resulting samples should be stored at –20 °C until RIS analysis can be performed. The entire sample should be used.

Also, genome amplification of samples with exceedingly low amounts of DNA (picogram quantities) is possible (Kiem laboratory unpublished data), but the impact of this additional, non-provirus-specific amplification step on efficiency of RIS integration profile identification is potentially dramatic. This is especially true in samples containing low levels of gene-modified cells (<5 %), or in samples with large numbers of low-frequency RIS. While commercial methods of unbiased whole genome amplification are available, the efficiency of amplification of specific genomic loci, and thus specific RIS, can vary significantly.

3.2.2 Fragmentation and Polishing of gDNA

To reduce inherent bias of DNA fragmentation by restriction enzyme digest in traditional LAM-PCR [19], most studies now use some form of non-restriction (nr) LAM-PCR or transposon integration coupled with a high-throughput sequencing approach [15, 17]. The method described here is a modified, whole genome sequencing approach. Specifically, the linear PCR in LAM-PCR techniques has been eliminated and instead the first step is to acoustically fragment the DNA along with other modifications outlined below [1, 18]. The advantage of initial gDNA fragmentation is that the resulting variation in DNA fragment lengths allows for quality control of PCR bias in RIS detection and additional population analysis not possible with LAM-PCR (*see* Subheading 3.4.1 longitudinal analysis of gene-modified clonal contribution).

The desired fragment length for MGS-PCR is 1,500 bp to assure that the maximum number of fragments contain a portion of the retrovirus LTR and sufficient flanking genomic sequence. With the development of focused acoustic DNA shearing by Covaris, it is possible to generate distinct size range DNA fragments reproducibly without thermal damage or GC bias in a high-throughput workflow [20]. A range of instrumentation is available to suit a variety of workflows. Following fragmentation, DNA ends may be damaged or recessed, requiring a "polishing" step to repair ends for blunt-end ligation of linker cassettes.

1. Between 100 ng–3 μg of gDNA is transferred to the appropriate vessel for fragmentation using a Covaris apparatus.

2. Following focused acoustic fragmentation, dry samples completely using a SpeedVac™ or similar method, and then resuspend dried pellets in 30 μL of QIAGEN's Buffer EB.

3. Damaged or recessed ends should be repaired using the End-It DNA End-Repair Kit (Epicentre) or similar DNA fragment

end repair kit following the manufacturer's instructions. The entire 30 µL sample of the fragmented DNA is polished for downstream manipulation.

When processing multiple samples, a DNA end-repair "Master Mix" is useful. The amount of Master Mix per sample should be as follows:

4 µL H$_2$O.

5 µL 10× End-Repair Buffer.

5 µL dNTP.

5 µL ATP.

4. For the final reaction mixture, 19 µL of the Master Mix is added to the entire 30 µL of fragmented DNA sample.

5. Following DNA polishing the sample is purified using a QIAGEN *MinElute* PCR Purification Kit following the manufacturer's instructions with the following exceptions: For each sample, (50 µL total volume), plus 250 µL (5× the sample volume) buffer PB is applied to one column. During the drying centrifugation spins the tube is rotated 180° to remove residual ethanol. After the final drying step the sample is eluted with just 11 µL of QIAGEN's Buffer EB, incubated at room temperature for 20 min, and then spun in a microcentrifuge for 1 min.

3.2.3 Ligation of Linker Cassette and Small Fragment Removal

Blunt-ended, double-stranded (ds)DNA linker cassettes are ligated onto the purified, polished, and fragmented DNA using the Fast-Link DNA Ligase system (Epicentre) or similar ligation system following the manufactures instructions with the specific reaction mixtures noted. When processing multiple samples a "Ligation Master Mix" is made that contains the following components in a per-sample volume:

1.5 µL of 10× Fast-Link Ligation Buffer.

0.75 µL 10 mM ATP.

2.5 µL of the dsDNA linker cassette (described in Subheading 2).

1. To make the final mixture, 4.75 µL of the Ligation Master Mix is added to ~9.25 µL of the polished DNA (the entire 11 µL sample with minor losses from the column above) and 1 µL of Fast-Link DNA Ligase.

2. Ligation reactions are incubated at room temperature for 15 min before stopping the reaction by incubated at 70 °C for 10 min.

3. After ligation, samples are purified using a QIAGEN *MinElute* PCR Purification Kit according to the manufacturer's protocol with the following exceptions: 75 µL (5× the sample volume) buffer PB plus the sample volume (15 µL) is applied to

a single column. During the drying centrifugation spins, the tube is rotated 180° to remove residual ethanol. After the final drying step the sample is eluted with 20 µL of QIAGEN's Buffer EB, incubated at room temperature for 20 min, and then spun in a microcentrifuge for 1 min.

To further purify the desired DNA fragments, small "contaminating" fragments are removed using solid-phase reversible immobilization beads such as AMPure XP.

4. Bring the purified/eluted DNA samples from above to 100 µL with QIAGEN Buffer EB and add 65 µL of room temperature AMPure XP beads.

5. Vortex the samples thoroughly and incubate for 5 min at room temperature.

6. Using a magnetic particle concentrator (MPC), pellet the beads against the wall of the microcentrifuge tube.

7. Remove the supernatant slowly, being careful not to disturb the pellet.

8. While leaving the tube on the MPC, wash the beads twice for 30 s each wash with 500 µL of freshly prepared 70 % ethanol.

9. After removing the last of the ethanol wash allow the pellet to air dry.

10. Once dry, add 33 µL of QIAGEN's Buffer EB and pulse-vortex the sample until the AMPure XP beads are resuspended, then vortex continuously for 15–20 s to dislodge the desired DNA fragments from the AMPure XP beads.

11. Briefly pulse-centrifuge the sample and then place the sample back on the MPC and transfer the supernatant containing the purified DNA fragments to a fresh, labeled microcentrifuge tube. Purified fragments can be stored at −20 °C for later use.

3.2.4 Standard Exponential PCR to Amplify Target RIS

Initial amplification of RIS is done using a standard PCR procedure with a biotin-tagged, LTR-specific primer that allows for enrichment of LTR-containing PCR products. Briefly, for multiple samples a "First PCR Master Mix" is formulated containing the following per-sample component volumes:

26.5 µL H_2O.

5 µL 10× PCR Buffer ($MgCl_2$).

1.5 µL 10 mM $MgCl_2$.

1 µL 10 mM dNTP.

1 µL standard Taq polymerase.

2.5 µL 10 µM linker cassette-specific primer #1 (*see* Subheading 2).

2.5 µL 10 µM biotin-tagged LTR-specific primer (*see* Subheading 2).

Note: If processing samples containing different retrovirus backbones, separate First PCR Master Mixes will have to be prepared for each backbone.

1. For each sample, 40 μL of the First PCR Master Mix is transferred to fresh PCR tubes with 10 μL of the SPRI bead-purified DNA fragments. For each sample, triplicate PCRs are set up and run in parallel to maximize amplification of RIS.

2. Samples should be subjected to PCR in a thermocycler. The PCR program consists of in initial denaturation step, 95 °C for 5 min, and then 30 cycles of the following program:

    (a) Denaturation at 95 °C for 1 min.

    (b) Annealing at 60 °C for 45 s.

    (c) Extension at 72 °C for 90 s.

The program should also include a single final extension at 72 °C for 10 min. PCR products can be held at 4 °C if immediate removal from the thermocycler is not convenient.

3.2.5 Solid-Phase Capture of Products from the First PCR

The biotin tag on the LTR-specific primer used during First PCR amplification of RIS is used to enrich for LTR-containing PCR products.

1. Triplicate PCR products for each individual sample are pooled into one tube with a total volume of ~150 μL.

2. Sufficient quantities of Dynabeads M-280 Streptavidin or other streptavidin beads should be prepared for capture of all samples with a two-sample excess using standard procedures.

3. 20 μL of streptavidin beads (per sample) should be transferred to a fresh microcentrifuge tube.

4. Place the tube on a MPC and discard the supernatant.

5. Rinse the beads twice with a wash volume equal to the initial bead volume of 1 mg/mL bovine serum albumin (BSA). During each rinse remove the tube from the MPC, gently resuspend the pellet, pulse-centrifuge to collect all material at the bottom of the tube, place back on the MPC, and remove the supernatant.

6. To prepare the streptavidin beads for binding, rinse once with 1× Binding Solution (*see* Subheading 2). Use a volume of 1× Binding Solution equivalent to the initial bead volume. Gently resuspend the pellet, pulse-centrifuge to collect all material at the bottom of the tube, place back on the MPC, and remove the supernatant.

7. Resuspend beads in 50 μL of 1× Binding Solution per sample and transfer 50 μL of the 1× Binding Solution/streptavidin bead mixture to the tubes containing the combined PCR product for a total volume of ~200 μL.

8. Incubate the samples for 1 h at room temperature on a shaking thermomixer to assure that the beads remain in solution.

9. After incubation remove the samples from the thermomixer and pulse-centrifuge to collect the entire sample at the bottom of the tube.

10. Place the tube on the MPC and remove the supernatant.

11. Remove the tube from the MPC and resuspend the pellet in 100 μL H_2O.

12. Place the tube back on the MPC and remove the supernatant.

13. Remove the tube from the MPC and add 10 μL formamide solution (*see* Subheading 2) and incubate at 65 °C for 5 min in the thermomixer.

14. To remove the formamide and prepare the samples for the second PCR remove the samples, pulse-centrifuge briefly, place the tube on the MPC, and transfer the supernatant containing the DNA fragments to a fresh tube.

15. Bring the samples to 100 μL with H_2O and purify the DNA using the *MinElute* PCR Purification kit according to the manufacturer's protocol with the following exceptions: 500 μL (5× the sample volume) buffer PB plus the sample volume (100 μL) is applied to a single column. During the drying centrifugation spins, the tube is rotated 180° to remove residual ethanol. After the final drying step the sample is eluted with 15 μL of QIAGEN's Buffer EB, incubated at room temperature for 20 min, and then spun in a microcentrifuge for 1 min.

3.2.6 Nested PCR

The nested PCR will further amplify RIS and also add necessary sequences required for various high-throughput sequencing platforms. For multiple samples a "Nested PCR Master Mix" is formulated as follows using per sample volumes:

34.5 μL H_2O.

5 μL 10× PCR buffer ($MgCl_2$).

1.5 μL 10 mM $MgCl_2$.

1 μL 10 mM dNTP.

1 μL standard Taq polymerase.

2.5 μL 10 μM linker cassette-specific primer #2 (*see* Subheading 2).

2.5 μL 10 μM LTR-specific barcoded primer (*see* Subheading 2) Note: Different Master Mixes need to be prepared if processing samples with different retrovirus backbones or multiplexing samples using barcoded primers.

1. For each sample transfer 48 μL of the Master Mix to fresh PCR tubes with 2 μL of the purified DNA fragments from **step 7**. Triplicate PCRs should be set up and run in parallel to maximize

amplification of RIS. The PCR program is the same as outlined above in Subheading 3.2.4.

2. After the nested PCR, the triplicate samples are combined and purified using the MinElute PCR Purification kit according to the manufacturer's protocol with the following exceptions: 250 μL (5× the sample volume) buffer PB plus the sample volume (50 μL) is applied to a single column. During the drying centrifugation spins, the tube is rotated 180° to remove residual ethanol. After the final drying step the sample is eluted with 18.5 μL of QIAGEN's Buffer EB, incubated at room temperature for 20 min, and then spun in a microcentrifuge for 1 min.

3.2.7 Agarose Gel Resolution and Purification of Products for Next-Generation Sequencing

Samples are gel purified to remove small fragments that can dramatically reduce the number of useable sequencing reads.

1. To visualize samples, 3 μL of Orange G Loading dye (Invitrogen/Life Technologies) is added to each sample.

2. The samples are loaded and run on Invitrogen's E-Gel pre-cast 2 % agarose gels containing ethidium bromide using standard settings for 52 min. A 100 bp ladder is run in parallel to visualize approximate fragment lengths.

3. The gel is removed from the plastic casing and visualized under a UV light. The PCR amplified products will appear as a smear.

4. Using a clean razor blade, the portion of the gel corresponding to products ~400–800 bp in length is isolated for final purification.

5. DNA fragments are purified from the agarose gel using the QIAquick Gel Extraction Kit and eluted in a final volume of 50 μL QIAGEN Buffer EB.

3.2.8 Final Purification of Products for Next-Generation Sequencing

To further remove small fragments to improve sequencing results, samples are purified using SPRI AMPure XP beads as in Subheading 3.2.3, with the exception that the final resuspension volume should be 15 μL of QIAGEN Buffer EB. The purified supernatant is the product collected for high-throughput sequencing. Further quality control and testing will be dependent upon the sequencing platform of choice.

3.3 High-Throughput Sequencing and Analysis of Sequencing Data

High-throughput sequencing methods and downstream processing considerations are described in Subheadings 3.3.1–3.3.3. This includes massively paralleled sequencing of RIS (*see* Subheading 3.3.1), computer-aided processing of resulting sequence reads for RIS identification and quantification (*see* Subheading 3.3.2), and expected sequencing results of RIS from complex samples (*see* Subheading 3.3.3).

3.3.1 Massively Paralleled Sequencing of RIS

Due to the rapid evolution and improvement of sequencing platforms, offering improved density, sequence accuracy and turnaround time, it is advisable to use commercial facilities that will provide and maintain the latest equipment and protocols. Most adaptations of current sequencing platforms and development of new sequencing platforms still rely on specific sequences flanking the DNA of interest, and thus, simple changes to primer design (*see* Subheading 3) can assure that samples processed by the methods outlined above are compatible with the most advanced sequencing systems available. Also, to reduce costs and overall sample processing time multiple samples can be barcoded (*see* Subheading 3), combined on single chamber sequencing hardware, and sorted post-sequencing using simple scripts often available as standard procedures from commercial facilities. FASTA format sequence reads are a computer script-friendly method for extraction and processing, and most sequencing facilities will deposit this file type or the desired sequencing format on a secure server that can be accessed remotely for downstream processing.

3.3.2 Processing Resulting RIS Sequence Reads

With the increasing number of sequencing reads possible using current platforms, automated processing to identify usable sequence reads is absolutely required to handle such large data sets. Any number of coding scripts can be adapted for these purposes (i.e., PERL), but specific scripts will not be addressed here. Instead, we will discuss more general considerations and potential issues to address early when developing your unique automated system. The method described here is a stringent criterion for identifying valid RIS sequences, as it is necessary to assure confidence in the resulting datasets and to avoid erroneous localization of RIS, which can complicate downstream studies (see below). For PCR-based long terminal repeat (LTR)-chromosome junctions, DNA sequences are processed as previously described [21–23]. Briefly, for sequences isolated using PCR-based methods, valid integration sites are scored after locating retrovirus LTR, intervening genomic DNA, and linker cassette sequence. Identification of the retrovirus LTR is absolutely required, but a computer script should be written to allow for mutations (additions or deletions) in the LTR that can result from common sequencing error. When the LTR and linker cassette sequences are identified, those sequences are "trimmed" off and the resulting junction sequences, the flanking genomic sequence of interest referred to as the "query," can be aligned to the appropriate genome using a stand-alone version of BLAT [24] that generates a BLAST alignment score. This score is generated using a heuristic approach with an approximation of the Smith–Waterman algorithm that relies on searching K-mers in the database rather than a linear search [24, 25]. These steps eliminate unusable sequences where no BLAST alignment score and usually occurs because the sequences are too short or the sequences correspond to the internal retrovirus vector that occurs because of the direct repeat structure of the virus backbone. Even with stringent steps taken to remove short fragments and the unavoidable consequence of internal retrovirus vector sequences, these undesirable groups will account for the majority of the sequencing reads (~70–90 %).

The next steps outline methods for quality control of the remaining, valid RIS sequences to retain only RIS that can be aligned to the appropriate genome with confidence.

1. *Removal of query sequences with >3 bp gaps between the LTR, identified by the computer script, and the start of the genome alignment.* These sequences are discarded because of the short "unknown" intervening DNA sequence.

2. *Removal of RIS with more than one genomic alignment wherein the secondary query alignment BLAST score is >98 % of the primary query alignment.* When aligning sequences to the appropriate genome, in most cases, there will be multiple potential alignments of varying quality. These sequences are eliminated because it is not possible to confidently align the query to a specific, single locus in the genome. For computer programming purposes, rounding up is not allowed. For example, if a single query produces two alignments and the second alignment BLAST score is of 97.5 % of the primary alignment BLAST score, the primary BLAST score would be retained and considered valid.

3. *Discarding RIS with unacceptable query identity to the appropriate genome.* An acceptable percent identity between the query sequence and the appropriate genome should be ≥95 %. Any query with an identity <95 % is discarded. For computer programming purposes, rounding up is allowed. Thus, a query with an identity of 94.5 % would be retained.

4. *Grouping of "like" RIS by accounting for sequencing errors (wobble).* High-throughput sequencing methods currently available have specific, inherent issues that result in a low percentage of mutations being introduced in the resulting sequence reads. As a result, the LTR genomic junction is often not perfect, therefore a best guess, referred to as "wobble," should be allowed. Briefly, all of the valid RIS identified after processing using the criteria outlined above are grouped by chromosome and in ascending order based on the genomic start position in the chromosome. Once this is completed, a 5–10 nucleotide window is used to identify "like" RIS. For example, an RIS identified on chromosome 8 at position 1,000,001 will be grouped with all subsequent valid RIS identified from a particular dataset on chromosome 8 at positions 1,000,001–1,000,010. This wobble grouping should be inspected by qualified personnel to assess whether the groupings actually involve distinct RIS.

5. *Identifying unique RIS in a sample.* In a full-scale sequencing run, millions of sequencing reads can easily be achieved. Some RIS will be sequenced thousands of times whiles others may only be sequenced once. In the event that there are a number of RIS in a wobble group, a representative unique RIS for each genomic locus is then used to represent the entire group. This processed list is the final list of unique RIS for the sample with no repetitive RIS sequences aligned to the same genomic loci.

6. *Query population data to retain for downstream interpretation.* The number of times a particular RIS is sequenced is one important piece of data that should be retained in final unique RIS lists and for our program is referred to as "hit count." Also, specific to MGS-PCR, the number of different query lengths that make up the total hit counts is also a critical component for population studies. The first step of MGS-PCR is DNA fragmentation with focused acoustic shearing. Since the DNA is randomly sheared before any PCR amplification, the sequencing product length is informative because every different query length corresponds to a different input cell in the starting sample carrying that same unique RIS (a "clone"). This sequenced fragment length data for our purposes is referred to as the "span count." The hit count and span count can be combined to answer more informative gene-modified contribution questions (*see* Subheading 3.4.1), to rule out PCR bias as the reason for high frequency of RIS capture (i.e., high hit count with low span count), and also indicates the quality of an MGS-PCR processing experiment.

3.3.3 Expected Sequencing Results of RIS From Complex Samples

A typical sample with >10 % gene marking and 3 µg input DNA will yield half a million sequencing reads on the Ion Torrent™ sequencing platform if the total number of samples per chip is 25 or less. Of these, 30,000–100,000 sequence reads per sample are typically usable, with identifiable genomic alignments based on the criteria described above. Depending on the clonal diversity within the starting cell pool, these RIS sequences can account for anywhere between 1 and 10,000 unique RIS.

3.4 Assessment of Dominant/ Overrepresented Gene-Modified Clones

Characterization of dominant/overrepresented gene-modified clones is critical to understand the fundamental underlying biology of virus vector systems and is described in Subheadings 3.4.1–3.4.3. This includes longitudinal analysis of gene-modified clonal contribution (*see* Subheading 3.4.1), RIS-specific clone amplification (*see* Subheading 3.4.2), and quantitative assessment of clonal contribution (*see* Subheading 3.4.3).

3.4.1 Longitudinal Analysis of Gene-Modified Clonal Contribution

Longitudinal analysis of gene-modified cell populations is a simple and effective method to identify a progressive, selective outgrowth advantage as a result of provirus integration. To accomplish this, employ a capture frequency analysis as a semi-quantitative assay to identify potentially dominant gene-modified clones [26].

1. The genomic location of the RIS is typically used as a unique identifier of particular clones. Within each experimental group, unique RIS that are sequenced most frequently (largest "hit counts" from Subheading 3.3.2, **step 6**) should be ranked. The top 10 most frequently sequenced RIS will be grouped together as the dominant clones at that particular time point. To calculate the relative contribution of each RIS to the clonal pool, each unique RIS hit count is divided by the total number of hits identified in the sample. This value is expressed as a percentage. For example, we process a sample and find the total number of RIS sequences is 100,000, representing a total of 100 unique RIS. One of these unique RIS accounts for 50,000 RIS sequences out of the 100,000 total RIS sequences (hit count = 50,000). Thus, the relative capture frequency of this clone within the detectable gene-modified cell pool would be 50 % [(50,000/100,000) × 100 %].

2. If a particular unique RIS is identified as one of the top 10 ranked sites at two or more time points analyzed and contributes to more than 20 % of the total sequence reads in either sample (as recommended by Dr Daniel Takefman, PhD, Division of Cell and Gene Therapies, Center for Biologics Evaluation Research (CBER), FDA; "Monitoring for Insertional Mutagenesis: FDA Recommendations"; Retroviral and Lentiviral Vectors for Long-Term Gene Correction: Clinical Challenges and Trial Design meeting, Washington DC, December 2010), this is considered a dominant clone and should be tracked in a RIS-specific fashion as outlined below. This type of data is most commonly presented as consecutive, segmented bar graphs. To verify true clonal representation as opposed to PCR bias, the hit count data is combined with the span count data for the same sample (Subheading 3.3.2, **step 6**) and presented as a scatter plot. For example, let us assume that the clone discussed above (50 % relative capture frequency), was falsely called out due to PCR bias. If this is the case, the span count for this unique RIS would be very low [1, 27]. However, PCR bias was not responsible for the high capture frequency, the span count would be high (e.g., 200). This comparison is not a direct ratio since current limits in sequence read length (~400 bp) limit the number of detectable fragment lengths to ~200 bp or less depending on how far receded into the LTR the provirus-specific primers anneal. Also, if a particular sample contains many difficult-to-amplify genomic sequences, PCR bias can be more prominent, and thus, this quality control should be assessed on a case-by-case basis. We have included a graphical representation of hit count and span count comparisons in Fig. 2.

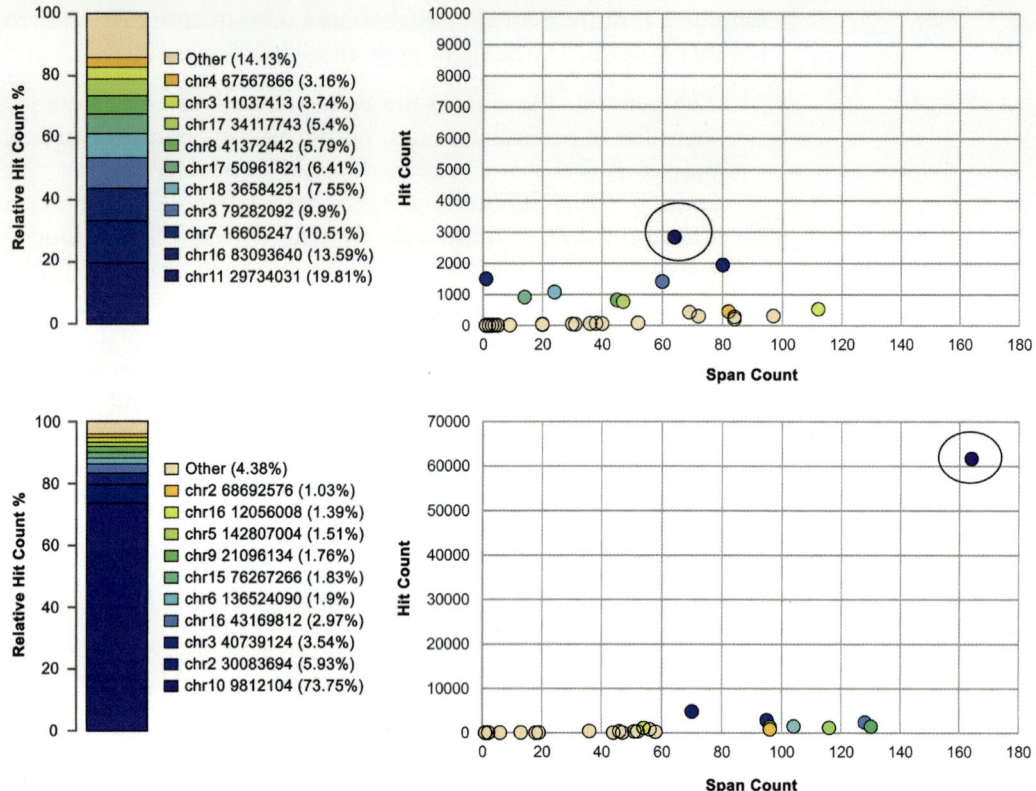

Fig. 2 Combining hit count and span count data for quality control of relative capture frequency analysis. Here we demonstrate how hit count (*y*-axis in both graphs and bar charts) can be quality controlled by examination of resulting genomic fragment lengths (span count; *x*-axis in graphs only). The *top panel* bar graph represents a clonal dominance (~20 % relative capture frequency) of an RIS on chromosome 11 at position 29734031 by relative capture frequency analysis alone. However, tracking of span count data in the corresponding graph demonstrate this capture frequency to be heavily influenced by PCR bias, as there are multiple, less dominant clones (e.g., yellow highlight), which were captured far less frequently, but displayed more genomic fragment lengths (i.e., higher span count). Conversely, the bottom panel represents a true clonal dominance as both the hit and span counts of the most dominant RIS observed by relative capture frequency analysis (chromosome 10, position 9812104) are significantly elevated compared to all other RIS identified

3.4.2 RIS-Specific Clone Amplification

Even with careful assessment and processing of valid RIS, specific RIS-associated genomic loci should be verified to assure accuracy in genomic alignment.

1. The simplest initial assessment is a BLAST alignment of the query sequence using a comprehensive database such as the National Center for Biotechnology Information's (NCBI) Nucleotide BLAST tool. Sequences that are BLAST aligned should return a similar sequence within the genome of interest as the top scored homologous regions.

2. An additional assessment for cross-contamination between samples should also be made (*see* **Note 2**). This is especially

important when considering longitudinal analysis of individual time points are processed and analyzed simultaneously, as cross-contamination can result in unique RIS being falsely identified in two separate samples. This is also important when considering experiments, such as mouse studies, where cell products for infusion may be cultured in bulk and transplanted into multiple recipients. Here, identical RIS detected in two or more recipient source samples cannot be distinguished from bulk-transduced populations or cross-contamination between samples.

3. For initial sequence verification of potentially dominant RIS, DNA from the original sample in which the RIS was identified should be used as template in subsequent PCR reactions. Primers used previously during the MGS-PCR process that are specific to the LTR are used for the "forward" primer. "Reverse" primers to designated RIS should be designed between 300 and 900 bp downstream from the genomic position identified immediately proximal to the RIS.

4. Standard PCR can then be used to specifically amplify a particular unique RIS. PCR products are visualized on 2 % agarose 1× Tris/acetate/EDTA (TAE) gels. The proper size band is excised, cloned into an appropriate blunt-ended plasmid vector, transformed, plasmid-purified, and sequenced to verify that the proper loci have been amplified. After the RIS has been verified, the clonal contribution over time should be analyzed as described below.

5. Once a specific genomic locus of the RIS has been verified by standard PCR, the surrounding genomic architecture can be analyzed to glean insight into what could be causing an overrepresentation. This can include transcriptome analysis to assess changes in local gene expression or presentation of alternative transcripts resulting from alternative splice events associated with provirus insertion.

3.4.3 Quantitative Assessment of Clonal Contribution

More information regarding the contribution of a particular dominant gene-modified clone may be required to further understand the underlying biology or, in the case of clinical settings, to address potential concerns for the patient. RIS-specific quantitative tracking is less labor intensive than carrying out large scale RIS analysis in a large number of time points and possibly different distinct cell populations. Thus, tracking specific RIS using methods such as real-time quantitative PCR with SYBR green can be extremely informative.

1. To prepare standards for SYBR green qPCR analysis, genomic DNA can be prepared using any appropriate kit for large-scale DNA isolation according to the manufacturer's protocol.

2. RIS-specific standards are generated by standard PCR amplification of the target insert with specially designed primers (from above Subheading 3.4.2, **step 2**).

3. For RIS quantitation, RIS-containing plasmids are generally used at concentrations of 0.005–500 fg. DNA concentration standards are determined using DNA isolated from appropriate cell sources with β-actin specific primers and used at concentrations of 5 pg–100 ng. Quantitative PCR is typically performed using SYBR® green dye and primers designed to the specific RIS of interest. SYBR® green qPCR kits are available from multiple manufacturers and should be used following the specific manufacturer's instructions. Longitudinal samples for analysis should be run in at least duplicate and preferably triplicate to normalize sample-to-sample variation. Due to different amplification characteristics, profiles of individual RIS are internally normalized, so the fluctuations will be reported specific to that RIS, not between different RIS.

4 Notes

1. *Primer design note*: General primer design is no different for PCR-based RIS analysis than for general PCR. The main consideration in designing primers for gene therapy vectors is that a portion of the vector is required for identification—long terminal repeats (LTR) in the case of retrovirus vectors. The primers have to be within approximately 30–100 nucleotides of the end of the LTR in order to also adequately sequence a reasonable portion of the genome for alignment purposes. Moving the second, nested primer fewer than approximately 20 nucleotides from the vector–genome junction should be avoided because erroneous annealing of the primer to analogous sequences in genomic DNA would result in a false vector–genome junction.

2. *Cross-contamination note:* With the ever increasing sequence reads available from a number of commercial sources the unintended consequence is that even very minor cross-contamination between samples is a significant issue. Thus, it is imperative that great care be taken to avoid any intersample mixing, even from initial gene transfer experiments. This should be considered throughout the protocol, from culturing cells, isolating DNA, processing samples for PCR, setting up PCR, and purifying samples for subsequent RIS analysis. Until the sequences are barcoded (second, nested PCR), even diligent technical work will still likely have some unavoidable cross-contamination that will need to be addressed [28].

Acknowledgements

The authors would like to thank Allie Evans and Sum-Ying Chiu for their technical assistance in conduct of this work. We also acknowledge the assistance of Grace Choi in preparing the manuscript. This work is supported in part by grants DK076973 (B.C.B.), HL115128, HL116217, HL098489, CA114218, HL085693, and AI097100 (H.P.K.) from the National Institutes of Health, Bethesda, MD. H.P.K. is a Markey Molecular Medicine Investigator and the recipient of the José Carreras/E.D. Thomas Endowed Chair for Cancer Research.

References

1. Adair JE, Beard BC, Trobridge GD et al (2012) Extended survival of glioblastoma patients after chemoprotective HSC gene therapy. Sci Transl Med 4:133ra57

2. Hacein-Bey-Abina S, von Kalle C, Schmidt M et al (2003) LMO2-associated clonal T cell proliferation in two patients after gene therapy for SCID-X1. Science 302:415–419, erratum appears in Science. 2003 Oct 24;302(5645):568

3. Kustikova O, Fehse B, Modlich U et al (2005) Clonal dominance of hematopoietic stem cells triggered by retroviral gene marking. Science 308:1171–1174

4. Ott MG, Schmidt M, Schwarzwaelder K et al (2006) Correction of X-linked chronic granulomatous disease by gene therapy, augmented by insertional activation of MDS1-EVI1, PRDM16 or SETBP1. Nat Med 12:401–409

5. Bender MA, Gelinas RE, Miller AD (1989) A majority of mice show long-term expression of a human b-globin gene after retrovirus transfer into hematopoietic stem cells. Mol Cell Biol 9:1426–1434

6. Cavazza A, Moiani A, Mavilio F (2013) Mechanisms of retroviral integration and mutagenesis. Hum Gene Ther 24:119–131

7. Nolta JA, Dao MA, Wells S et al (1996) Transduction of pluripotent human hematopoietic stem cells demonstrated by clonal analysis after engraftment in immune-deficient mice. Proc Natl Acad Sci U S A 93:2414–2419

8. Devon RS, Porteous DJ, Brookes AJ (1995) Splinkerettes–improved vectorettes for greater efficiency in PCR walking. Nucleic Acids Res 23:1644–1645

9. Tisdale JF, Hanazono Y, Sellers SE et al (1998) Ex vivo expansion of genetically marked Rhesus peripheral blood progenitor cells results in diminished long-term repopulating ability. Blood 92:1131–1141

10. Wu T, Kim HJ, Sellers SE et al (2000) Prolonged high-level detection of retrovirally marked hematopoietic cells in nonhuman primates after transduction of CD34+ progenitors using clinically feasible methods. Mol Ther 1:285–293

11. Kim HJ, Tisdale JF, Wu T et al (2000) Many multipotential gene-marked progenitor or stem cell clones contribute to hematopoiesis in nonhuman primates. Blood 96:1–8

12. Schmidt M, Hoffmann G, Wissler M et al (2001) Detection and direct genomic sequencing of multiple rare unknown flanking DNA in highly complex samples. Hum Gene Ther 12:743–749

13. Schmidt M, Glimm H, Lemke N et al (2001) A model for the detection of clonality in marked hematopoietic stem cells. Ann NY Acad Sci 938:146–155

14. Wu X, Li Y, Crise B, Burgess SM (2003) Transcription start regions in the human genome are favored targets for MLV integration. Science 300:1749–1751

15. Paruzynski A, Arens A, Gabriel R et al (2010) Genome-wide high-throughput integrome analyses by nrLAM-PCR and next-generation sequencing. Nat Protoc 5:1379–1395

16. Ronen K, Negre O, Roth S et al (2011) Distribution of lentiviral vector integration sites in mice following therapeutic gene transfer to treat -thalassemia. Mol Ther 19:1273–1286

17. Wang GP, Garrigue A, Ciuffi A et al (2008) DNA bar coding and pyrosequencing to analyze adverse events in therapeutic gene transfer. Nucleic Acids Res 36:e49

18. Beard BC, Trobridge GD, Ironside C et al (2010) Efficient and stable MGMT-mediated

selection of long-term repopulating stem cells in nonhuman primates. J Clin Invest 120: 2345–2354

19. Uren AG, Kool J, Berns A et al (2005) Retroviral insertional mutagenesis: past, present and future. Oncogene 24:7656–7672

20. Quail MA, Kozarewa I, Smith F et al (2008) A large genome center's improvements to the Illumina sequencing system. Nat Methods 5: 1005–1010

21. Trobridge GD, Miller DG, Jacobs MA et al (2006) Foamy virus vector integration sites in normal human cells. Proc Natl Acad Sci U S A 103:1498–1503

22. Beard BC, Dickerson D, Beebe K et al (2007) Comparison of HIV-derived lentiviral and MLV-based gammaretroviral vector integration sites in primate repopulating cells. Mol Ther 15:1356–1365

23. Beard BC, Keyser KA, Trobridge GD et al (2007) Unique integration profiles in a canine model of long-term repopulating cells transduced with gammaretrovirus, lentivirus, and foamy virus. Hum Gene Ther 18:423–434

24. Kent WJ (2002) BLAT–the BLAST-like alignment tool. Genome Res 12:656–664

25. Smith TF, Waterman MS (1981) Identification of common molecular subsequences. J Mol Biol 147:195–197

26. Cartier N, Hacein-Bey-Abina S, Bartholomae CC et al (2009) Hematopoietic stem cell gene therapy with a lentiviral vector in X-linked adrenoleukodystrophy. Science 326:818–823

27. Biffi A, Montini E, Lorioli L et al (2013) Lentiviral hematopoietic stem cell gene therapy benefits metachromatic leukodystrophy. Science 341:1233158

28. Schmidt M, Schwarzwaelder K, Bartholomae CC et al (2009) Detection of retroviral integration sites by linear amplification-mediated PCR and tracking of individual integration clones in different samples. Methods Mol Biol 506:363–372

Barcoded Vector Libraries and Retroviral or Lentiviral Barcoding of Hematopoietic Stem Cells

Leonid V. Bystrykh, Gerald de Haan, and Evgenia Verovskaya

Abstract

Cellular barcoding is a relatively recent technique aimed at clonal analysis of a proliferating cell population of any kind. The method was shown to be particularly successful in monitoring clonal contributions of hematopoietic stem cells (HSCs). An essential step of the method is retroviral or lentiviral labeling of the hematopoietic cells. The unique feature of the method is the generation of a vector library containing specific artificial DNA tags, generally known as barcodes. The library must satisfy multiple essential requirements. Importantly, considering the number of possible variations within the barcode sequence, the actual size of the barcoded vector library, and the number of clonogenic (stem) cells in the given experiment should be in ratios far from saturation. Excessive bias in barcodes frequencies must be avoided, and the library size must be assessed prior to the sequencing analysis. The final sequencing results must undergo statistical filtering. If all requirements are met, the method ensures profound sensitivity and accuracy for monitoring of the clonal fluctuations in a wide range of biological experiments.

Key words Barcode, Vector, DNA, Random, Stem cells, Clonality

1 Introduction

The problem of monitoring clonality in hematopoiesis is descending from early theoretical and experimental studies published first in the 1960s [1–3]. With the realization of the existence of hematopoietic stem cells and the clonal nature of hematopoiesis, questions of the extent of heterogeneity of HSCs were raised. Such heterogeneity includes the variation in the level of contribution to differentiated blood cells in time, and to each particular blood cell lineage [4].

Several approaches have been suggested which allow to obtain this kind of information. This included both theoretical efforts [5, 6] and experimental assays (in vitro assays, spleen colony forming assay, limiting dilution and single cell transplantation, chemical labeling, viral integration site analysis). At present it is widely accepted that HSCs comprise a small fraction of all hematopoietic

Kevin D. Bunting and Cheng-Kui Qu (eds.), *Hematopoietic Stem Cell Protocols*, Methods in Molecular Biology, vol. 1185, DOI 10.1007/978-1-4939-1133-2_23, © Springer Science+Business Media New York 2014

cells, which are quite heterogeneous in repopulating activity, and often are highly variable with respect to lineage contribution. Thus, determining the behavior of HSCs at the single cell level has become an important goal in stem cell biology.

To improve quantitative aspects of clonal monitoring, recently others and we have suggested an upgraded version of the retroviral or lentiviral HSCs labeling technique, using a specially barcoded vector library [7–9]. This approach considerably simplified the technique of the clonal monitoring, since it only required accurate DNA isolation from any cell sample of interest, and sequencing of the barcoded region [7, 8, 10–12]. Although straightforward in principle, the quality of the library and the data (sequence) analysis are crucial factors for the adequate interpretation of the collected results. Ignoring essential steps of the protocol might jeopardize the successful clonal counting. Multiple technical nuances of the cellular barcoding are the subject of the current report.

2   Materials

2.1   DNA

2.1.1   Retroviral Vectors

MIEV (originally obtained from Prof. Craig Jordan), 633 (Open Biosystems, USA), SF91 (originally obtained from Prof. Christopher Baum), and their derivatives. Lentiviral vector pGIPZ (Open Biosystems, USA) derivatives. All vectors must be adapted for barcoding in the following way: there must be at least 2 sticky unique restriction sides close to the end of CDS of the enhanced green fluorescent protein (eGFP) or turboGFP (tGFP) transcript, usually located upstream of the 3' long terminal repeat (LTR). This vector locus is relatively nonfunctional and sequence variation in this locus should be selectively neutral.

2.1.2   Adapters

Custom-made barcode DNA adapters (forward and reverse complement).

2.1.3   Additional Vectors

For lentiviral transductions, packaging construct (pCMV Δ8.91), glycoprotein envelop plasmid (VSV-G) (both obtained from Dr. Hein Schepers [13]).

2.2   Enzymes

Restriction endonucleases: *BsrgI*, *BamHI*, *ClaI*, *BstBI*, DNA ligase (Thermo Fischer Scientific, USA; New England BioLabs, USA). Note that these enzymes are all supplied with buffers and instructions for use.

2.3   Commercial Kits

Mini-prep kit (Fermentas, USA), PCR cleanup kit (Roche, Switzerland), DNA isolation kit (Sigma-Aldrich, USA) or their analogues. All kits are supplied with detailed instructions, which must be followed.

2.4 Mammalian Cells	Packaging cell lines: for retroviral transduction—Phoenix-ECO cells (originally obtained from the Nolan lab); for lentiviral transduction—293T human embryonic kidney cells. Appropriate murine cells (for hematopoietic stem cell purposes we routinely use flow cytometry-purified stem cells).
2.5 Culture Media and Additional Chemicals	StemSpan culture medium (StemCell Technologies, Canada) supplemented with penicillin/streptomycin (P/S) and cytokines: interleukin-11 (R&D Systems, the USA), stem cell factor, Flt3 ligand (both Amgen, the USA), Dulbecco's modified Eagle's medium (DMEM) culture medium supplemented with P/S, fugene (Promega, the USA), retronectin (Takara, Japan), polybrene, 0.1 % gelatin solution (Sigma, USA).
2.6 Flow Cytometry	We perform cell sorting on MoFlo XDP and MoFlo Astrios (Beckman Coulter, the USA). Fluorofore-conjugated antibodies for cell sorting are acquired from Biolegend (the USA) and BD (the USA).
2.7 Disposables for Cell Culture and Transduction	T75 culture flasks, 6- and 12-well plates, 7.5 % BSA solution (PAA, for blocking purposes), 2 % BSA solution in PBS, 0.2 % BSA solution in PBS, 10 % FCS solution, +10 % FCS medium, Eppendorf tubes, sterile low protein binding filters (Millex HV sterile syringe filter with Durapore PVDF membrane, 0.45 µM, Millipore, the USA), 5 and 10 ml syringes, 25 ml syringes with luer lock, cryogenic vials (preferably with screw wire inside tube to prevent spills), 15 and 4.5 ml centrifuge tubes.

3 Methods

3.1 Barcode Linker	Two complementary barcode adapter sequences (forward and reverse complement) are ordered for synthesis at a company of choice (in our case BioLegio, the Netherlands). Adapters should carry the barcode sequence in its middle part. The barcode sequence consists of repeated random fragments interspersed with definite sequences. The barcode is usually flanked by 6–21 base long perfectly complementing sequences on both sides. At each end the adapter contains sticky overhangs compatible with the sticky ends of the cut vector (section below). For instance: if the vector is cut with *BsrgI-BamHI* enzymes, then the barcode should carry GTAC 5′-overhang for forward adapter and GATC for 5′-overhang of the reverse complement adapter. For this particular case the structure of the complete barcode adapter sequence will be as shown in Fig. 1a (from Gerrits et al. [8]). Note that this version preserves intact restriction sites. This original version of barcode was modified into a slightly reduced version (first NN pair was replaced by GG, Fig. 1b [8, 10]). Alternatively, the barcode shown in the figure was used in lentiviral vectors (Fig. 1c [10]).

A GTAC*AAGTAA*NNATCNN*G*AT*SS*AAAANN*GGT*NNAAC*NN TGTAAAACGACGGCCAGTGAG*
        *TTCATT*NNTA*G*NNCTA*SS*TTTNNCCANNTTGNN ACATTTTGCTGCCGGTCACTCC*TAG

B GTAC*AAGTAA*GGATCNN*G*AT*SS*AAAANN*GGT*NNAAC*NN TGTAAAACGACGGCCAGTGAG*
        *TTCATT*CCTA*G*NNCTA*SS*TTTNNCCANNTTGNN ACATTTTGCTGCCGGTCACTCC*TAG

C GTAC*CAGTAA*GGNNNACNNNGTNNNCGNNNTANNNCANNNTGNNN GACGGCCAGTGAC*
        *G TCATTCC*NNNTGNNNCANNNGCNNNATNNNGTNNNACNNN CTGCCGGTCACTGC*TAG

Fig. 1 Structure of barcode linkers. Details of linker design are described in the text

The barcoded adapters can be further modified according to the designer's preferences. Essentially, the number of N positions define the total number of possible combinations in the library, namely, $4^N \times 2^S$ (where N stands for any base, S is either G or C). Upon adapter ligation into the vector the latter version of the barcode adaptor eliminates original restriction sites (TGTACA became TGTACC), therefore once inserted the barcode cannot be removed by repeated restriction with *BsrGI-BamHI*.

3.2 Barcode Annealing

Synthesized barcode oligonucleotides usually arrive as a dry pellet. This is diluted with sterile pure water to a concentration 10 μM, and incubated at 37–40 °C for 1 h in a closed tube to dissolve it completely. For annealing forward and reverse complementary adaptors are mixed at a 1:1 ratio and 3 volumes of water are added. Further, a 10 mM Tris pH 7.4 buffer with 1–5 mM MgCl₂ (final concentration) and 50–100 mM NaCl is added. Alternatively, a ligase (no polyethylene glycol (PEG) added) or restriction buffer (for instance *EcoRI* buffer) can be added from 10× stock at dilution 20-fold. The buffer composition is approximate, the pH might vary in a range 7.3–7.8. Similarly, MgCl₂ (or other salts of Mg) should be present in sufficient concentrations, which can vary from 1 to 10 mM. Excessive amounts of ethylenediaminetetraacetic acid (EDTA) must be avoided since DNA is naturally a Mg^{2+} salt, and EDTA will chelate Mg^{2+}. PEG must be avoided because it will disturb DNA melting and force annealing at higher temperatures.

The mixture is placed in a closed tube on a dry thermostat with a sufficiently massive metal rack (to preserve heat for a few hours). The tube is heated to 90 °C, the thermostat is then turned down to the room temperature to allow for *passive* cooling of the mixture during 2–3 h. The mixture can be directly used for ligation to the vector of interest; alternatively aliquots of annealed adapter are stored in the freezer at −20 °C for later use.

3.3 Vector Preparation

Before barcoding the researcher must decide which vector to use and which locus of the vector is most suitable for barcoding. Our preferred locus is a region upstream of 3′LTR (usually flanked by eGFP/tGFP from other side). This site is nonfunctional, therefore integration of the barcode should not affect vector properties. The locus of interest must contain two unique sticky-end forming restriction sites. If the locus of interest does not reveal a suitable pair of restriction sites, they must be made by site-directed mutagenesis and/or custom made adaptors (in case only one available site is present, more sites can be added via adapter ligation).

Vector DNA is usually prepared from *E. coli* stocks, namely, single colony seed into 2 ml Luria broth (LB) medium supplemented with proper antibiotics, incubated with rotation overnight at 37 °C. Antibiotic concentrations can be found for instance online at the Roche LAB FAQs book. Plasmid DNA is isolated using available mini-prep or midi-prep kits (Thermo Scientific mini-prep kit, protocol according to the commercial provider of the kit) and stored in at −20 °C in small aliquots.

Routinely we use a pair of sticky end restriction endonucleases which cut the vector near the end of eGFP CDS upstream of 3′LTR. The MIEV vector is usually cut with *BsrgI* and *BamHI* [8]. Restriction is performed in a buffer compatible with both enzymes. Both NEB and Thermo Scientific offer a range of common buffers (NEB Double Digest Finder, Thermo DoubleDigest pages). A similar scheme is adapted for the properly modified SF91 vector [8], as well as for the pGIPZ modified vector [10]. Usually 20–40 µl of the standard mini-prep (0.2–0.4 µg/µl, 60–80 µl total) is taken for the restriction. Details of the restriction protocol (buffer, duration) must conform to recommendations of the enzymes provider, and can be found in the accompanying instructions and on the companies' Web sites (NEB or Thermo Scientific). After restriction is completed (usually 6–8 h is enough), the reaction is terminated by heat-inactivation at 80 °C for 20 min and chilled on ice. The cut vector can be either used directly or cleaned from reaction components using PCR cleanup kit (according to the recommendations of the company). In practice, 10 µl of the mini-prep (0.2–0.4 µg/µl) is sufficient to proceed with construction of the vector library. The rest of the prep is stored for later use.

3.4 Ligation

When barcode adapters and vectors have been prepared (barcodes annealed, vector cut with appropriate restriction enzymes generating mutually compatible sticky ends) the ligation can be carried out. Routinely, freshly cut vector is incubated at 65–85 °C for 20 min to inactivate restriction enzymes, diluted three times with sterile water, and barcode adapters are added to the linearized vector. Ligation buffer (usually 10× concentrated) and T4 DNA ligase is added to the mixture in amounts recommended by

Table 1
Calculation of linker to vector ratio

Component	Length, bp	µg/µl	Final 1:1 ratio	1:10 ratio
Plasmid DNA	9,000	0.1		
Linker	60	0.7		
Dilution	150	7	1,050	105

the company protocol. In case it is a regular DNA ligase, the initial temperature is set to 19 °C for 30 min, then gradually decreased to 17 °C and kept overnight. Fast ligase protocol can be performed as well, temperature conditions are defined by the enzyme provider. Essential for this step is mixing the adapters and vectors at a correct ratio. Unlike regular cloning purposes aimed on obtaining only few vector constructs of the desired configuration, vector barcoding is aimed on maximal efficiency of the barcode integration into the vector (i.e., preventing the formation of non-barcoded vectors, *see* **Note 1**). For this purpose an approximately 10–50-fold excess of the barcode adapter over the vector concentration is essential. In practice, the calculation can be done as follows. First, consider the length of the barcode and the vector. Second, consider concentrations of both (vector concentration can be measured, barcode adapters arrive with exact description how much µg is synthesized). From these two measurements we fill out the table above and find a final volumetric mixing ratio of the vector and barcoded linker. To illustrate, in our case we had the numbers specified in Table 1. This means that for a ratio 1:10 (vector to barcode) we take 20 µl of vector and 0.2 µl of a barcode adapter. Note that further excess of the barcode concentration (above indicated range) will inhibit efficiency of the ligation. It will lead to the drastic decrease of the successfully ligated product. In addition, the risk of a barcode concatemer forming will increase. Additionally, a titration of the vector to barcode ratio can be performed to ensure optimal conditions for ligation.

3.5 Post Ligation Handling

After ligation is completed, the sample is usually heat-inactivated at 80 °C for 20 min. Further it is cleaned up (we use Roche PCR cleanup kit). This is not a strictly obligatory step, yet cleaned DNA usually gives better efficiency for *E. coli* transformation. If the adapter is made in a way that eliminates the original restriction site (like lentiviral adapter above for *BsrGI* site), a control cut with corresponding enzyme (*BsrGI*) is recommended. This will further minimize the risk of the concomitant presence of unbarcoded vectors in the barcoded library.

3.6 E.coli Transformation

Competent *E. coli* cells (NEB, Life Sciences, or analogue) are transformed with a ligated vector mixture according to the protocol provided by the company. It is usually 5 µl of the vector DNA (0.05–0.1 µg/µl) per 50 µl of the *E. coli* cell suspension. At the end of the protocol the suspension is diluted to 300 µl by SOC medium. Cells are plated in variable dilutions (usually up to 50 µl per plate, totally 6 or more plates per one ligated vector prep) to ensure single colonies growth on petri plates. Note that barcodes consisting of completely variable uninterrupted N-mers (which are used by some groups) cannot be used in this protocol; such barcodes have a high risk of strand misalignment, which will be eventually modified by the DNA-repair system of *E. coli*.

3.7 Analysis of the Vector Barcoding Efficiency and a Library Size

At the optimal conditions all colonies on the plates should be well separated from each other, so one can count them reliably. This number is usually under 100–200 colonies per plate (at the vector concentration mentioned in the previous chapter). At this stage it is important to take 20–40 colonies from the plates, perform vector isolation (mini-prep from Thermo Scientific, Qiagen, or any other company) and test for the presence of barcodes. Insertion of the barcode in the vector causes a slight but detectable increase in the size of eGFP-LTR fragment of the vector. Using appropriate restriction enzymes around the barcoded site or PCR with barcode flanking primers will allow to detect the frequency of barcoded vector preps and to count the efficiency of barcode integration. Note that both unbarcoded vectors and aberrant barcodes will be detected at some frequency. It is inevitable that a fraction of the vectors will be not properly barcoded. It is important to ensure that the total percentage of such aberrant vectors is kept low (usually under 5 %). The quality of barcoding will determine the further strategy of preparing vector prep library. If the efficiency is high, colonies can be pooled in 5–20 colonies per flask and cultured as one midi- or maxi-prep. As a rule 1–2 ml of LB medium is used in liquid cultures per single colony in the prep. Culturing in pools saves labor and timing costs. Yet, the quality of such prep can be reduced by differences in barcode concentrations and by presence of non-barcoded vector. Additionally, the whole prep can be used only once, because preparing an *E. coli* stock from such a prep is not an option. Alternatively, each colony can be cultured separately and mini-prep is performed with each cultured clone individually. This approach requires considerable labor and time efforts. However, it provides the best quality result at the end since each clone is confirmed and stored separately, and all aberrant clones can be discarded. In this case *E. coli* stocks must be made for each barcoded vector. This will allow to re-culture and regenerate the very same barcode library whenever it is needed, so the library become endlessly reusable (*see* **Note 5**).

Table 2
Calculating the risk of identical barcodes and barcodes differing by 1 nucleotide based on the barcode structure and library size

Type	Barcode	Max complexity	Library size	Probability to pick the same barcode	Probability for barcodes to differ by 1 nt
1	NN...NN...SS...NN...NN...NN	4194304	100	0.000091	0.001
			500	0.00008	0.004
2	NN...SS...NN...NN...NN	262144	100	0.001	0.009
			500	0.0026	0.052
3	NNN..NNN..NNN..NNN.. NNN..NNN..NNN	4.398E+12	500	<0.0005	<0.0005
			2,000	<0.0005	<0.0005

At the end, all isolated barcoded vector preps are mixed at equal amounts (equal μg DNA per each barcoded vector), which essentially becomes a barcoded vector library of known size/complexity. Importantly, the library size is counted at this stage for the first time (*see* **Note 4**). It is based on the actual number of colonies collected from the plates. Note that the library size should be far below the total number of possible combinations of the barcode sequence (*Library size* $<< 4^N \times 2^S$). This important condition is required for two reasons:

1. To avoid the risk of obtaining the same barcode more than once (this will eventually decrease the real size of the library).

2. To make sure all barcodes are sequence-wise significantly different from each other [14], therefore the chance of having two barcodes differ in only one nucleotide will be negligible. This is important for sequence data filtering.

More exact solutions can be found using random simulations, Poisson and binomial distribution models. For detailed analysis (if not available in the lab) we recommend to refer to specialist in bioinformatics (*see* **Note 3**).

For the three types of barcodes we routinely calculate the risk of obtaining the same barcode twice or receiving two barcodes that differ by one nucleotide (with a minimal distance (D_{min}) of 1). The calculations are shown in Table 2.

The statistics are obtained from a random simulation script in Python (taking into account the complexity of the barcode and the library size). Each parameter was tested in 50–500 simulations.

One can see from the table that barcode type 2 can be used for medium size libraries. If the size of a barcode type 2 library is increased to 500 barcodes, the library reaches its maximal size as the odds of including barcodes that are identical becomes critical.

Barcode type 1 suits well for a 500 barcodes library, while barcode type 3 can be used for much bigger libraries, which can count into the thousands.

Low saturation of the library is necessary for library validation by deep sequencing; however, it is not entirely critical in experiments with relatively low number of clones. If the number of clonally expanding cells in the biological experiment is 50–100 times below the size of the library, then problems of low barcode sequence distance or high redundancy in the library will be compensated by low sampling frequency in the real experiment. However, if the expected number of clonally expanding cells is close to or higher than the library size, then the size of the barcode library is critical. In both cases, it is beneficial to measure the number of clonal cells by at least two independent methods, one of those being barcoding.

3.8 Creating Viral Particles and Transduction of Hematopoietic Stem Cells

For retroviral vectors an ecotropic packaging cell line is routinely used. Excellent description of protocols is available at the Nolan lab page (http://www.stanford.edu/group/nolan/protocols/pro_helper_dep.html). Some variations in the method can be dictated by specifics of the cell line used. Our routine protocols have been described recently [8, 10] both for retroviral and lentiviral [13] barcoded vector libraries.

1. Day 1: Cell plating
   - Plate Phoenix-ECO cells: 2.5×10^5 cells per 1 well of 6-well plate in 2 ml DMEM + 10 % FCS. Incubate overnight at 37 °C, 5 % CO_2 to allow cells to attach.

2. Day 2: Transfection
   - Prepare transfection mix. Amounts needed for one well:

DMEM (no serum)	ad	100 µl
Fugene		3 µl
Barcoded vector construct		1 µg

   - Calculate the volume of the medium to be added, pipette in microcentrifuge tube.
   - Add dropwise fugene (do not touch walls of the tube), then vector preparation.
   - For mock transduction, add no vector.
   - Tick gently, no vortexing, allow complex formation for 15 min at room temperature.
   - Add 100 µl transfection mix dropwise to Phoenix-ECO cells, swirl really gently, incubate at least for 12 h (optimally 24 h) at 37 °C, 5 % CO_2.

3. Cell sorting

- Collect the bone marrow cells and FACS-sort desired populations (in our case lineage negative Sca1$^+$ c-Kit$^+$ CD150$^+$ CD48$^-$ cells (LSK150$^+$48$^-$) and control LSK CD150$^-$ CD48$^-$ cells).

- Collect cells directly from cell sorter into microcentrifuge tubes coated with 7.5 % BSA to avoid cell losses during transfer of cells to the plate.

- Plate 10,000–20,000 sorted cells into 1 ml StemSpan + cytokines in 1 well of 12-well plate (block wells with 7.5 % BSA, remove it before plating).

4. Medium change

- Carefully remove medium of Phoenix-ECO cells by suctioning.

- Replace medium with 1.5 ml medium of eventual target cells (Stemspan, no serum should be added to prevent inactivation of the virus, no cytokines to avoid stress to Phoenix-ECO cells), incubate O/N 37 °C, 5 % CO_2.

5. Day 3: Coat wells with retronectin

- Dilute retronectin stock (as received from a company) 40 times. This solution can be reused 4–5 times.

- Add 1 ml retronectin solution to desired number of wells of 12-well plate, gently rotate to cover the whole bottom of the plate.

- Leave at room temperature for 2 h.

- Remove retronectin solution by suctioning.

- Block retronectin by adding 1 ml of 2 % BSA solution in PBS. Leave for 30 min at room temperature.

- Remove BSA solution (optional—wash with 0.2 % BSA solution in PBS before cell plating).

6. Virus harvest:

- Very carefully collect supernatant from Phoenix-ECO cells using 5 or 10 ml syringe.

- Pass through filter (Millex HV, low protein binding) to remove residual Phoenix-ECO cells.

- Add 1 ml of PBS solution to Phoenix-ECO cells.

7. Cell transduction:

- Collect LSK48$^-$150$^+$ and control (LSK CD48$^-$ CD150$^-$) cells that have been prestimulated in StemSpan + IL11 + SCF + Flt3L for 22–24 h (14 ml screw cap tubes).

Wash every well at least two times with 1 ml of PBS + 0.02 % BSA. Check under the microscope that all the cells were collected. Repeat the washing step if necessary.

- Spin cells down for 5 min, discard the supernatant.
- Add:
  - Viral supernatant in desired volume.
  - Polybrene solution.
  - Cytokine solution.
- Transfer cells to retronectin-coated wells, centrifuge for 15–25 min at 22 °C.
- Incubate O/N at 37 °C 5 % CO_2.

  Check fluorescence of Phoenix-ECO cells to control efficiency of transfection.

8. Day 4: Control whether transduction was successful

- Collect transduced and mock-transduced control (LSK48⁻150⁺) cells in 14 ml tubes.
- Add 10 % FCS up to 14 ml to wash (and partly inactivate) the virus and spin down.
- Remove supernatant by suction.
- Resuspend in PBS + 0.02 % or propidium iodide solution.
- Bring on ice to FACS-analyzer and check fluorescence in GFP-channel. %GFP in control cells indicates whether transduction was successful. It is usually lower than in LSK CD150⁺ CD48⁻ cells, and additionally not all cells express GFP at this time point, so the value underestimates the real transduction frequency and has to be used only for indication.
- If transduction was successful, collect LSK CD150⁺ CD48⁻ cells (as described above). Count the number of cells in a small aliquot using hemocytometer.
- Leave a small aliquot in culture with StemSpan + 10 % FCS + cytokines to check final transduction efficiency.

9. Day 6–7: Check final transduction efficiency

- Collect remaining cells, analyze by FACS.

3.9 Transduction Protocol with Lentiviral Vector Library

1. Day 1: Cell plating

- Coat T75 flasks for 2 h with 0.1 % gelatin (4 °C). Use 3 flasks per transduction.
- Plate 293T cells: 5 (1.5–5) × 10⁶ cells per flask in 10 ml DMEM + 10 % FCS. Incubate O/N at 37 °C, 5 % CO_2.

2. Day 2: Fugene transfection

- Prepare transfection mix. Amounts needed for one T75 flask:
  - Tube 1:

DMEM–FCS	100 µl
Packaging construct (pCMV Δ8.91)	3 µg
Glycoprotein envelop plasmid (VSV-G)	0.7 µg
Vector construct (pGIPZ derivative)	3 µg

  - Tube 2:

DMEM–FCS	400 µl
Fugene	21 µl

  (Add Fugene in medium, not against the edges of the tube).

- Add content of tube 1 to tube 2, flick gentle and allow complex formation for 20 min at RT.
- Add 500 µl transfection mix dropwise to 293T cells, swirl real gentle, incubate O/N 37 °C, 5 % CO_2.

3. Day 3: Medium change

- Carefully remove medium of 293T cells by suctioning.
- Replace medium with 5.0 ml medium of eventual target cells (e.g., HPGM, Stemspan, etc), incubate O/N 37 °C, 5 % CO_2.

4. Day 4: Virus harvest

- Very carefully collect 5 ml medium from 293T cells into a 12 ml tube.
- If there are a lot of floating cells, consider centrifugation first.
- Pass over two filters (Millex HV, low protein binding) to remove residual 293T cells. Nb: first time, the filter might clog and snap, since a lot of 293T cells are also in this medium, hence the second filter step with new filter.
- Freeze virus-containing supernatant in aliquots of 500 µl in cryotubes in –80 °C and use when necessary.

3.10 Critical Parameters

Number of target cells. To address the research question appropriately, it is important to approximate expected number of barcoded target cells in the population to be studied. For our experiments with HSCs, we performed limited dilution transplantations to estimate the stem cell frequency (*see* **Note 6**). Ex vivo culture and gene transfer can functionally influence transduced cells, and these factors have to be considered in experimental planning.

Barcode per cell ratio. A viral transduction event is a random Poisson process [15]. For any given transduction efficiency one can assess probabilities of single, double, triple etc. viral transduction per cell. While at low transduction efficiencies only one barcode is incorporated in most cells, number of vectors per cell might dramatically increase when reaching transduction efficiencies higher than 75 %. Researchers must be able to assess those probabilities experimentally and theoretically.

Limitations of blood sampling. For studying kinetics of clonal contribution to hematopoiesis, blood samples can be repeatedly taken from the same mouse. Since the blood volume should not exceed ~100 µl within intervals of 1–2 months, it creates some limitation for the resolution and statistical power of the experiment. If cells of different lineages are sorted from the blood sample the number of sorted GFP⁺ cells defines the upper limit of the maximum number of barcodes that can be found in a sample. The sensitivity of barcode detection should be tested in samples containing known number of cells and barcodes. Work in clean environment free of high copy DNA handling is critical to prevent cross-contamination in the process of DNA extraction and preparation for PCR. It is a good practice to test all samples in duplicate to allow testing for consistency of reads if problems are detected. For DNA isolation standard commercial kits are routinely used. In our lab we combine RedExtract (Sigma) and Blood and tissue DNA isolation kit (Macherey-Nagel, Germany), which allow for extraction of DNA from cells in a range 1,000–1,000,000. For highly proliferative cells available in small numbers (HSCs), it can be recommended to perform monoclonal expansion in vitro to allow for robust barcode detection [10].

3.11 Barcode Sequencing and Primary Data Handling

3.11.1 Primary Processing

Genomic DNA is subjected to PCR with specifically tagged (indexed) primers, allowing for multiplexing of multiple samples in a single sequencing run. This obviously greatly reduces the cost of sample analysis. Details of primer tag design we use have been previously described [10, 14]. Unlike barcode sequences, primer tags for multiplexing are custom made 8–9 nt long DNA oligonucleotides added to the 5′ end of the PCR primers sequences. Importantly, tags must differ from each other by at least three bases, in other words they must conform to the minimal distance, $D_{min} > 2$, to resist at least 1 sequencing error. Various barcoded DNA samples can be pooled together and sequenced in one lane. We use Illumima HiSeq machine for barcode detection. In this configuration 200–300 samples per run can be routinely sequenced and generate sufficient number of reads per sample.

When sequencing is completed a few essentials steps must be done.

1. Quality controls. In addition to machine filtering, we routinely discard all sequence reads with low quality reads (or N) if these occurs within the expected barcode region.

2. Data compression. All identical reads are compressed as one single sequence, and the frequency value (number of reads a particular sequence occurs) is kept next to the sequence.

3. Data demultiplexing. Experimental samples are separated on the basis of the exact match to the sample multiplexing tag and PCR primer fragment, 13 nt total length. Note that at this stage more sophisticated algorithms can be applied, allowing for instance single error correction and therefore improving recovery of reads.

3.11.2 Filtering of Sequencing Noise

When all steps mentioned above are completed a barcode sequence is routinely identified using matrix-scoring method (for our scripts in Python and Perl we used MOODS package [16]). Note that limiting the analysis to only exact matches to the barcode backbone sequence, contains the risk of missing slightly aberrant barcoded structures which occasionally occur (*see* **Note 2**). The initial set of reads is further reduced to the set of unique barcode sequences (with all frequencies counted and stored), ignoring all other errors/mutations outside of the barcode region. The first and the most powerful step of sequencing noise filtering is applying the rule that in a barcode data set, sorted in descending order by frequency, barcodes that occur less frequently and that have a $D_{min} = 1$ with respect to the most abundant barcodes, must result from PCR and/or reading errors, and are removed. Failure to remove these spurious barcodes results in greatly inflated clonal counts. Note that this rule can be applied if barcode library size and clonality of the analyzed population make a probability of any pair of barcodes in the population on such distance negligible (see chapter barcode library construction above). Authors must be able to assess those risks or consult a statistician.

Some additional noise filters to consider are:

Number of reads per cell. Suppose 30,000 barcoded cells were sorted, from these cells DNA was extracted and PCR of the barcode sequence performed. Subsequent sequencing returned 100,000 reads per whole set. These reads will be unequally distributed among different unique barcodes. Barcodes with frequencies less than 1/30,000 of total reads, in this case 100,000/30,000 = 3.3, will be discarded since their relative representation is less than one barcode per cell, per definition indicating presence of technical noise in the sample.

Biological significance. Suppose in a time series a barcode has been detected which never reached a frequency above 0.5 % of the population. Even though this barcode might be real, its contribution to overall blood regeneration remains of low biological significance (keeping in mind the routine standards for acceptable limits for HCS contribution). On this basis this barcode could be rejected as it non-robustly contributes to blood cell formation.

Consistency of barcode spectra in time series (outliers). Our experience shows that the great majority of clones can be consistently detected in a time series, which allows concluding that overall blood contribution is stable, although gradually drifting in a time [8, 10]. If this assumption is accepted, incidental outliers in time series are suspect to result from technical errors or contaminations. It is advised that samples organized as time series are checked for consistency and all outliers should be taken with caution. Performing barcode analysis in duplicate helps resolving this issue.

4 Notes

1. Synthesis of the barcoded vector via a ligation protocol suffers from several imperfections, such as the occasional truncation of the backbone, concatenation of barcode sequences (when used in excess), and some fraction of vectors remaining unbarcoded. Thorough experimental optimization of the protocol should keep those side products at minimum.

2. Retrieving barcoded sequences must allow for detection of those aberrant vectors, and they must be considered as real and included in the analysis.

3. Increasing the barcode library size is a labor intensive process. Therefore, researchers must consult specialist in statistics on a reasonable library size for planned experiments.

4. Library size cannot be determined solely on the basis of sequencing data, because sequencing alone does not discriminate between real barcodes and false barcodes which are generated due to PCR/sequencing/reading errors.

5. Bias in barcoded vector representation inevitably reduces effective library size, and therefore must be kept at minimum.

6. It is desirable to have a parallel, independent assessment of the initial frequency of the barcoded clonogenic cells. Counting clones using barcodes is probabilistic by nature. It relies on specific probabilities of D_{min} values in the barcoded vector libraries; also it is used in combination with technical thresholds and distinct levels of biological significance. This creates certain risks of obtaining false positive and false negative types of errors. Those risks must be approached using proper statistical analysis.

Acknowledgements

This study was supported by the Netherlands Organization for Scientific Research (TopTalent grant to E.V.), and the Netherlands Institute for Regenerative Medicine (NIRM).

References

1. Till JE, McCulloch EA (1961) A direct measurement of the radiation sensitivity of normal mouse bone marrow cells. Radiat Res 175: 145–149

2. Becker AJ, McCulloch EA, Till JE (1963) Cytological demonstration of the clonal nature of spleen colonies derived from transplanted mouse marrow cells. Nature 197: 452–454

3. Kay HE (1965) How many cell-generations? Lancet 2:418–419

4. Copley MR, Beer PA, Eaves CJ (2012) Hematopoietic stem cell heterogeneity takes center stage. Cell Stem Cell 10:690–697

5. Harrison DE, Astle CM, Lerner C (1988) Number and continuous proliferative pattern of transplanted primitive immunohematopoietic stem cells. Proc Natl Acad Sci U S A 85: 822–826

6. Abkowitz JL, Linenberger ML, Newton MA et al (1990) Evidence for the maintenance of hematopoiesis in a large animal by the sequential activation of stem-cell clones. Proc Natl Acad Sci U S A 87:9062–9066

7. Lu R, Neff NF, Quake SR et al (2011) Tracking single hematopoietic stem cells in vivo using high-throughput sequencing in conjunction with viral genetic barcoding. Nat Biotechnol 29: 928–933

8. Gerrits A, Dykstra B, Kalmykowa OJ et al (2010) Cellular barcoding tool for clonal analysis in the hematopoietic system. Blood 115:2610–2618

9. Schepers K, Swart E, van Heijst JW et al (2008) Dissecting T cell lineage relationships by cellular barcoding. J Exp Med 205:2309–2318

10. Verovskaya E, Broekhuis MJ, Zwart E et al (2013) Heterogeneity of young and aged murine hematopoietic stem cells revealed by quantitative clonal analysis using cellular barcoding. Blood 122:523–532

11. Grosselin J, Sii-Felice K, Payen E et al (2013) Arrayed lentiviral barcoding for quantification analysis of hematopoietic dynamics. Stem Cells 31:2162–2171

12. Naik SH, Perie L, Swart E et al (2013) Diverse and heritable lineage imprinting of early haematopoietic progenitors. Nature 496:229–232

13. Schepers H, van Gosliga D, Wierenga AT et al (2007) STAT5 is required for long-term maintenance of normal and leukemic human stem/progenitor cells. Blood 110:2880–2888

14. Bystrykh LV (2012) Generalized DNA barcode design based on Hamming codes. PLoS One 7:e36852

15. Fehse B, Kustikova OS, Bubenheim M et al (2004) Pois(s)on – it's a question of dose. Gene Ther 11:879–881

16. Korhonen J, Martinmaki P, Pizzi C et al (2009) MOODS: fast search for position weight matrix matches in DNA sequences. Bioinformatics 25: 3181–3182

INDEX

Kevin D. Bunting and Cheng-Kui Qu (eds.), *Hematopoietic Stem Cell Protocols*, Methods in Molecular Biology,
vol. 1185, DOI 10.1007/978-1-4939-1133-2, © Springer Science+Business Media New York 2014